AT HOME ON AN UNRULY PLANET

AT HOME ON AN UNRULY PLANET

FINDING REFUGE ON A CHANGED EARTH

MADELINE
OSTRANDER

HENRY HOLT AND COMPANY

NEW YORK

Henry Holt and Company
Publishers since 1866
120 Broadway
New York, New York 10271
www.henryholt.com

Henry Holt® and Ⓗ™ are registered trademarks of Macmillan Publishing Group, LLC.

Library of Congress Cataloging-in-Publication Data is available.

ISBN: 9781250620514

Our books may be purchased in bulk for promotional, educational, or business use. Please contact your local bookseller or the Macmillan Corporate and Premium Sales Department at (800) 221-7945, extension 5442, or by e-mail at MacmillanSpecialMarkets@macmillan.com.

First Edition 2022

Designed by Meryl Sussman Levavi

Printed in the United States of America

1   3   5   7   9   10   8   6   4   2

I want you to act as you would in a crisis. I want you to act as if our house is on fire. Because it is.

—Greta Thunberg

Because we have not made our lives to fit
our places, the forests are ruined, the fields eroded,
the streams polluted, the mountains overturned. Hope
then to belong to your place by your own knowledge
of what it is that no other place is, and by
your caring for it as you care for no other place, this
place that you belong to though it is not yours,
for it was from the beginning and will be to the end.

—Wendell Berry

But the ethereal and timeless power of the land, that union of what is beautiful with what is terrifying, is insistent. It penetrates all cultures, archaic and modern. The land gets inside us; and we must decide one way or another what this means, what we will do about it.

—Barry Lopez

CONTENTS

AT HOME ON
AN UNRULY
PLANET

PROLOGUE

To contemplate the meaning of *home* isn't some kind of scholarly undertaking. It's more like sifting through a cardboard box of old photos and keepsakes, riffling through memories and images. In an instant, the word conjures the most vivid associations, the most visceral pieces of personal history, thoughts that wrap around me like a warm blanket, nostalgia so bittersweet that I can taste it.

The first image that arrives in my mind is of rain tapping the windowpanes of my early childhood house while outside, intricate tongues of lightning streaked and jagged across a purple sky. Here I could watch in sheltered awe, no matter what the sky unleashed.

Then, the midwestern city where I finished high school, surrounded by fields of corn and soybeans—the summer air hot and damp, the streets near the local college foot-worn and leading to a shaggy coffee shop and a take-out restaurant that sold crab dumplings and fried rice. Rings of shopping malls and big-box stores at the outskirts. I was never comfortable here, but it became a part of my identity, like some bit of clothing you wear because someone gave it to you, pulling it on again and again until it softens and fits to your particular frame.

Then, the hundred-year-old Seattle house where I live now. The desk where I write—cluttered with books and mementos, a photograph of my grandmother, the wooden bookends that used to belong to my grandfather, a red stone I plucked from the trail on an arrestingly lovely nature walk. From here I gaze out the window at a pair of crows squabbling in the backyard magnolia tree. Above me the slant-ceilinged attic bedrooms where multiple generations of people slept before I arrived here. In the distance, the groaning undersong of the highway and the port nearby and its sounds, a train whistle, metal shipping containers cracking loudly against one another in the distance, the moan of a cargo boat, the roar of a jet plane above. The sheen of the blue estuary that circumscribes my city, not visible from here but always present.

Home is "not a house for sale or a site for 'development' but the place by which one is owned, year after year loved and known," writes poet and essayist Wendell Berry. For now, my house, this city, this frayed and beautiful bit of urban ecosystem, are mine, and I am theirs—until we part ways.

Home is also a negotiation between the essentials you need, such as food and shelter, the life you construct, and the rhythms of your surroundings. More than anything, home offers safe refuge and a means to create stability, both physical and psychological. I have been privileged to always have a home with sturdy walls. I am aware—with the occasional quivering feeling of hyperalertness—that my sense of safety could rupture in an instant, perhaps if the fault line that lies beneath Seattle decides to quake.

Other invisible but insidious threats lurk just beyond my walls. Down the hill from my current house lies the industrial corridor that has lashed itself to the edge of the Duwamish River—named for the Indigenous people of this area but also known as a Superfund site, among the most toxic places in the United States. The lower river valley has some of the worst air pollution in Seattle. Not long ago, a group of high school science students collected tree moss in the valley and found it laced with arsenic, lead, and chromium. I assume that these threats do not intrude noticeably into my personal space, my air, my body, but I don't know for certain. Safety is partly a story we tell ourselves.

And my house, like the majority of dwellings in Seattle, was built for a certain set of conditions. A maritime climate—long wet winters alternating with crisp blue summers when the thermometer's mercury rarely used to slide above 80 degrees Fahrenheit. My city was made with mountains in mind: the water that flows through my household taps was drawn from the high-elevation meltwater that resupplies the city reservoirs all spring and summer. Mountain snows and glaciers recharge the rivers that spin the city's hydropower turbines and keep my lights on, fire up my electric teakettle, and power the laptop computer on which I write this.

As years pass, these conditions are no longer as reliable. Some years, the snows are scant. Summers here are increasingly hotter, smokier—as wildfires in the surrounding mountain ranges encroach on the city. The air becomes more difficult to breathe. One achingly dry, hot summer could create the right conditions not just for smoke but also for a catastrophically large fire much nearer to my home, even in this damp place of cedar and Douglas fir. I am in a more precarious position than I used to be. We all are.

Everywhere, the weather, the sky, the water, even the terrain on which we have built our homes is becoming unruly. It is literally unsettling—causing the unsettlement of some places that used to be more livable. As I write this, the American West is parched, millions of acres in extreme drought or worse. From Western Canada to California, wildfires are driving people from their homes. Burning down houses and neighborhoods and communities. Destroying belongings of both physical and emotional value—old pianos and guitars, wedding dresses, furniture, knitting needles, cars, garden tools—while also devouring human shelter and livelihoods. At the same time, the first hurricanes of the season are heading toward Mexico and the Gulf, the Caribbean, the American Southeast and Mid-Atlantic. How many thousands of people will flee to escape the next round of storms? How many more will ride them out, hunker under fragile roofs while heavy winds shriek and pound? Who will dodge the blows and who will lose everything, the homes they've built, the lives they've created?

It is too easy to recite a list of ongoing calamities. In the summer of 2021, one in three Americans experienced some kind of weather disaster. Elsewhere, the Italian island of Sicily reached what may have been the hottest ever recorded temperature on the Continent—nearly 120 degrees Fahrenheit. Villagers on the Greek island of Evia organized a brigade to try to stop wildfires from burning down their homes. At the same time, a vicious drought hung over Angola, and thousands fled their homes for nearby Namibia.

The world has always been stormy. Some of these events would have happened at any time. But each year, the likelihood of larger calamities creeps up. According to the calculations of hundreds of scientists, the whole planet is about 2 degrees Fahrenheit warmer than when the foundation of my house was laid. Because of this addition of heat energy, events that were once by definition anomalous—a catastrophic flood, a megafire, a severe drought—are becoming almost routine. The warming of the planet has also caused the level of the sea to rise about eight inches since the beginning of the twentieth century. Because of the emissions human societies have already sent into the atmosphere, over the next twenty to thirty years—as my school-age niece and half-sister journey into the middle of their lives—the planet's temperature is guaranteed to continue climbing. And unless we choose collectively to prevent it by transforming the way we live—from our economies to our politics to the built environments of our homes and cities—the world will become hotter still, until it is perhaps too unreliable, too dangerous for people to occupy many places that once held thriving communities and histories and cultures.

But not all of this is inevitable, and I am not giving you a book of doom. I don't want to ruminate on all the ways we might be evicted or displaced. This is a book about home. I want to consider how we settle in. I want to think about how we choose to live now, so that we may continue to have safe places in the future.

What happens when the rhythms, the seasons, the known patterns within which we have built our homes, our lives, our towns, our places, go off-kilter? How do we meet our needs? How do we negotiate the weather, the water, and the terrain when things turn hotter, stormier, and more unpredictable?

These aren't hypothetical questions. This is now the dilemma of our time. How do we make a home on this unruly planet?

·ᛝ·

This book dwells in the particulars of personal experiences—it offers a gathering of stories about people who are facing calamities at home and the ways they confront new risks and disruptions, often with courage and insight. It is about the hope and imagination that come from a sense of connectedness to a place and a community.

But in the backdrop of each of these stories is a global crisis known far too prosaically as climate change—the alteration of the fundamental rules, seasons, temperatures, tides, and weather cycles on which we have based our lives. Before I tell you these stories of home, it's worth spending a moment thinking about this planet we all occupy. This may seem like a pairing of opposites—a mismatch in scale. It isn't possible to squint at the whole blue Earth out someone's kitchen window. But that is part of the great failure of perception that has placed us all in this crisis: we haven't recognized how the intimate spaces of our lives and the workings of the planet are tied together.

The connection between home and Earth is inherent in our language. The word *ecology* originates from the Greek word for home, *oikos*, and the suffix *logia*, meaning "the study of something." The modern meaning refers to the scientific discipline that connects us to other species and to our natural surroundings. The word *economy* grows from the same root and *nomos*, meaning "to manage"—the management of home. In common parlance, we discuss ecology and economy as if they are opposites. But an economy should be a means "to organize our relationships in a place, ideally, to take care of the place and each other," according to Movement Generation Justice and Ecology Project, an environmental justice organization. For decades, the environmental justice movement has been vocal about the connections between home and ecosystem and the need to defend home from pollution. But our increasingly globalized economy is becoming ever more disconnected from home, place, and planet—managing our planetary home as if the whole Earth is controllable by humans without limit.

In simple terms, coal, oil, and gas—energy sources formed from ancient fossil carbon—fuel the global economy and power most of our homes. As we have burned that carbon, we have altered the physics and chemistry of the Earth's atmosphere on a grand scale. We have known about this problem for more than three decades, since scientists first began drawing international public attention to the subject of climate change, but we have failed to reorganize ourselves in order to take care of it. As a result, the planet is becoming ever more unruly. *Unruly* has several meanings, including "uncontrollable"—and also "stormy, tempestuous, characterized by severe weather or rough conditions." Climate change is fundamentally a crisis of how we relate to the world around us—it's a crisis of home.

Nearly fifteen years ago, when I began writing about climate change, the threat felt close to the bone to me—as it did for many fellow journalists, scientists, and activists I met who were also following this crisis. In the space of my own home place, I noticed little shifts in, say, the opening of spring flowers, the first ripening of tomatoes in my summer garden, and how short and damp the cross-country ski season in the mountains was becoming. I was also aware of the massive disasters, such as Hurricane Katrina, that were predicted to keep arriving on a more frequent schedule. I understood that all of these were harbingers of a much bigger crisis, which would not just affect the world at large but also disrupt even the most fundamental aspects of life at home.

It is obvious now to most people: climate change has arrived. And to make a home anywhere in this moment is to reckon with a problem that could easily blow our doors down, flood or smoke us out, or erode the ground beneath our feet. Everything that we took for granted is now in question. We have to reexamine how to live.

ßī

This book offers four tales about people who are confronting crisis at home and seeking answers. They represent four faces of twenty-first-century calamity—wildfires in a northwestern community, floods in

Florida, collapsing permafrost in Alaska, and an accident in a refinery town in California. In each, the people who rise up to respond are also deeply committed to their communities. All four tales are set in my home country, the United States. I felt it was important for Americans to recognize themselves in these stories: people in this country have been slow to understand their roles in this crisis and have only just begun to realize the personal risks they face. The stories I offer here parallel crises happening all over the world. They are an invitation to consider how our own lives are changing and what we all need to be doing to prepare and respond.

In form, they are also ancient narratives. Many cultures recount stories in which heroes defend their homes against a threat—from the Greek hero Theseus, who slays the Minotaur so that Athens won't have to keep sacrificing its youth to the monster, to blockbuster movies like *Avatar* or *Guardians of the Galaxy*, in which the protagonists rescue, in the first case, an otherworldly forest, and in the second, the entire universe. Humans have often feared that a shadow might be cast over their homeland and that they would be called to defend what they love.

In real life, climate change is forcing people to cope with unprecedented circumstances at home, and no one can yet claim a clear narrative of either triumph or tragedy. Invariably, the solutions are never individual—they often involve enormous community efforts to plan for threats in the present and future, to reengineer a place, to reshape the choices that are available, to reexamine what matters. Sometimes these efforts require conjuring aspirations that are so beautiful and optimistic they sound almost absurd—until they begin to bear fruit. They require confronting powerful institutions, swimming through a deluge of misinformation, fierce brawling over the smallest matters, stumbling and then standing up again. I have tried to render the moments of mess and misstep, grace and creativity, as honestly as possible.

This book is also a sort of quest—for a new sensibility about home and place in challenging times. Interspersed among the four narratives is a series of chapters that meditate on the nature of home and rootedness. These are structured like essays, and you could think of them as quiet spaces to take a breath—little refuges from the intense winds of calamity where we can ponder more reflective questions. I have long admired deeply rooted souls such as Wendell Berry and bell hooks, people who know their places, have stitched their identities into the fabric of a particular landscape, and can gather wisdom and insight from there. But I have wondered, in a time of climate change, in an increasingly globalized society and economy, is this kind of rootedness still even possible? Can we really find our way back to an old home or enduringly attach ourselves to a new one?

To make or remake a home is also to change your identity. What kind of people will we become on this unruly planet? Who should we strive to be?

What follows is an effort to find out.

PART ONE

CHAPTER 1

—

THE FIRE

home, n. *the place where a person or animal dwells. . . . a collection of dwellings; a village, a town.*

home truth, n. *1. an unpleasant fact that jars the sensibilities. 2. a statement of undisputed fact.*

Susan Prichard grew up in the 1970s and 1980s in a green, placid place—on an island in the estuary that curlicues around Seattle and into Canada. She and her family were avid hikers. So as a kid, she spent a lot of time among trees, along the coast and in the mountains, and she loved especially the old-growth forests—the gnarled and mossy stands of centuries-old trees that inhabited the coastal Pacific Northwest.

But at a young age, she was also haunted by a pair of big-world worries. One was global, about the cold war. *Nuclear holocaust was a really big topic when we were teenagers,* she recalled later. *And then the other one was much more local; clear-cut logging was everywhere.* She worried as she watched the old trees fall.

Slim-framed with straight cedar-colored hair, Susan is levelheaded and also passionate about the things she cares most about. And at the age of thirteen—partly as an act of rebellion against her dad, who suggested once that she might not be cut out for it—she decided to become a scientist, so she could equip herself with hard evidence that would help people understand how to keep this place of forest and water and mountains safe and good and healthy.

Her college years coincided with the peak of the "Timber Wars," when environmentalists lashed themselves to trees to stop clear-cutting. But she found the activists' views strident and sometimes distorted, and devoted herself instead to science—searching for solutions in evidence gathered from the forests themselves.

Then, in graduate school at the University of Washington, in the early to mid-1990s, Susan realized that there were even larger forces of upheaval at work here than the logging crews. She began to immerse herself in what is called *disturbance ecology*—the study of volatility in nature. She learned how to extract buried lake sediments, which could collect and preserve bits of ash and pollen over millennia. Under a microscope, she could sift through the grains of plant matter and particles in order to reconstruct a picture of the forests. And she also studied climate change, poring over the predictions about temperature, rainfall, snow, fire, and tree habitat that climate scientists and ecologists had spun with computer models.

Much of the American public had first become aware of climate change in the 1980s, especially after NASA scientist James Hansen testified on the subject before Congress. (Scientists had predicted a possible crisis in the Earth's atmosphere from carbon dioxide emissions since at least the 1950s.)* By 1992, the concentration of carbon dioxide in the atmosphere had risen to nearly 360 parts per million—just above the level that Hansen and his colleagues would later identify as the crucial threshold for keeping the planet safe for humans, 350 parts per million. The planet had already warmed by less than half a degree Fahrenheit, though some places, like the Arctic, were feeling more heat. (Moreover, even a small uptick in temperature can alter

---

* Swedish scientist Svante Arrhenius is often credited for first proposing, in 1896, that excess carbon dioxide in the atmosphere might warm the Earth. According to the book *All We Can Save*, though Eunice Newton Foote, the first female climate scientist, theorized about the connection between atmosphere and temperature decades earlier, she never received much recognition.

or upend natural systems. Consider the difference between 32 and 33 degrees Fahrenheit: one is ice, and one water.)

But the impacts the world witnessed then were still relatively mild. In that moment, climate change was mostly a problem of the future, and scientists had already predicted, with a great deal of accuracy, what it would mean. In 1990 and 1992, the first reports by the Intergovernmental Panel on Climate Change—the IPCC, the most authoritative international scientific body studying climate change—made it clear that global warming was already happening and that it could lead to drastic changes in the world as we know it. One of the many impacts these reports described was to forests worldwide: some tree species and forest ecosystems would "decline during a [hotter] climate in which they are increasingly more poorly adapted," read the report, and "losses from wildfire will be increasingly extensive." So scientists like Susan and her colleagues were trying to figure out with greater precision how such losses would actually play out: What would happen to the trees and the humans living in areas affected by wildfire?

There had always been, and there would always be, wildfires in the American West. In both the historic and the fossil record were major seasons of fire, sometimes burning millions of acres. Through her lake-sediment detective work, Susan's doctoral dissertation offered a sort of arboreal history: an account of how different tree species advanced and retreated as various blazes burned around a mountain lake. And fires were often ecologically good—wildflowers and vigorous tree seedlings springing up in their wake. Some plants crave fire almost as much as they need water and nutrients. Some trees, like lodgepole pines or the giant sequoias of the Sierra Nevada, even make cones that will only open and drop their seeds after they are heated and broken open by flames.

But as climate change grew into a more significant crisis, the future would bring worse fires than anyone had experienced in living memory. The new age of fire would be amplified by the carbon the world was emitting from burning even more ancient forests—the forests that had rotted and become compressed into coal, oil, and natural gas. Fires would become more massive, more destructive, and more difficult to control—"megafires," as they would later be called.* Susan could envision this happening in a theoretical way—but the idea belonged to a set of abstract trend lines running into the murk of the future. A great distance away, glaciers and sea ice were already shrinking in the Arctic. But in that moment, climate change—the gorgon of fires and hurricanes and droughts—wasn't yet looming over her roof, wasn't yet menacing the lives of anyone Susan knew. It was a thing of decades to come. In that moment, Susan was stalwart and optimistic. She believed that good information would persuade people to take steps to fix the problem. So she set to work studying wildfires with the hope that she could help people prepare.

She couldn't then imagine how soon that hotter, fierier future would arrive.

In her early thirties Susan took a research job working with both the University of Washington and a federally funded laboratory focused specifically on wildfires. Her employers agreed to let her telecommute, and she moved with her wife, Julie, from Seattle to the Methow Valley, a river basin on the east side of the Cascade Mountain Range that enfolds a collection of tiny rural communities. Their new house was nestled against pines, aspens, and cottonwoods just south of the Pasayten Wilderness that stretches northward to Canada—just the sort of forest Susan wanted to study. It was several miles outside Winthrop—a town of a few hundred whose city planning decisions are governed by a "Westernization ordinance" that requires storefronts and signs to appear as if part of a Clint Eastwood movie set, wood siding in rustic frontier colors, for the purposes of attracting tourists.

By this time, Susan was pregnant with the couple's second child, and they wanted to raise their family in a place close to nature. The pair wondered how the conservative community would regard them. *We*

---

* Jerry Williams, the former national director of fire and aviation management for the U.S. Forest Service, coined the term *megafire* and convened a scoping group on such extreme fires in 2003. "We're seeing . . . a new type of fire . . . in the U.S., Russia, Australia, Greece, South Africa. . . . It seems like every year we see a 'worst one.' And the next year we see a worse one yet," Williams said in 2011, in a talk recounted in the book *Megafire* by Michael Kodas. The meaning has been hard to pin down, with some experts insisting the term should refer to the intensity of the fires. The most common definition, according to many sources, is any wildfire larger than one hundred thousand acres.

definitely worried a little bit about being two women and having kids in the valley; that was very uncommon. But when their daughter was born, neighbors brought them casserole dinners and stews for two weeks thereafter. The valley community turned out to be more accepting than dogmatic. People were just used to taking care of one another.

The Methow Valley had witnessed plenty of fires. Many were kept under control and quickly snuffed. But some had been deadly: in 2001, four firefighters died trying to put out a more-than-nine-thousand-acre blaze called the Thirtymile, and many in the valley were still mourning their loss.

In 2006, a heat wave radiated across the United States, sometimes called the "Great Heat Wave," described as "epic" and "epochal" by the *San Francisco Chronicle*, though others would turn out to be grander and deadlier in years to come. Record-breaking temperatures struck California, and sweltering weather caused heat-related deaths from the West Coast to Missouri, all the way to New York City. In late July in the Methow, after the temperature broke a hundred for three straight days, a towering, toadstool-shaped column of smoke rose from the darkly forested mountains above Susan's house. A wildfire had lit the wilderness.

She hoped it could be a beneficial fire, the kind that birthed seedlings and regrowth. And she and her wife collected their son, then age four, and their eighteen-month-old daughter and drove down the road to watch the red glow in the distance, as if it were a performance. *And the fire at first was really exciting. I remember, we just watched it and were just stunned by the smoke plume.* But the power of the fire was also ominous. The smoke plume *resembled the aftermath of a bomb explosion,* Susan wrote later in an article about living with wildfire.

On the mountainside, the blazes leaped into the crowns of lodgepole pines and mountain spruce—each becoming a roaring torch, spraying embers onto the next. Two fires merged to become the Tripod Complex, named for a nearby peak. The Tripod burned east and crossed the divide between two rivers, then fanned out and merged with a second wildfire. The flames tore rapaciously through the trees between the valley and the Canadian border, less than twenty-five miles from where the fire started—through stands of pines already weakened by infestations of beetles—and eventually grew to more than 175,000 acres, one of the largest fires in the state in more than half a century (and bigger in land area than the 150,000-acre Camp Fire that would destroy Paradise, California, in 2018). Finally, the smoke became so suffocating and dense that it drove Susan and her family out of their home; they evacuated to her parents' house on the west side of the mountains. A combination of luck and firefighting confined the fire's physical impact to unpopulated parts of the forest. The Tripod smoldered in the wilderness for the rest of the season. More than $82 million was spent on efforts to contain the fire. In October, snowfall finally quenched the last of it.

A year later, Susan and some of her colleagues in the Forest Service began searching the burned forest for clues to help them interpret what the fire meant for the forest ecosystem. Over the next three summers, she devoted several weeks to wrapping her arms around burned trees in the wilderness, winding a measuring tape around their trunks to record their diameter (sometimes a proxy for tree age), and she developed a fondness for the charcoal scent of charred trees postfire.

Later, in the reports and images of blackened acres that she spread across her desk, the theoretical and the immediate seemed to merge. The Tripod Complex was a megafire, and it had happened next to her home. Climate change was arriving in her valley, and there would be more fires like this.

But Susan was an optimist—she studied images from before and after the fire, captured by satellites, and found therein a story about hope. And her optimism would never really waver, not even after the much larger catastrophes that would come.

🐾

I lived in Seattle for nearly a decade before I understood that it was not just a place of rain—but also surrounded by forests that could go up in flames.

Older locals who had grown up here mostly boasted about the dampness—don't expect a real summer like in other parts of the world, they would tell you. Keep your rain jacket on for Memorial Day. And some scientists and public figures predicted that the Pacific Northwest, its western edges buffered by the Pacific Ocean and shrouded for eight or nine months in rain and cloud, wouldn't feel climate change as quickly as other parts of the country. This region would be a refuge, they said, and millions would come here to escape the hotter conditions elsewhere.

But the Pacific Northwest is cleaved into two halves by the Cascades, the jagged volcanic peaks that slice north to south through the region. To the west is the rainy, green coast, and to the east, the rain shadow, the arid zone that forms on the leeward side of mountains—making the region where Susan Prichard lives a not-so-mild place of searing, dry summers and deeply snowy winters. The two halves of the Northwest are knit together by the forests that stretch from one set of mountain flanks onto the other, transitioning across the miles from stands of dark-barked Douglas firs and redolent cedars into rust-colored ponderosa pines, lodgepole pines, and larches. Large fires are far more common on the eastern side, but all of these forests can and sometimes will burn.

In recent years, Pacific Northwest summers have often arrived early and sprawled languidly across the long days of June well into September. A too-hot summer sun seems sometimes to glare down on the mountain peaks—which are no longer capped with as much snow as in years past. The Seattle of the past rarely sweltered, but now hot weather is becoming more routine. And on the eastern, rural side, the heat waves grow ever longer, more extreme, and more savage.

In July 2014, when the high temperatures rose into the mere 80s and low 90s Fahrenheit—which used to be considered sweltering for Seattle—several days in a row the *Stranger*, the city's famously snarky alternative newspaper, named it "HOTPOCALYPSE 2014" and offered a "survival guide," including a list of the rare establishments with air-conditioning. But though parts of the region burned fiercely that summer, the characteristic dampness of the western wet side spared it from major fire, and wind patterns kept most of the smoke from eastern wildfires out of the city.

The following summer, however, was not as lucky for Western Washington. The smoke arrived abruptly, blowing south from wildfires in Canada. Fourth of July weekend, 2015, as I boarded a ferry and crossed between two of the San Juan Islands off the coast, I stared into a mustard-colored, acrid sky. The air clung oppressively to my skin like an itchy blanket, so that by the end of a daylong excursion riding the seaside hills on my bicycle, I felt like I had a fever, like my body itself was ablaze. That year, the *Seattle Times* posted a series of images of smoky sunsets—the sun an orange bulb behind a curtain of ash.

When the smoke came again in late summer 2017, I felt with a stomach-clenched certainty that this was a sign of climate change—the crisis I had written about for years—showing up to leer at me from above. The sky dropped ash, which clung to garden spiderwebs and the leaves of plants. It clotted onto the surface of my fresh cup of tea. The whole experience felt eerily like being cupped inside a terrarium, like there was no way to step outside and breathe freely. "The fires' impact—the claustrophobia, the tension, the suffocating, ugly air—feels like a preview (and a mild one) of what's to come if we don't take immediate and drastic steps to halt and mitigate climate change," wrote Seattle-based columnist and author Lindy West.

July and August 2018 were even more oppressive. Rivers of smoke gushed from both the interior of British Columbia and the center of Washington state across hundreds of miles to merge in the skies above my house. Seattle's air quality was suddenly among the worst in the world, worse than smog-choked Beijing's, and drifted into the U.S. Environmental Protection Agency's "very unhealthy" range, vicious enough to threaten the well-being of anyone who was breathing. The city skyline appeared as a smudged pencil etching through the ash. When I walked the streets, I wore an N95 respirator mask purchased from the hardware store—similar to the kind that healthcare workers would later use as protection from the spread of coronavirus. But my throat stung anyway.

A year later in August, I drove over the mountains and visited the North Cascades Smokejumper Base in the Methow Valley, run by the Forest Service. The valley is also the "birthplace of smokejumping," which is something like the firefighters' equivalent of the Army Rangers, an elite style of firefighting that involves parachuting into the wilderness. The campus—which included a runway; a brown house that functioned as an office; a metal-sided warehouse; and what looked like a barn but was labeled the "Lufkin Parachute Loft"—lay halfway down the valley between Winthrop and Twisp, an unpretentious town with a family-run lumber store on one end and on the other, the old Idle-A-While Motel, in a set of cabins originally built by the U.S. Forest Service. In the summer and early fall, the base offered free tours to any curious person who might stop in, and I had long wanted to get a better handle on the reality of fire.

A fire called the Williams Flats was burning about seventy-five miles to the southeast, and there had been seven thousand lightning strikes in Washington and Oregon that weekend, though any flames they'd ignited hadn't yet grown into big fires in this part of the region. The firefighter who led me through the base had his eye on some more thunderheads brewing above us. He was wiry, with sturdy arms like tree limbs, in a red T-shirt, talking in a low voice, monitoring my face to make sure I was catching on.

"Storm chasers," he said, referring to himself and his colleagues. "We're going to monitor anywhere there's lightning." He pointed to the sky. It was full of scraps of cloud, like torn fabric, blue gaping between them. "Look right here. See the little bumps, the swirls? We're right on the edge of that thunder cell." He pointed again, drawing my attention to pebbly and divoted surfaces in the matrix of clouds. "If you look just under there, where the blue meets that wisp, that's part of that cell that's trying to push up. You can see how it's kind of cauliflower-looking." There were generally only two proximate causes of fire: lightning and humans.

He led me to a small but muscular airplane, white with a maroon stripe, a military craft built in the 1980s. "She's a savage," he said, giving the plane an admiring look. "It's not uncommon to see this in Afghanistan flying for our troops." But this aircraft's intended purpose here was to allow up to eight people to parachute to a location near the edge of a fire, hike in, and try to get the flames under control. It was a hazardous job, more so in this region of the world, where the weather could swing suddenly. The topography is also tricky to navigate and could trap a firefighter in dangerous, even lethal conditions. "In the fire world, there's a lot of us that call this R5 times two, as kind of a slang," explained the firefighter to me. The Forest Service had defined nine regions, and

California, a place of extreme wildfires, was number five or "R5." But this part of eastern Washington had similarities to that more southerly location. "A lot of guys will not come up here. It's a very extreme place."

The parallels to the military were not coincidental. The U.S. Forest Service had battled against wildfires from the early twentieth century until now, in order to allow people like me to live in a tamer West. Beginning in 1921, U.S. Army planes were used to patrol for wildfires. The history of smokejumping was closely entwined with World War II paratrooping, and the only all-Black airborne unit in military history had been tasked first with responding to the threat of Japanese "fire balloon bombs" and then with fighting actual wildfires via smokejumping. By 1935, the Forest Service's official policy was that every wildfire should be extinguished by 10:00 A.M. on the day after it was first reported. (The policy was reconsidered and changed in 1978. But in the last decade, the Forest Service has invested more than a trillion dollars every year in fire suppression and still describes itself as "the world's premier firefighting agency.") Engineers eventually developed specialized firefighting airplanes—such as air tankers, which can drop flame retardant on a blaze; water scoopers, which can pull freshwater from a nearby water body and drop it on a fire; and smokejumper aircraft like the one I was now looking at. But the Forest Service still sometimes uses military planes during the height of fire season.

Places like the North Cascades Smokejumper Base had waged a war against wildfire for decades so that western communities could be safe. Home here had developed almost an adversarial quality—the presumption was that you had to fight fire to live here.

But now climate change had arrived, and fire couldn't be held back any longer. The flames would come again and again, and in some battles with even the best-trained firefighters, the fire would win.

🔥

In the years after the Tripod Complex Fire, Susan Prichard became more vocal about both climate change and wildfires.

In 2006, in response to a community member who had expressed denial, she wrote a defiant though polite letter to the editor of the local paper explaining the science of climate change. A few days later, when she stopped to pick up some cuts of meat from a local lamb farm, the farmer—a tall man in suspenders, quizzical—asked her to explain further. *I didn't feel like he necessarily completely believed me*, she thought later. *I knew he was probably pretty conservative.* But the conversation was respectful, and she began to have more of them, tentative but factual and levelheaded.

She, Julie, and the kids moved closer to the center of Winthrop—to a barn-red house surrounded by an ample lawn of green grass and beyond it, patches of wild grasses and bitterbrush, a spiky shrub with flowers like tiny wild yellow roses, and a scattering of ponderosa pines.

From her new home office, Susan continued to study the behavior of the Tripod Fire. Parts of the forest that had burned in 2006 were old-growth. Parts hadn't seen another fire in decades. But some others had previously been clear-cut, some just cleared of small trees and brush and then selectively logged, and some charred by past wildfires. And some had been deliberately burned in a management practice known as "prescribed fire," in which low ground fires are lit, usually during mild weather, in order to consume "fuel"—the brush, grass, little trees, and understory that might feed a hotter wildfire. She searched for answers in this landscape: What could the Tripod and the experiences of this forest tell us about the more severe fires that would come in the years ahead?

In the distant past, before European Americans inhabited the area, there had been much more fire in this place, which had generally been a good thing, ecologically speaking. She knew this partly from her studies of ancient pollen. Those past fires had been on average smaller, more frequent, and less severe than the Tripod. They would have turned the landscape into a collage or a mosaic of little patches—different burn ages, different mixes of trees, diverse habitats for many kinds of animals and plants, including those that preferred deep tree cover and those that sought open spaces. Each of these little fires would also have consumed some fuel and helped keep the next wildfire from spreading as quickly or growing as hot. But Susan's sense of how and why past fires had happened also changed over time. She used to think a few were set by humans, but most were the product of nature and randomness—which is to say, they were ignited by lightning.

Then came a paradigm shift in how Susan and many other scholars understood the history and ecology of North America. When early anthropologists and historians tried to estimate the population of the continent at the time that Christopher Columbus arrived, they had often turned to incomplete colonial census data gathered much later by Europeans, and they had failed to account for the scale of the devastation that occurred when Old World diseases, combined with colonial violence, wreaked havoc on Indigenous communities. These omissions produced lowball calculations of the pre-Columbian Indigenous population. Some evidence of this distortion lay in plain sight—in Indigenous oral histories and records that had never been given their due by most academics. But the dubious assumption that Indigenous societies were tiny and American landscapes largely untouched has lingered in many fields, including ecology. In the 1990s, the scientists who became Susan's mentors were still sometimes leaning on these assumptions to try to understand the historical forces that had shaped the Pacific Northwest.

But over at least the past couple of decades, some historians, geographers, Indigenous scholars, and other researchers have called for a reckoning with evidence that had been ignored. Before Europeans arrived in North America, Indigenous communities were not "living with minimal impact on unspoiled nature," wrote Indigenous scientists Robin Wall Kimmerer and Frank Kanawha Lake in a 2001 paper, but strongly influencing North American landscapes. Advances in fields like molecular anthropology, which uses genetics to answer historical questions, allowed for new reassessments of the past. In 2011, for instance, researchers from the University of Washington and Yale analyzed ancient and modern DNA and surmised that the number of people living in the Americas had dropped by more than half around the time European diseases were introduced.

If you revise the pre-Columbian population upward to account for such a catastrophic loss and you correct outdated and biased assumptions about Indigenous knowledge and ecological expertise, the whole continent becomes a radically different place, managed by large and influential societies long before Europeans arrived. This means that North American landscapes were never just wild, not just the design of nature; for many generations, they were managed deliberately and skillfully by Indigenous people.

Susan's next-door neighbor in Winthrop is a historian and an expert on legal issues affecting North American tribes. He had learned via oral histories that tribes such as the Wenatchi and the Methow—both now part of the twelve Confederated Tribes of the Colville, a 1.4-million-acre reservation about thirty miles east of the Methow Valley—had used prescribed burning, also sometimes called "Indigenous burning" or, more recently, "good fire," every year to increase habitat for wildlife and berries. Later, a California ecologist named Will Harling reached out to Susan with even stronger views on how profound a role Indigenous communities have in managing both ecosystems and fire. Good fire is part of a process of Indigenous "world renewal," he explained. "It's humans' responsibility to manage fire on our landscapes. It's how we manage for future generations. We don't leave them a fuel-choked tinderbox. We leave them a landscape where there's diversity and abundance, and there's safety and security."* In other words, Indigenous communities had given the West a legacy and a strategy for how to care for the land and keep communities safe—a set of traditions and knowledge base that scientists and government land management agencies had largely neglected or pushed aside.

The practice of deliberate fire-setting had endured in some places over the decades. In southern states and in many tribal reservation communities, including at the Colville Reservation, the tradition of setting controlled fires to manage land never went away. In the 1950s, the famous wildlife biologist and policy advisor A. Starker Leopold began making the case that fire could have a healthy role in ecosystems; his work led directly to the creation of some of the first federally run prescribed fire programs in the national parks in the 1960s. But government agencies have budget constraints, tribes only have authority over their own landholdings, and both often have to contend with thorny legal hurdles. In the West, efforts to bring back good fire have been tiny compared to the vastness of the landscape and the fires of the past.

---

* In the words of Kimmerer and Lake, "Indigenous practice and philosophy offer us an alternative view of the 'natural' fire regime, in which humans regain their role as 'keepers of the fire' and the symbiotic relationship between humans, forests, and fire is reestablished for mutual benefit."

The question remained, could these kinds of practices still be relevant in an era of megafires and climate change? Susan and her colleagues ran an analysis based on field research, the count and measurement of trees they'd done in the area that had burned in the Tripod Fire. Then she and another scientist used satellite data to search for patterns over an even larger area of forest. What factors determined whether trees lived or died during the severe conditions of the Tripod Complex? They saw a pattern emerge—the Tripod had spared more trees in the places that had experienced prescribed fire in the recent past.

Despite the threat of climate change, Susan hung on to a vision of the future in which her valley, her home, could return to the kind of collage-forest that had existed in the past, a landscape that would be better able to tolerate the heat of the future. But old policies of fire management were slow to change, and the new, warm summers didn't leave the West much lead time for such preparations.

☙

In mid-July 2014, as HOTPOCALYPSE came to a close in Seattle, the heat on the east side of the mountains rose into the upper 90s Fahrenheit, then into the triple digits.

On Monday, July 14, lightning lit four fires around the valley near Twisp, Winthrop, and another tiny community, called Carlton. As with the Tripod, Susan and her family wanted to see it for themselves. So they drove to a spot south of Winthrop where it was possible to look across the valley and watch the fire glow in the distance. Susan found the flames to be mesmerizing at night, almost like an astral phenomenon, the aurora borealis or an eclipse. A couple days later, she left for Seattle for a set of professional meetings.

By Thursday, the day before her birthday, Susan heard that the fires had grown. Eighteen thousand acres, not yet a megafire, but menacing in the torrid winds that were then heaving across the valley. Collectively, these fires would be called the Carlton Complex.

She was determined to head home over the mountains to her family that night and arranged a ride with a lawyer friend who also lived in Winthrop. She and the friend made a stop at a hardware store so he could pick up a generator and cram it in the backseat of his Volkswagen Golf. Susan's belongings were a tight squeeze around the four-foot-wide appliance, and there was no room in the little sedan for a second generator for Susan's house. *Damn it*, she thought. *We're probably going to lose power.* Her house, like many in the valley, relied on well water supplied by electric pumps, and when the power went out, so did water for drinking and for sprinkling the yard (which could actually provide a buffer against a traveling wildfire). As Susan and the friend set out on the interstate, Julie called to say that the power was already down at their house.

They drove north to Highway 20, then headed east—up through North Cascades National Park—past one of the hydropower dams that feed Seattle's power supply, winding up through about eight hundred square miles of wilderness. One of those impossible roads that swoop across a valley and sidle up to cliff edges, curving and arcing up and up, with gravel pull-offs for gawking at jagged granite mountain faces, wide green lakes, churning rivers, and plunging waterfalls. The road offered an almost mythic transition between one side of the mountains and the other, from city up into sky then back to home.

It wasn't until they exited the wilderness and hit the valley floor that they could pick up static-fuzzed radio and the first cell phone signals. The news was worse than Susan had imagined. *We were getting radio reports of major evacuations of the communities of Carlton and Pateros,* about 15 and 30 miles south of Winthrop, respectively—and *no one had ever predicted* the fire would spread so far.

The day had reached dusk, and as the road snaked back down, they entered the valley full of pines and finally saw it. *It was just horrifying to look down-valley and see just this wall of black smoke. And then at the base of it was this glowing red.*

Susan began to cry. *It just hit me that I had no idea how all those people living in the backwoods were going to survive it. When I saw the smoke plume—and the winds were easily over thirty-five miles per hour—I just assumed not everyone would get out.*

Unmistakably, the Carlton Complex had grown into a giant, a megafire. And unlike the Tripod, this one was coming for people's homes.

Susan couldn't sleep that night. The official government-run website for tracking wildfires wasn't updated yet, and the only reports she could find were anecdotal, on Facebook. In the wee hours, she left the house and drove the country roads, *a safe driving tour that would not be in the way of any emergency vehicles, just to look at the fire. The initial front of the fire had already passed. There was this huge column of smoke in the distance. But then the flanks of the fires were still burning. Individual trees were torching. Bushes were burning.*

On Friday, the cell phone lines went down and would be out for more than a week. Susan and her family drove to the post office, which had suddenly become a primitive form of social media, sprouting a series of paper notices about whose houses had burned down.

That evening, they went to the house of some friends for dinner. It was one of their regular gathering places: they were all part of a group that held rotating soup nights through the winter. The friend ran her house on solar panels, and therefore still had electricity. At the gathering, they swapped stories, shreds of information, and rumors people had heard about what was happening, what had burned, who was safe, and who was not.

Then they sang to Susan. *People kind of ruefully lit the candles on my birthday cake. They're like, "Yeah, a little bit more fire for you. How are you feeling about this, fire ecologist?"*

❧

The battle against the Carlton Complex fires began as a routine operation.

Carlene Anders was a volunteer firefighter who ran a daycare more than thirty-five miles down the valley from Susan and Julie in the town of Pateros on the Columbia River. She had deep roots here. Her grandfather had been the Pateros mayor in 1952, and her mother had graduated high school in a neighboring town. Carlene had served for seventeen years as the ski school director at a nearby mountain pass called Loup Loup (French for "wolf, wolf"). She had noticed how feeble some winters were becoming—across the West, the ski season has shortened by about a month since the 1980s—and she found it heartbreaking.

She also knew fires—by firsthand experience. She became a wildland firefighter for the state in 1984, and two years later, at the age of twenty, had served at the North Cascades base as one of the first two women smokejumpers in the state. Early in her career, she had also worked on some prescribed burns for the Forest Service. By 2014, she had spent seventeen seasons fighting fire in one context or another. *Fire is alive*, she felt. *It's a live thing, and it takes its own shape. And you have to have seen enough of it to understand what it's going to do.*

Carlene had seen big fires before, including a devastating cluster of fires in Oregon in 1987. But when the Carlton Complex arrived in July 2014, it was the first time Carlene witnessed fire conditions that were so thoroughly dangerous and intractable.

When the first lightning strikes hit the valley on that Monday, she left the kids at the daycare under the supervision of her staff and rode out with Engine 1511, Pateros's yellow fire truck, crisscrossing the area to respond to different reports of fire starts. Four separate fires lit in different parts of the valley. They were still small, but they could not be quelled on that dry, hot day.

On Tuesday, Carlene and her crew drove up-valley. The largest of the four fires, called the Stokes, stood at about six hundred acres that day. Seven rural households evacuated as the Stokes raced across the grass, and more than a hundred firefighters tried to rein in the blaze, including a Canadian crew manning an air tanker.

On Wednesday, the torrid winds picked up, and the Stokes fire grew from about 1,700 to 7,000 acres, and there was no time to rest. *We had a meal on Wednesday at four o'clock, and I think I laid down for twenty minutes right there on a trampoline while we were protecting somebody's house.*

But it wasn't until Thursday that Carlene fully realized how different this fire was from anything she had seen in her almost three decades of experience. That day, gale-force winds surged through the valley. The fire spread monstrously—from 18,000 acres in the morning to 45,000 in the afternoon to 168,000 by the end of the day. Embers flew: not the moth-size bits of campfire ash you would normally associate with that word but chunks of burning debris bigger than a person's forearm, up to two feet long and suspended in the air. When they landed, they lit more fire, a phenomenon called "spotting."

Normally you have a head of a fire and you can flank it and pinch it off and actually fight it and have control of it. But this, there was no control. We would hit a house. We would think it was taken care of and then come back—it's on fire again.

The number of firefighters battling the Carlton Complex grew to about five hundred. But it wouldn't be enough.

That afternoon, Carlene fought fire on one side of the Methow River. Across the valley on an opposite hill stood the house where she had spent her adolescence, from eighth grade onward. The whole family had built it together; they'd spent twenty-three Christmases there. Her mother now lived there and rented out a cabin on the same property. It was just beyond Carlene's line of sight. Then she watched the wind tilt a column of smoke toward her family's land and saw fire dashing up the hill and over the ridge. Fire is propelled by wind, running in the direction of its own smoke plume, like an animal chasing its tail. *I knew when that column up there laid down that there was a good chance that my family home was getting burned down. And then it was going to reach my mom.*

Time seemed to distort for a moment. *There was a cabin right by the water. I just remember the helicopter dropping water on that cabin and just everything was slow motion, like the rotors. I remember doing this 360 around as I was trying to move this hose off the highway and thinking, we are not going to contain this. We're not going to be able to handle this.*

She could have left right then, she thought. If she'd had an oxygen tank ready, so she could breathe through the smoke, maybe she could have rushed to her mom's property and most likely saved at least that much—the house, the cabin, her family's history. But firefighters have a sort of creed, like a military ethic, not to leave their crew. So she stayed with the engine.

Then the fire lieutenant got a call that the fire had charged toward Pateros and was endangering the city water supply. Carlene and her crew piled into the engine and took off down the serpentine highway that followed the Methow River. For a moment, Carlene's cell phone had reception, and she was able to reach her husband. They made a plan to meet at the Pateros fire hall—Carlene, her husband, her son, and her mother.

But when her engine arrived that evening, still before dark, none of her family could be found, her cellphone battery had died, and the whole center of her hometown was ablaze. Fire had spotted over the crest of a hill and spread to the grass around water towers, the cemetery, the parsonage on the hill. The entire landscape seemed to glow and roar eerily, as if a volcanic eruption were spreading across the ground. Fire on the edges of the highway, fire along the railroad tracks. Carlene was with a fire engine crew of three people, and, at first, there was only a handful of other firefighters defending the downtown—including a second engine crew and someone manning a tender, a fire truck equipped with a water tank, and a couple of additional privately contracted firefighting rigs. She saw her daughter's car parked at the fire hall. Her kid had just turned eighteen years old, had signed up to fight fire, and was probably also out battling the Carlton Complex somewhere.

There was no time for Carlene to contemplate whether her family was okay. The assistant fire chief and the firefighters made a quick decision to defend the center of town. They didn't want the whole community to wither. *If we didn't keep downtown, nobody would come back.* First, they had to evacuate the residents. A growing line of cars was already trundling slowly onto the highway, but *there were people everywhere still.* The fire trucks drove the streets, looping through the center of town and up the hillside, shouting over the engine PA, telling people to evacuate. *This isn't a drill. Please leave town now,* Carlene repeated. The summer sun hadn't set, but the smoke grew so impenetrable and dark that the streetlights came on.

The district fire chief had put out a call for help to any firefighting crews in the state, and a series of engines from Chelan, a town about fifteen miles to the south, showed up at the edge of Pateros. The crews doused the flaming yard of the hardware store, the shrubs and trees at the edges of the streets, the strip of grass between the railroad tracks. Carlene drove the yellow truck as they advanced toward the Methodist church. Her companions sat on top of the engine and tried to blast the adjacent foliage. The whoosh and gasp of the fire was so loud that none of them could hear, so the other crew members would thump the roof of the engine to signal to Carlene, *one tap to stop and two to go. It was so hot that they could barely handle it. I mean the fire was blowing right into them. And the transformers were exploding. So you'd hear these huge pops. And then all the propane tanks were releasing.*

The crews blasted the town with so much water that they drained the water towers by 10:30 P.M. Some of the firefighters drove to the neighboring town of Brewster to fetch more water, but they were drawn into a battle against flames to the north and never returned. The remaining crews pumped water from the Columbia River. At nearly 11:00 P.M., Carlene got a call from her mother and teenage son from an evacuation center in Chelan; they were safe. But she would not see them again until Saturday, when she would also sleep briefly, her first chance to rest in four days.

The hours stretched out into an endless moment, across day and night, outside the normal movement of time. Everything Carlene had was on the line: *I knew we lost mom's home, and my grandma's home in Brewster was threatened, and our house was threatened here, and I thought, Oh my gosh, our whole family's gonna lose all their homes at some point.* By luck, her own house and her late grandmother's were spared. But her mother lost her home and her rental cabin; moreover, she had owned an orchard down the valley, and there, two houses, eight fruit pickers' cabins, a bathhouse, and a cookhouse were all consumed by fire.

Still, the battle wasn't over. The firefighters pursued the fire south of town. *The fire moved, jumped this ridge, went down and jumped the highway all the way down to the river. We lost homes out there.*

That weekend the winds changed direction, pushing the fire toward Twisp, Winthrop, and Chelan. The governor sent a hundred members of the National Guard to assist.

The following Tuesday—a week and a day after the Carlton Complex began—President Barack Obama arrived in Seattle, a place that felt farther away from this valley than usual, for a pair of campaign fundraisers. He agreed to send emergency federal aid to the burning communities. By then, an estimated 150 or more homes had been destroyed in a region of just forty thousand residents. (After all the damage had been assessed, the number of razed homes was more than 300.)

The emergency dragged on inexorably through the summer. *We were going back out almost every night for fires. For weeks and weeks, we were out at night till two, three in the morning. A lot of times, we'd get calls that people just were scared to death because there was outlying fire that was going the whole time.* On Wednesday, July 23, about ten days in, rain fell on the hillsides and dampened the Carlton Complex Fire but also triggered some mudslides.

In mid-August, another fire lit—directly across the river from the smokejumper base—when a car was pulling a trailer with a broken wheel along the highway, and the friction of metal against pavement sent sparks into dry grass. A firefighting crew dashed across the river to defend a house on a hill above the valley and prevent a second Carlton Complex–size fire from cutting loose. That fire was under control within four days. But several other residences and outbuildings were razed. (One of these belonged to one of Susan Prichard's soup-night friends: a house, two cats, eighteen chickens she kept, an art studio, and a vegetable garden were all gone.)

At the end of August, rains arrived again, triggering even worse mudslides that trapped cars on the roads, knocked a house off its foundation, covered a highway with a five-foot-thick layer of debris, and swept a firefighter off the road, trapping him in his truck, though he was able to escape.

On August 24, the Carlton Complex Fire was finally declared fully contained. It had scorched 256,000 acres, or 400 square miles, a footprint larger than the five boroughs of New York City. It was the largest fire on record in the state.

In Pateros, the water towers were scorched. Several children in every grade were houseless, along with about one-fifth of the school district staff and one-third of the fire department. Many months would pass before the kids in town would stop "playing fire" (like playing house, except your house is burning down and you have to pack up and evacuate).

Afterward, Carlene and her neighbors looked at the burned-up town and felt that they could never let this happen here again. But another year would pass before Carlene realized she had crossed over into a new sort of reality. They would all meet this sort of fire again, and they would all have to learn how to protect themselves, how to recover, and how to help one another through.

🐦

So many stories about disaster close the curtain before you see what happens afterward. Especially in this era of catastrophe, we can always distract ourselves with the next fire, the next flood, the next tragedy—ride the crest of the drama without asking what happens in the years after a place burns. But it seemed especially important to me to understand what makes it possible for people to recover in this era of more common disasters. I wanted to know what had become of the Methow communities now that the smoke was gone. So I made repeated trips there.

On the day that I visited the town of Pateros, five years after the Carlton Complex, two museums were open. The first occupies a part of city hall, a building that formerly served as a combination jail, court, police department, and fire station. In rows of dusty glass cases lay an array of various artifacts and mementos, including a series of black-and-white photographs that were like a love letter to the commercial apple growing that began here around the turn of the twentieth century—a cider-making party, well-known orchardists of the era, the irrigation pipeline bridge that brought water to the fruit trees. Mounted to a set of wooden panels was an outdated-looking painting of Paleolithic people with wooden clubs pursuing a saber-toothed cat. But I was really here for a meeting with Carlene Anders, who had become the mayor of Pateros, and when she finally found me, studying a sign about the region's first inhabitants next to an array of what looked like cowhides, she said I had come to the wrong place. It was the second museum that told the story she wanted me to hear, and she led me out the door into the sun-drenched afternoon.

She wore a loose-fitting purple tunic over a pair of jeans, and her hair, the dark-yellow color of end-of-season grass, was pulled into a ponytail. She had robust hands, the kind you might expect to find building houses or kneading dough, and the well-defined lines in her forehead gave her a perpetual expression of earnestness. As she marched me through the tiny downtown riverfront, she recited a list of facts and figures about the town without pause. Behind us to the east, the Columbia River shone wide and pale blue, the same river Woody Guthrie exhorted to "roll on" in his famous anthem to hydropower, and there was a dam about seven miles downstream. We began to walk along the path that the Carlton Complex Fire had traveled. As I squinted into the dazzling sky to the northwest, Carlene pointed to a pair of round, squat cylinders with colored tiles decorating the sides of a golden bluff that rose above Pateros. "Those are our water towers that got burned up there." Both had been repaired. But in the fire's aftermath, manganese, a heavy metal, had rushed into the water supply and filled the pipes, a common trouble after wildfires loosen earth and liberate certain minerals from the soil. The problem had never really gone away. "Over the last three years we've been working on replacing the water system here for the city," she explained. "It's a $7.6 million project."

Immediately in front of us lay the bare ground of a construction site and a partly assembled concrete wall. The city was building a new well here after its previous ones failed—and was incorporating a stage and an interpretive center into the design of the pump house.

I couldn't observe any evidence of the wildfire. But Carlene gestured to the many things that had burned down and were now gone, as if conjuring ghosts. "It was all on fire. You used to have trees and all kinds of stuff between the railroad tracks and the highway right there."

We walked through a parking lot and followed a sidewalk, made a stop at her office to collect some keys, then crossed the street to a retail building with a red metal roof and dark windows. It was only temporarily a museum. She and her husband owned the building, she told me, and were planning eventually to open a restaurant there called Fire and Ice, in recognition of Carlene's history as a firefighter and ski instructor. The name was also, as her mom would point out, an inadvertent Robert Frost allusion ("Some say the world will end in fire, / Some say in ice"). When she opened the glass door and switched on the lights, we were facing a large square mirror, mounted in a white wooden frame and propped against a cloth-draped stool. Carlene grinned sheepishly as I took a photograph of our reflection. Hand-painted across the mirror were the words WELCOME TO THE SMOKE AND REFLECTIONS EXHIBIT.

Stepping farther into the room, we encountered a series of room dividers covered in black cloth with displays of images mounted on them, then a table set with an array of burned and warped metal and glass. Some of the items laid out here were recognizable. A glass bottle with a curved and distended neck. A shovel end with no handle. But some had liquefied and re-formed into the sort of bizarre shapes candle wax can make when it drips. "Everything melted, all the radiators, all the cars; all the wires melted in place," Carlene said. Then quietly, "This was my mom's stuff."

Above the table hung a photograph of the metal frame that once sat beneath a modular home, warped and sunken and covered in bits of ash. "This was my mom's." It had stood on the orchard property. "It was her retirement plan. She had rentals. She lost all of our homes except for one." The house that remained, also at the orchard, had been equipped with woodpecker-resistant cement siding, which had also turned out to be fire-resistant. The thirty-acre orchard that her mother owned had been uninsured. "She had gotten mad," Carlene explained. "A year and a half before, she had had welding equipment stolen, and the insurance company wouldn't pay for the welding equipment. So she canceled her insurance. She lost the shed, the tractors, the eight picker sheds, the kitchen, everything, lost it all." There were other similar stories throughout the community. Three-fourths of the people living in Pateros and surrounds had been uninsured or underinsured at the time of the fires. Some people had believed they were covered only to discover loopholes and exemptions in their policies.

The entire exhibit had a handmade feeling, laminated photographs pinned on black fabric. An image of a brick chimney still standing while the rest of the house it had belonged to was nothing but ash. An ATV so warped it looked like folded cloth. Some images were donated by community members, including a local photographer. Some were Carlene's. A picture of a young woman and an older man clearing a yard full of ash beside a concrete wall. "This is my daughter. This is the house that we built when I was young. This is my husband," she said and gestured to the image. It occurred to me that she had been reciting these same details to people for years.

"Is it hard to keep telling this story?" I asked.

"It depends. People told me that you had to tell the story eight to twelve times before you start to lose that emotional piece of it, and so telling a story probably helps."

Plus there were reasons to keep reminding people. As she and I walked through the exhibit together, she worried aloud about the complacency that can set in even after a crisis. "The problem is, five years down the road, are we still going to remember?" she reflected. "And it's going to get worse. There's no way it's not going to get worse. So we better be prepared, better do as much as we can while we can."

"I'm scared to death for the other side of the mountains." She looked at me meaningfully. "The earthquake when it comes will be tough on everybody too, be lots of fire then." The west side of the mountains, my side, had active and dangerous tectonic fault lines, and while the forests there were damper, they could certainly burn if warmer weather dried them out enough.

"Where do you live? Just saying," she said.

I mumbled something about my house in Seattle, realizing that I had no particularly solid plan for any of the situations she was alluding to.

🐌

How do you write a story about this era of disaster that doesn't end in tragedy? How do you make a life and a community in a recurring set of crises and still offer any kind of stability or safety? I sensed it would require some combination of preparedness and pragmatism—the kind I had seen from Carlene Anders—and imagination, the variety Susan seemed to cultivate in her work.

So on a different day, an autumn morning of overcast and diffuse light, I went to visit Susan up-valley from Pateros. I wanted a dose of her optimism. And she volunteered to take me in her silver pickup truck to see the scars of the Rising Eagle Fire, the one that had burned through her friend's neighborhood.

As we turned off the highway, she pointed to the weedy roadside. The couple whose broken-down trailer had sparked the fire had later been sued by nine insurance companies that were trying to recoup money from property loss. The claims were ultimately dismissed. "It was the most ridiculous thing because, for one, the state had obviously not maintained the weeds along the highway," she said. "And you could blame all of us for this exceedingly dry summer. We're all complicit in climate change." (Arguably some, I thought, like the CEOs of top carbon-emitting companies, carried far more blame than an elderly couple with a rickety trailer, but that was a subject for another time.)

We drove into the Rising Eagle neighborhood and trudged around a bare lot where a house had once been. By then, the ash and debris and mess had been cleaned up so thoroughly—no more walls, no old windows, no scarred pipes or other debris from human habitation—that it seemed as if the house had been sucked straight into the heavens. But a band of gravel remained and some concrete blocks, probably marking the edges of what used to be landscaping, along with a wilted lilac bush. Beside these, in the sparse shadows of charred pine trees, the land itself was reclaiming the space. Thick and bushy willows, serviceberry bushes, and buckbrush (often called by its Latin name, *Ceanothus*) had sprouted, their branches gnawed short by local mule deer. Across the ground lay the feathery leaves of yarrow, a wildflower and old herbal treatment for toothaches, and the golden bunches of a native wheatgrass. Aspen seedlings lifted delicate branches, their pendulous leaves winking in the breeze. "They do such an amazing job resprouting after fire," mused Susan.

"Everything's so dry," I said.

"I know, everything's pretty dry," she echoed a little wistfully. "This summer, I felt like we were going to get a break because we had such a late spring and a lot of snowpack. But one of the things that people have been talking about is that now it's not just about an early snow melt that kind of makes for a long fire season. It's also these kind of punctuated hot, dry periods. We just had an exceptionally dry and hot series of days."

We drove up the hill, where an abandoned plastic newspaper delivery box still hung from a tree. A local wildlife biologist had converted it into a birdhouse by mounting a wooden board to the front with a round hole drilled in the center.

Then we headed to the crest, where two brand-new houses stood, western and modern with siding roughly the same color as pine bark. Their wide, shining windows gazed out at the river and the leather-colored hills—and also at the burned snags of dead ponderosa pines jabbing into the sky.

The whole tableau before us fascinated Susan, she said. Perhaps this was the new aesthetic, the beauty of fire, an implied acceptance that one lived in a volatile sort of world. "These snags, you know, are going to get more beautiful over time," she remarked, "because they'll become silver and drop their fine branches. So there is a certain beauty to them if you can embrace it."

Moreover, the burn scar would protect the residents. It was the same phenomenon Susan had studied after the Tripod Fire. Anything in the path of an old fire was less likely to burn again for a while.

But not everyone could afford to construct a fancy home like this. Not everyone could rebuild or recoup quickly. Not everyone could easily move past trauma.

"I came out the other side of these fires still feeling like one of the primary answers for our community is to embrace fire, not try to put a wall up and avoid all fires in the future. And that's such a difficult message," she told me. But the lessons of ecology told her that home and ecosystem would always be intertwined. Fire was as much part of this place as the houses and towns, the trees and sagebrush. To make a home and persevere in this warmer and unrulier era, people would need to make more space for flames.

But to do this would require unflinching realism and tremendous effort and creativity.

And there would be plot twists and, tragically, more disasters down the road.

CHAPTER 2

HOMESICK

What do we call this unprecedented moment of home-instability, widespread displacement, unease with our surroundings, the feeling that we can no longer trust what is familiar?

In such a moment of collective distress, the act of naming can carry a lot of power—to acknowledge and legitimize disquieting and traumatic experiences and to create a sense of shared struggle and aspiration. In various moments in human history—especially in moments of unrest—people have tried to name the pain that comes from the disruption of home: a complex set of feelings, muddled together like paint colors on a palette, that includes longing, love, grief, existential angst, and even a lurking sense of dread. Loss of home can also evoke the pain of dispossession, profound cultural and personal disorientation, and righteous anger, all of which can haunt a society for generations.

After the English invaded Wales in the thirteenth century, the word *hiraeth* (pronounced like *here-eyeth*) became a fixture in the Welsh language, to express the societal disruption of living under colonial rule. The feeling remains a defining part of Welsh identity. "*Hiraeth* is a protest," writes essayist Pamela Petro. "It's a sickness [that comes on] because home isn't the place it should have been."

In 1688, Johannes Hofer, then a medical student at the University of Basel in Switzerland, assembled a set of case studies to document the pain of home-disruption. Hofer was born in southern Alsace two decades after the Thirty Years' War—a conflict that turned the region into "a smoldering land, amputated of half its population," writes historian Thomas Dodman. The war's long aftermath left this part of Europe in a state of economic stagnation and political instability. Later, his hometown, the independent city-state of Mulhouse (which wasn't part of France proper until 1798), became a sanctuary to refugees fleeing religious persecution in France. At the time, it was also common practice in Western Europe to hire Swiss mercenaries to provide muscle, firepower, and military expertise to local militias. And these young soldiers reportedly suffered a common, chronic heartbreak, *la maladie du pays*, literally "the disease of the country" in French, or *Heimweh*, "home-woe" in Swiss German. In the thesis Hofer produced as part of his studies, he gave a scientific name to the pain of home loss that he had witnessed throughout his life. He called it *nostalgia*, derived from Greek, "composed of two sounds, the one of which is *Nosos*, return to the native land; the other, *Algos*, signifies suffering or grief," he wrote. In English, a century later, this emotion would also receive the name *homesickness*. But to Hofer, nostalgia was also a medical condition whose symptoms included fever, nausea, sleep disturbance, fatigue, and respiratory problems, along with "palpitations of the heart, frequent sighs, also stupidity of the mind." Untreated, it could be fatal, and there were documented deaths among Swiss soldiers attributed to this malady. By the nineteenth century, the symptoms of nostalgia included "tachycardia, skin rashes, hyperhidrosis [excessive sweating], hearing difficulties, convulsions, heartburn, vomiting, diarrhea, and any rales or wheezing that a stethoscope might pick up in the chest," according to Dodman. We would probably now attribute many of these symptoms to other psychological ailments, such as post-traumatic stress disorder.

In the twentieth century, the meaning of *nostalgia* became more detached from home and homesickness. Instead, it was about time, the longing for the real or imagined comforts of the past. Modern nostalgia can be pleasant or cloying but also dangerous—a desire for a moment that has never truly existed, a longing for a home or homeland that shelters some and exposes others to violence and displacement.

The German word *Heimat*, which translates roughly as "homeland," can conjure a "beautiful and terribly violent past," writes cultural theorist Julia Metzger-Traber. For some Germans, the word simply evokes belonging or safety or connection with nature, but *Heimat* was also used in the 1930s and 1940s to romanticize Nazi ideology. That meaning still haunts Germany, and in 2018, when the German Ministry of the Interior formally renamed itself to include the word *Heimat*, a group of journalists of color published an anthology called *Eure Heimat ist unser Albtraum* ("Your Homeland Is Our Nightmare"). To them the word still evoked a "white majority society which is being threatened by people who are not allegedly part of it," reflected Fatma Aydemir, and not the "radical diversity that is the reality in Germany."

But in discarding the original notion of nostalgia, we may have underestimated the impact that place and home—our connectedness to community and ecosystem and planet—have on the human body and our ability to navigate our lives. Having a home is part of human well-being; when home is disrupted, it can make us literally sick. It is a kind of trauma. Social psychiatrist Mindy Fullilove has described the pain felt by displaced communities—especially Black communities uprooted because of gentrification, discrimination, and urban development—as "root shock," or "the traumatic stress reaction to the loss of some or all of one's emotional ecosystem."

In an era of climate crisis, we will have to reckon with new complexities in our relationships to home, and even more people will experience the shock of being uprooted. In the long run, if we fail to address the crisis that is disrupting our planetary home, there will be hardly any safe refuge left.

🐦

Australian scholar Glenn Albrecht has devoted decades of his career to searching for language that might define the home-woe of the twenty-first century, the grief of climate change.

For Albrecht, home has always, in some way, involved birds. As a child in the 1950s and 1960s, he kept a secret aviary in his backyard in Western Australia. As a university student, he considered becoming an ornithologist. But Albrecht was also a dreamer—a self-described "tree-hugging hippie"—and he was eventually drawn to the humanities, to pondering meanings and societies, language and human relationships to the natural world, and to a decades-long career studying what might happen to our collective sense of home in an era of climate disruption.

In the early 1980s, as a Ph.D. student, Albrecht and his wife made a home near Newcastle, the most urbanized part of a bushy and idyllic section of southeast Australia called the Hunter Region, a place of dairy farms, wineries, and wallabies. For birds, the Hunter River Valley offers a stopover along the vast migratory route called the East Asian–Australasian Flyway, which runs from Alaska and Siberia all the way to New Zealand. Albrecht's enthusiasm for the avian world led him into a set of leadership roles with groups involved in protecting wetlands, and then he began to learn what was truly threatening the well-being of both the feathered and the human residents of the Hunter—coal mining.

Coal mining in the Hunter dates to the late eighteenth century, but its visible impacts on the landscape were relatively modest until the mid-twentieth century, when changes in mining technology and a booming export market made it viable for the industry to expand radically. Between 1981 and 2012, the amount of land occupied by open-cut mines in the Hunter increased almost twentyfold. There are global implications, of course: coal produces the highest carbon emissions* per unit of energy of any fossil fuel. From the perspective of the atmosphere, it is the dirtiest fossil fuel. Physically, at the scale of a local and regional landscape, open-cut or open-cast is a kind of strip-mining, akin to the mountaintop removal mining that has devastated Appalachian landscapes. The process leaves a permanent and raw scar, devoid of topsoil. Such mines can also discharge high levels of arsenic, lead, mercury, and nickel into nearby water supplies. They fill the air with toxic dust, and people of the Hunter have learned to shut their doors and windows

---

* There are multiple gases that contribute to the warming of the Earth when humans emit too much of them into the atmosphere. Carbon dioxide is by far the most prevalent, but methane, nitrous oxide, and fluorinated gases are also part of the problem. The term "greenhouse gases" refers to all of these and harkens back to when climate change was called "the greenhouse effect" in the late twentieth century. Throughout this book, I'm using "carbon emissions" as a generic term for greenhouse-gas emissions.

during strong winds. Children living in the region have high rates of asthma and chronic respiratory diseases.

From above, an open-cut coal mine looks like some geological aberration, a sort of man-made desert or a recent volcanic eruption. Albrecht was confronted with such a scene one day while driving along a highway through the Hunter. "I stopped and got out of my car and looked west up the valley at the desolation of this once beautiful place," he wrote in his book *Earth Emotions.* "In front of me lay hundreds of square kilometers of open-cast black coal mines feeding two large power stations. . . . The air was thick with dust and there was the acrid smell of burning coal from the spontaneous combustion that was erupting in various actively mined locations. . . . A dull roar went up in the far distance as a mine detonated a panel of overburden, and a cloud of orange smoke drifted over that end of the valley." To Albrecht, just to behold these mines felt like an "acute traumatic event."

Meanwhile, he began fielding calls from friends, neighbors, and other conservationists in the Hunter. "In their attempts to halt the expansion of open-cut coal mining and to control the impact of power station pollution, individuals would ring me at work pleading for help with their cause. Their distress about the threats to their identity and well-being, even over the phone, was palpable," he wrote later in an academic article. To Albrecht, these conversations were both personal and a subject of scholarly inquiry, because there seemed to be so few satisfactory efforts, at least in Western society, to characterize the agony and angst people experience when their surroundings become marred.

In the 1990s, a group of psychologists, including Alan Kanner, Theodore Roszak, and Mary Gomes, established the field of ecopsychology to recognize the connection between people's emotional health and the natural world. But even now, the field remains marginalized, says Kanner. "Many clinicians treat their client's grief, anger, and fear about the environmental crisis as somehow peripheral to what truly matters, as either symbolic of human-centered concerns or even a kind of avoidance of their core personal issues," he explains. Of course, volumes of research demonstrate that health—mental health, especially—is not just an individual pursuit but heavily influenced by what is happening in society, in our communities, and in the environment. But Western medicine is not really designed to treat the collective causes of medical troubles, and we are only just beginning to talk about the ways that the climate crisis is also a mounting health catastrophe. (In September 2021, more than two hundred medical journals released a joint statement that climate change is now the greatest threat to public health worldwide, in part because of its impact on mental health.)

Like Hofer, Albrecht thought it would be useful to name the experience of watching one's home environment unravel. (He wasn't the first to name such a feeling. There are words in Indigenous languages that could have filled the gap, but none had yet migrated into the English language.) Around the turn of the millennium, he decided to coin his own word. "With my wife Jill, I sat at the dining table at home and explored numerous possibilities. One word, 'nostalgia,' came to our attention as it was once a concept linked to . . . homesickness," he wrote. Hunter Valley residents were homesick, but they hadn't gone anywhere—the place they lived in just no longer offered the kind of comfort, solace, or safety one would expect from home. Albrecht came up with the word *solastalgia*, using the suffix -*algia*, meaning "pain," and the same Latin root as the words *solace*, *console*, and also *desolation*. (In Latin, *solacium* means "comfort" and *desolare*, "to leave alone," so the word *solastalgia* suggested the loss of comfort, the loneliness of being estranged from home.) He published the first academic paper on the idea in 2005.

Some neologisms never make it out of the realm of private conversation, and some molder in the corners of academic journals as useless jargon. But occasionally a word like this catches a bit of zeitgeist, like wind, and gets borne aloft into the culture at large.

Over the next several years, it seemed Albrecht's mellifluous word had tapped into a kind of angst about life on a warming planet. Artists and creatives picked up on the idea. A British trip-hop band produced an instrumental track called "Solastalgia," and a Slovenian artist recorded an album, also called *Solastalgia*. At the beginning of 2010, the *New York Times Magazine* ran a profile of Albrecht and commissioned sculptor Kate MacDowell to create a porcelain representation of solastalgia—a brain full of delicate trees and Australian wildlife.

The neologism also offered a useful means of describing and studying how the impacts of climate change reach beyond tangible, physical, and economic damages. A team of social scientists identified feelings of solastalgia among people from rural northern Ghana, a region devastated by climate change–related drought and crop failure. A collaboration of environmental scientists and public health researchers observed solastalgia in communities affected by hurricanes and oil spills in the Gulf of Mexico. A Los Angeles physician named David Eisenman stumbled across the idea of solastalgia when interviewing survivors of the 2011 Wallow Fire, the largest wildfire on record in Arizona. "All around them was this blackened, charred landscape," he remembered. Over and over, he heard them express "the sense that they were grieving [its loss] like for a loved one." He and his team designed a special set of survey questions to gauge the problem and asked fire survivors to rate their responses to statements such as "I feel like I have been grieving for the loss of the forest affected by the Wallow Fire" (71 percent agreed or strongly agreed) and "seeing the forest affected by the Wallow Fire has been stressful" (93 percent concurred). The more uneasy they felt about the landscape itself, the more at risk they were for other kinds of psychological distress.

It's now becoming more common for writers, artists, and scholars to talk about the emotions of the climate crisis. We have even more ways to name the experiences of people living through megadisasters and the slow attrition of beloved places—including "climate grief," "ecological grief," and "environmental melancholia." In a 2020 survey by the American Psychological Association, more than two-thirds of American adults said they'd experienced "eco-anxiety."

We are moving into an era defined by homesickness.

⁊

Does it matter if we name or even notice this kind of angst?

In recent years, it's become abundantly clear how much our actual, physical homes and lives are at risk, all over the world. In 2019 alone, 24.9 million people around the globe were effectively evicted from their homes by natural disasters and climate change impacts, including 1.5

million in the Americas. A record-breaking monsoon season in India displaced one million people that year and killed 1,600. Communities that survive disasters, both large and small, face damage that is hard to even tally. Various economists have tried to estimate the harm of climate change to our societies in monetary terms—up to sixty-five billion euros per year (or around seventy-three billion dollars at the time of this writing) in one estimate.

Others have made calculations of potential economic losses based on factors such as wage-earning potentials and gross domestic product (methods that could easily devalue the impacts on vulnerable people and on less tangible aspects of human well-being). Still, the world could lose up to 18 percent of GDP by 2050 if nothing is done about climate change. But such calculations strike me as profound underestimates of a phenomenon that could easily tear apart the basic fabric of our societies, economically and physically.*

Former vice president Al Gore has called climate change a "challenge to our consciousness and our moral courage." The journalist David Wallace-Wells—in his sobering book on the climate crisis titled *The Uninhabitable Earth*—lambasted our society for "an incredible failure of imagination."

Imagination and empathy, feelings and morality are closely intertwined. And I have long wondered if we have failed in some more personal realm. Have we cultivated a tendency to shut off our feelings and distract ourselves from what is, in truth, horrifying? Have we persuaded ourselves for too long that climate change is the problem of others and that the storms will never rattle our own roofs? And if we all faced our grief, would we find the collective will to take the kind of drastic action required to stanch the destruction?

---

* In the last several years, a whole field of study has emerged to quantify the intangible losses associated with climate change. Losses related to culture, identity, heritage, emotional well-being, and the sacredness or spirituality of people's relationship to a place or a community—not to mention experiences like the joy, love, beauty, or inspiration found in a cherished landscape—are nearly impossible to quantify in economic terms. So scholars of intangible loss are now trying to find other ways to account for them, formally, for the Intergovernmental Panel on Climate Change. "We have to find a better way to make visible what is often overlooked, ridiculed, dismissed as too personal, not generalizable, not quantifiable," says Petra Tschakert, who is a professor of geography at the University of Western Australia and who has also studied solastalgia in Ghana.

In 2013, I met Glenn Albrecht while I was on a writing fellowship in Perth, a Western Australian city of Spanish-style architecture, white-sand beaches, brightly colored wild cockatoos, and some of the most profuse biodiversity of any city in the world. He had moved back to the country's west coast in 2009 to take a post as a professor of sustainability at Murdoch University and was living with and caring for his elderly mother in a house about thirty miles outside Perth that they had nicknamed Birdland. He had also broadened his work beyond solastalgia and was creating an entire lexicon of polysyllabic words related to climate change. At a seminar at a local university, I watched him—a gangly, energetic man—urge the few dozen people in the room to pronounce several other neologisms, in the manner of a children's television program. Albrecht waved his arms like a drum major in a marching band, as we sounded out in unison "*SUM*-BI-*OS*-IT-Y," *sumbiosity*, which refers to a utopian-sounding state in which people live in balance with the Earth. (I had doubts about whether this word would catch on, though I could appreciate the sentiment.)

The environs of Perth were most definitely not in such a state. Because of climate change, the amount of regional rainfall has dropped by about 20 percent since the 1970s, and since 2000, the populace has been slurping most of its water from shrinking aquifers. I had twice been to Perth years before—once as a student and once shortly after graduating—and at the time, it had felt like a second home to me, a colorful place where I came of age and immersed myself in its wild and urban beauty. And even on my return, more than a decade later, I could see something strange had happened. Many of my favorite lakes and wetlands—from a small duck pond near a university to marshes and big expanses of open water full of black swans—had shrunk. Some had reportedly even caught on fire in recent years—the peat beneath them smoldering for weeks. In the exurbs, entire forests of jarrah trees, a species of eucalypt with lustrous red heartwood and lacy white flowers, were dying from a combination of stress and a fungal disease. "I'm witnessing the disruption of my own home biophysical environment, the things that I grew up loving," Albrecht told me.

Meanwhile, the idea of solastalgia had taken on a life of its own, and in the Hunter, the concept had been used in a 2013 court ruling to stop the expansion of a coal mine by the company Rio Tinto: the solastalgic pain of local residents was named as one of several reasons to halt the project. Albrecht had testified on the negative impacts on citizens and how the project would likely make one village unlivable. "Hence, approval for this project would set a precedent for all future cases where the intrinsic values of human culture (including agriculture), good human health (physical and mental) in nonpolluted environments, aesthetic and scenic beauty, amenity, a strong sense of place, and community cohesion will all be trumped by the dollar value of resources irrespective of the negative impacts imposed in obtaining them," he wrote in a supplementary report for the hearing. It was maybe the first time that something as intangible as love of home had nearly as much legal standing as pure economics. The decision was overturned again in 2015, but the community group there has continued to try to fight the mine.

After Albrecht's mother passed away, he returned to southeastern Australia full-time in 2014—to a place at the edge of the Hunter Valley that he and his wife named Wallaby Farm, where they could live nearly off the grid, with solar power and a farm full of fruits, herbs, and vegetables. But in 2019, wildfires raged around his property—part of the massive outbreak of flames called the Black Summer that would devastate much of Australia and draw international attention. One fire ignited about a mile from Albrecht's house. Albrecht wore a face mask much of that season to cope with the searing smoke. He kept watch for any embers that might drift through the air and alight on his property.

When I spoke to him not long thereafter, he said, "We're actually in the process of trying to sell it and move. We're being driven out by climate change." (Though I later heard he and his wife had actually stayed put.)

In our interview—just after the explosion of the Covid outbreak worldwide—he took the same slightly detached, professorial tone that I had always heard from him. I couldn't hear his emotion in his voice. "The bushfires were a massive psychoterratic experience," he observed, drawing on another Latinate word he had coined.

I asked him about a post he had placed on Facebook during the height of the fires. It was full of expletives and occupied some space between humor and rage. "The land that we love is being fried and a bunch of fuckwits in charge are doing nothing about it," he had proclaimed. "It is frying because the joint is getting fucking hotter. Our trees and gardens are fucking dying and frying because of fucking climate warming. People are being fried because of fucking fires."

This post was, he said, an expression of his anger in the Australian vernacular.

When I read it, it was like hearing a battle cry in the distance—a roar about everything we love, everything we are losing, everything we must try to defend.

—

CHAPTER 3

THE FLOOD

home, n. *A person's own country or native land. Also: the country of one's ancestors.*

Until she moved to Florida, Jenny Wolfe had never seen a hurricane and never imagined how storms like this would both upend her life and define her work.

She grew up in Iowa—a state often troubled by tornadoes but nothing as gargantuan and watery as a tropical cyclone. Then she moved to the Sunshine State in 1996 with a boyfriend who had set his sights on an out-of-the-Midwest college experience—specifically at the University of Florida at Gainesville.

Gainesville lies inland and thus is sheltered from the full force of tropical storms—though Jenny experienced a few alarming moments with *the edges of hurricanes* as they crossed over the central Florida area. In an especially ferocious storm in 2004, *I remember feeling scared enough to push up a mattress against our apartment window and sit in the closet because something sounded like an explosion, a nearby transformer. The hurricane had that train sound like a tornado.*

Her first academic major in Gainesville was architecture, but the field never felt like it fit her properly. Eventually she switched to political science. *Then I learned there was a thing called historic preservation, a* profession that involves saving, protecting, and documenting buildings, objects, and artifacts that offer up evidence of the long arc of the human story. And she realized she was drawn to historic places and old architecture. She enrolled in the University of Florida's graduate program and researched a thesis about Cedar Key, a town on a group of islands, also called the Cedar Keys, off the state's west coast in the Gulf of Mexico. These islands had in the nineteenth century been a source of cedar for pencil production, a port, and a railroad terminus. They held a pair of productive timber mills (one of which had briefly employed famed conservationist John Muir, who also had the misfortune of contracting malaria there). But the Cedar Keys were ravaged by an 1896 hurricane that killed about a hundred people and destroyed the mills. In the years that followed the community went through a period of economic decline. It is now mostly a tourist destination and fishing community with a historic site, museum, and a handful of seafood restaurants.

Jenny observed that Cedar Key, surprisingly, had no particular plan for protecting its historic structures from the next hurricane. Her thesis offered a nearly exhaustive review of the community's historic assets, along with a recommendation for the site's stewards to use mapping technology like Geographic Information Systems, GIS, to make a record of what was there before disaster struck again. *These disasters are predictable in terms of their nature and ability to have a devastatingly widespread impact*, she wrote.

But in 2011, when she was hired as the historic preservation officer of the City of St. Augustine, a burg of fifteen thousand people on Florida's northeast coast, she didn't think she would still be planning for floods and storms.

By disposition, Jenny is an organized person—in appearance, fine-boned, with long, blond hair and a heart-shaped face; in dress and manner, neat but not ostentatious, with a formal way of speaking, especially about her work.

But her feelings about her new job and home would eventually go far beyond formality. It was an overwhelming but rewarding job, and she would grow to love this storied place. Established in the sixteenth century by the Spanish, St. Augustine is the oldest continuously occupied European-settled city in the United States—sometimes nick-named the Ancient City. Her new desk stood inside city hall, which shares space with a quirky museum of antiquities and high art inside the old nineteenth-century Alcazar Hotel—a Spanish Renaissance Revival building capped with ornately carved terra-cotta and brick towers, like a reverse red-velvet layer cake.

A block away lies the old Spanish plaza, laid out in 1573, with archae-ological finds now interred beneath almost every bench and lamppost. And from there, the historic details spread out in all directions like chapters in an epic. The city's most iconic structure is a star-shaped seventeenth-century stone fortress called the Castillo de San Marcos—built from coquina, a stone formed by the compression of piles of tiny clamshells deposited here more than a hundred thousand years ago (when much of what is now Florida was underwater). North of city hall is a tight cluster of narrow eighteenth-century colonial streets full of shops. And to the south lies a neighborhood called Lincolnville—one of the key battlegrounds of the 1960s civil rights movement. In the streets beyond here you can find all of the charming poshness and kitsch of Florida—midcentury drive-in motels with neon signs and tur-quoise swimming pools and flashy tourist attractions such as a Ripley's Believe It or Not! museum and a pirate museum.

Jenny moved to St. Augustine during a spate of calm weather: for a handful of years, there were no major storms. But her arrival was also the beginning of a new reckoning with the oceans. Climate change has been a difficult subject politically in Florida. But a decade ago, at least some federal and state experts and Florida civic leaders claimed they were owning up to the troubles that were being unleashed on the state by warming air and rising water. In 2009, the U.S. govern-ment released the second National Climate Assessment (the first was in 2000). It contained a formal if banal acknowledgment that climate change had already arrived, that the storms brewing in the Southeast carried extra charge and force because of it, and that the sea had already risen and would continue to climb up onto the land and carry more water into yards and streets and highways. Meanwhile, the population of Florida had doubled in the past three decades, and "the quality of life for existing residents is likely to be affected by the many challenges associated with climate change," the report noted. A few months later, four counties down the coast from St. Augustine—Miami-Dade, Palm Beach, Broward, and Monroe—formed an entity called the Southeast Florida Regional Climate Change Compact. Their founding resolution acknowledged that, because of the unfortunate combination of low-lying land and rising seas, the region's five-million-plus residents would face "disproportionately high risks associated with climate change." The counties pledged to push for policies both to cope with these risks and to reduce carbon emissions. (This pledge was promising but didn't prevent Governor Rick Scott from purportedly banning the term "climate change" from official use in the middle of his time in office from 2011 to 2019. Nor has it prevented real estate tycoons from planning expensive housing developments in flood-vulnerable areas.)

At the federal level, the Obama White House also directed agencies to produce adaptation plans—something like what Jenny had written for the Cedar Keys but based on future projections, which were much more sobering than the experiences of the past. The National Park Service, which oversees the National Register of Historic Places and hundreds of historic sites—including the Castillo de San Marcos—released its first plans for managing climate-related troubles in 2010 and 2012.

Between 2012 and 2014, St. Augustine added a new seawall to the waterfront boulevard, Avenida Menendez, with money from the U.S. Federal Emergency Management Agency, or FEMA. This was intended as an improvement on two older seawalls, better protection against floods than the first one, made by the Spanish in 1696, and the second, built by graduates of the U.S. Military Academy in the mid-nineteenth century—and torn apart by a hurricane in 1846 and a tropical storm in 2008. But the wall is only high enough* for a Category 1 storm, the least severe. Two years later, the state of Florida launched a series of pilot studies, in partnership with the National Oceanic and Atmospheric Administration, on coastal flooding in three communities. One

---

* It is about seven feet above a standardized benchmark for sea and tide levels called the North American Vertical Datum of 1988.

would be St. Augustine: an evaluation of how this city might fare as the waters rose.

By this time, Jenny had gone through some personal disruptions and transitions. The Gainesville boyfriend had become a husband and then an ex. She moved alone to an apartment in a neighborhood called Davis Shores. It had been platted in the 1920s on Anastasia Island, a barrier island across the river from the city center, and built up during a series of mid-twentieth-century real estate booms. Jenny's place was a 1940s garage that had been transformed into a cottage. It was dainty and mint green with a metal roof, its inner walls covered in heart pine tongue-and-groove paneling. She dearly loved this place, a space of her own. She planted three raised garden beds full of vegetables and adopted a puppy. But Davis Shores is low-lying and susceptible to regular soaking by the tides. Already, even on a sunny day, a high tide could send water from the estuary into the creeks, to the storm drains, to the streets, dampening the underbellies of the cars parked there.

St. Augustine's sea level rise report came back from the state of Florida and NOAA in June 2016 and offered an alarming prognosis for the city. With one and a half feet of sea level rise, nearly a third of the city's road network would frequently be swamped by the highest high tides and small storms and a fifth to half of the buildings and structures in the downtown historic district would flood around once a month or more. This scenario could be reality as soon as the 2040s, according to the report. By about two and a half feet, the city would have passed a tipping point: twenty-four times as many buildings would be vulnerable to tidal flooding every day. This could happen as soon as the 2050s if global society kept gasping out carbon rapidly.

Then in late September 2016, as if to illustrate how powerful and raucous water and wind could be, Hurricane Matthew rose up from the Caribbean. It was a Category 5 hurricane, the strongest possible designation on the basis of wind speed, the first storm of this level of ferocity to spin through the Atlantic Basin since 2007. At its peak, off the coast of Colombia, Matthew huffed out 165-mile-per-hour winds. The storm weakened a little to a Category 4, then slammed into Haiti on October 4 (with winds still at 150 miles per hour), and here it produced a horrifying disaster, killing an estimated one thousand people, razing 80 percent of the buildings in the city of Jérémie, ripping apart crop fields, and flooding villages. Then it struck Cuba, furiously attacking the province of Guantánamo and tearing the roofs off houses. Then it continued northward.

An evacuation order for St. John's County, wherein lay St. Augustine, began early in the morning on Thursday while Matthew was pummeling the Bahamas, and Jenny left for a friend's house farther inland, west of the city. There, she anxiously watched news reports and Facebook videos of the rising waters.

As the hurricane approached Florida, it remained at sea: "the western edge of Matthew's eyewall barely clipped NASA's Cape Canaveral launch facility," read a report from the National Hurricane Center. By the time it traveled past St. Augustine and Jacksonville, it was a Category 3 (120-mile-per-hour winds) shifting toward a Category 2 (110 miles per hour). It never made landfall in Florida, striking South Carolina instead. St. Augustine could be considered, in the grand scheme of things, lucky.

But even good luck in a hurricane can still look rough. In video footage from the storm, St. Augustine resembled an underwater ruin, palms flopping limply like grass in the wind, streets full of turbulent rivers.

On Saturday, Jenny drove her car back into the city under a jarringly blue sky. *I was driving through—I don't want to say a war zone—but it felt like you're going to a site of devastation.* Downed fences. Fallen trees and debris. Signs wrenched off posts. Halloween pumpkins transported by floodwaters, heaped on the street.

At city hall, although someone had piled sandbags in front of the doors, water had seeped through some windows on the side of the building, drenching a ground-floor conference room and ruining the carpet and flooring. At the museum that shared the same building, it had flooded the hallway, the basement, and a historic pool that had been drained and was functioning as a café.

Later that day, Jenny crossed the bridge—the Bridge of Lions, built in 1926 and adorned with actual marble lions—to her own neighborhood. When she arrived at her apartment, she saw that Matthew had swept up miscellaneous belongings from the yards and storage units nearby and strewn the alley in front with garments, outdoor furniture, and bicycles. The floodwaters had soaked the rooms inside her place, ruining her old wedding album, shoes, kitchen appliances, a dresser full of clothes, thumb drives, and various other personal effects she'd left there. Throughout the island, the wastewater system had overflowed, and the city warned residents to boil their water (in case traces of human waste ended up in the tap). Her place emitted a powerful odor that she would remember vividly but never be able to describe, stranger and more potent than sewage.

After sizing up the mess, she donned rubber boots and gloves to salvage what she could of her home. In the next few weeks, she would turn her attention to the rest of the city, to help other St. Augustinians recover, to try to protect the place from the next flood.

The storm wasn't as bad as it might have been. But it carried a more powerful symbolism than some of the hurricanes before it. It was a moment of foreshadowing that would force tough questions about the city's future—and what everyone living there would do when the water arrived again and again at their doorsteps. Jenny would put her efforts toward helping the city she loved stay safe for as long as possible.

※

To have any kind of home, you must have at least a little history. If you reside somewhere only one day, it's not home yet. You have to accumulate stories and a sense of familiarity. Writ larger, to have a nation, you need a story about origins. Collectively, we must be able to write ourselves onto the landscapes we inhabit in some sensible way.

So many human histories start at coasts, and the United States is no different. People often arrive in a place by water and are drawn again and again to make homes near the edge of the sea. But you can never live in or near this littoral zone without reckoning with the power of the ocean. The sea will almost always have the last word on what happens to coastal land.

Beginning at least five thousand years ago, the ancestors of Florida's Indigenous groups (including the Timucua and Calusa, both of whom may have later integrated into the Seminole Tribe) built mounds along the Gulf of Mexico and the Atlantic coasts—comprised mostly of shells from oysters and clams, which they ate in prodigious quantities, along with other material like pottery, bone, and charcoal. Academics and archaeologists used to think shell mounds were just ancient trash heaps. But most scholars now believe they were far more sophisticated. Mounds are human-made landforms, a type of Indigenous engineering. They can take the form of ridges, rings, hills, even entire islands. Ken Sassaman, an archaeologist at the University of Florida and an expert on mounds, calls this *terraforming*, a term borrowed from science fiction, usually referring to the process of building a world from scratch. Shell mounds may have had various functions: they were perhaps shelters from wind and waves (homes were sometimes built on the leeward sides), elevated building pads for houses or monuments, ceremonial sites, or plazas. They were often oriented in the precise direction of the rising sun on the summer or winter solstice.

At some mounds, there is evidence of regular feasts and celebrations. "These were world renewal ceremonies—ritual practices to bring the world back to balance," Sassaman believes. To balance and renew the world, you had to connect with your ancestors and share stories about both the past and the future.

In some locations, people buried their dead inside the mounds. Sometimes the people of this era would even dig up and move entire cemeteries full of bones and relocate them to new places.

Sassaman thinks they did this to keep their ancestors from washing away in the tide. And keeping track of their forebears and their history was arguably part of survival. During the era of mound-building, the ocean rose slowly, perhaps an inch or two every fifty years, less than half to one-third of the current rate of sea level rise. But there were sometimes sudden disruptions—such as a hurricane tearing apart and drowning the land. The people who built Florida's mounds probably knew the sea could change, Sassaman contends, because they had a long memory and a record of history, kept alive both by oral tradition and by the structure and symbolism of the mounds. Knowing about the floods of the past could help them prepare for those of the future. Indigenous cultures often had a cyclical sense of time. "The future is more certain for people that look to the past," he says.

By contrast, to have an ahistorical world—to forget the past—can be dangerous. Paradoxically, it can make a society less able to perceive and respond to change.

The modern American relationship with history is inconsistent and fractured. Many of our most revered national memories, historic architecture, and monuments stand at the coasts. Yet we have had a tendency to knock down old buildings in favor of erecting new ones. And for the last few years, shimmering new real estate developments, glassy luxury condos, palatial beach houses, and boxy McMansions have flowered like weeds all along the Atlantic coast, from Boston to Miami, even in areas that are already prone to floods or will be in the near future. Estimates of the real estate damages wrought by rising seas vary enormously across sources, but they usually run into the billions. According to a study by the Union of Concerned Scientists, for instance, in about two and a half decades, roughly 136 billion dollars' worth of real estate on both the East and the West Coast could suffer from chronic flooding.

By the end of the century, if the world emits carbon at the highest and most disastrous rates, that figure could rise to more than a trillion dollars, and one million homes in Florida alone could be soaked by the sea more than twice a month on average. Every coastal community worldwide will bear the impact, and both the world's wealthier citizens, living in cities such as Miami, New York, and Amsterdam, and its poorer ones, in places like Lagos and Dhaka, will ultimately feel the losses.

Too often, the American approach to these eventualities is to look away. In 2012, real estate developers successfully pushed North Carolina legislators to ban the use of scientific predictions that foresee catastrophic sea level rise in any state and local agency coastal planning. Such a policy is akin to making it illegal for North Carolinians to imagine the future. Sometimes it feels as if the United States exists in a suspended state of now, as if we are pretending the past has never been and the future will never arrive.

Historical records tell us that the sea was lower when we first built our coastal cities and towns, our roads and bridges. And it keeps rising. We haven't readied our homes for the water of the future, and we haven't reckoned with what we are all about to lose.

ꝰ

There are two main ways that heating the planet causes the seas to rise. One, as water warms, it expands. So a warmer ocean simply takes up more space. Two, much of the planet's water is locked up in large masses of ice—such as the Antarctic and Greenland ice sheets and mountain glaciers. When this ice melts and turns to water, it runs to the sea, as water will do. Since at least the 1970s and 1980s, scientists have known that the extra heat that was being trapped in the atmosphere by our carbon emissions would also enlarge the volume of the oceans, and the rise of the seas has gathered speed in the decades since. This trend is not reversible on the time scale of a human life. You can't put the ice back, and the carbon that we have already sent into the atmosphere and the heat energy now stored in the oceans guarantees that they will keep rising.

The same set of circumstances also means that coastal weather will keep growing more intense. The now-agitated atmosphere unleashes its extra heat energy in the form of storms and hurricanes. So this much is certain: anyone who lives at the coasts will be confronted with more water, more wind, and more deluges, now and in the future.

But human society still gets to choose how severe the problem becomes. The rate and extent of future sea level rise depends on carbon emissions. Our choices will have a direct impact on St. Augustine: in just one lifetime, will its streets fill with calf-deep or waist-deep water during the highest tides? You could imagine the place coping with and adapting to the first scenario and falling to ruin in the second.

A few years ago, I began learning about sea level rise, hurricanes, and coastal adaptation through the eyes of those who had dedicated their lives to history and not just from the experts trained to prognosticate the future. The conundrums the historic preservation community is facing and the solutions they have tried to pursue—in old cities like Boston or New York or Newport, Rhode Island—seemed more realistic to me than, say, wild designs for futuristic floating cities or hurricane-proof houses (which might withstand a storm but wouldn't be terribly functional if the community around them was crushed). I couldn't imagine how we would face the future without taking stock of the past. I didn't think we should move forward into the twenty-first century without hanging on to some of what our ancestors had built or some of the stories they had handed us. How would we make meaningful and enduring homes and communities if we lost all of the trials and errors and lessons that came before us?

Many historic preservationists and planners look to Europe as an example of how to prepare for water to arrive in landscapes deeply imprinted by history—and often full of old, immovable architecture. In some cases, you try to save what's important; in others, you have to let it go. The goal is to choose deliberately and wisely. Some European approaches have involved letting the water in. For instance, a roughly decade-long Dutch project called "Room for the River," begun in 2007, bought out houses and properties along the Rhine, the Meuse, the Waal, and the IJssel Rivers in the Netherlands and helped residents relocate, and then opened up the floodplains so that rising waters had a place to go.*

Other approaches involve trying to keep the water out, but in a time of rapid water rise, this is a trickier proposition. For one thing, it is difficult to divorce a water-bound economy and culture from the ocean by cleaving the two with a wall or barrier. On a trip through Italy, I spoke to a few locals in Venice about the enormous tidal gates the city was building—a project that has become infamous because of cost overruns (with a total price tag of around $8 billion), decades of delays, and a corruption investigation. Some Venetians distrust and disapprove of the whole endeavor. The gates are designed to rise up and defend the city, barring stormwaters from entering, the way medieval castle gates kept out invading soldiers. But they are only intended for the highest of tides and storm surges, because to close them is to shut down boat traffic, temporarily isolating the city and the port. Nearly half of Venice can still flood before the gates rise up.

---

* "Room for the River," or *Ruimte voor de Rivier*, is described in vivid detail in one of the best-known accounts of climate change, Elizabeth Kolbert's *Field Notes from a Catastrophe* (2006).

I spoke to one environmental scientist in her brick-lined office overlooking one of the city's famous canals. She called it "propaganda" to suggest that the gates alone could save Venice. In October 2020, the project entered a trial phase, and the gates were deployed against a flood for the first time. They successfully kept a major tidal flood away from the famous San Marco Square and the cathedral. But they are manned based on the weather. Two months later, when the forecast failed to accurately predict the tides, the gates were not activated, and four and half feet of water rushed into Venice.

There will be a time in the future when the city can't keep raising and lowering the gates. Eventually the sea may have the last word on what happens to Venice.

Historic cities in the United States face the same sorts of quandaries. In the aftermath of Hurricane Sandy, Lisa Craig, then the chief of historic preservation of Annapolis, Maryland, launched a discussion about preparing for sea level rise. This seaport, first built by seventeenth-century Puritans and later expanded by oystermen, holds quintessential pieces of the American story and collective identity. George Washington resigned his commission as commander in chief of the U.S. Army in the Annapolis State House in 1783. From 1783 to 1784, the city served as the U.S. capital. Here the U.S. Naval Academy—full of granite-walled, slate-roofed, Beaux Arts–style buildings—has trained some of America's most important political and military minds: Jimmy Carter and John McCain, twenty-six members of Congress, fifty-four astronauts, and five state governors. But NOAA had recently revealed that Annapolis was also a hot spot of "nuisance flooding." Over about fifty years, the frequency of chronic flooding in Annapolis increased by 925 percent and, in about four more decades, it is predicted to occur about fourteen times more often again.

Craig held a major public engagement campaign called "Weather It Together" to push the city to plan better for coastal flooding disasters and to make decisions about which of its historic resources it would try to protect. But when a new mayor was elected in 2017, Craig felt the city's interest in the subject beginning to ebb, and she resigned.

A year after this, I visited the Annapolis City Dock—a quaint area of brick sidewalks, a marina, some historic inns, and a few shops and restaurants at the waterfront—with an architect named Michael Dowling, a longtime expert on historic preservation. Serious looking but congenial, with white, windblown hair, a close-trimmed beard, and a furrowed brow, he had worked to fix up a theater in downtown Annapolis after Hurricane Isabel drenched the place in 2003. He has since provided expertise to the city and to local and regional organizations on how to protect historic buildings from sea level rise. His theater restoration project had taken a decade. Meanwhile, protection efforts by the City Dock still felt piecemeal. The city planned to install a stormwater pumping system. Some property owners had raised their foundations. Others were using devices called "door dams" to try to keep tidal floods out. We stood in front of the theater, painted a stormy-sea blue with a steeply angled red roof, and looked up the street. Here Dowling had carefully tested and selected mortars and materials to rebuild the centuries-old brick and wood exterior after Hurricane Isabel. "High tide is projected to be at the top of their doorsill twice a day," he said mournfully. "That's just a regular tide. Let alone a storm surge or anything. And elevating historic buildings is difficult and expensive. And there's still a school of preservation that says you shouldn't do it."

He pointed to some buildings in the distance where the owners had raised the foundation. You could do this, property by property, he mused, all the way up the waterfront, though the city would still have streets that turned into canals on a high-tide day.

"To save us in 2050, we need a seawall that's several feet higher. Who's gonna pay for it?" he said. And big sea barriers, like the one in Venice, can create their own problems.

"Then you close off stormwater from getting out to the bay. You could cause flooding from the water that's coming down the Severn River and the creeks. And another approach is, let it happen. Abandon and retreat, and that'll just continue up Main Street."

The next year, a citizen advisory committee would come together to protect the City Dock, and Michael Dowling would continue to provide expertise. The community has since considered a series of strategies that would buy time and keep some of the flooding out of this water-prone place—including raising a walkway, building a seawall, and installing a (possibly retractable) barrier—but the price tag remains high, up to $50 million or more. Parts of the City Dock are now underwater fifty to sixty days per year.

People can try to hang on to places like Annapolis at a steep cost, or they can simply choose to neglect them and let the sea take them. These are the choices.

⚓

When she arrived in St. Augustine, Leslee Keys, also a preservationist by trade, thought she might be mostly done with hurricanes. The city is partially protected from storms by both coastal geography and a barrier island. So there was a sense that South Florida had the real hurricane problems. In 1999, after Hurricane Floyd, the *Tallahassee Democrat* ran an article that even suggested the city had some mystical form of protection: "Watched over by a 300-year-old Spanish Fort, a cross built after the last hurricane, and a mysterious statue of the Virgin Mary . . . there was little structural damage to the historic buildings and homes in this northeast Florida city."

But Leslee quickly realized the city wasn't altogether immune to tropical storms. *I've been through a lot of hurricanes. I tend to pay a lot of attention.*

Before moving to St. Augustine in the late 1990s, Leslee had spent four years working for the state historic preservation office in the Florida Keys (a random but somehow apt coincidence of place-name and personal name). Part of her job was to help historic properties there recover from Hurricane Andrew, which was in 1992 the costliest disaster in the United States (until it was surpassed by several others, including Hurricanes Katrina in 2005 and Sandy in 2012). The other part was to instruct them in how to prepare for floods still to come. She grew up in Indiana, but the Florida Keys had given her a fast education in hurricane preparedness. And the small necklace of islands that extends from the bottom of the state had a more matter-of-fact approach toward hurricanes than the mainland. *In the Keys, when there's a hurricane warning, everybody gets ready for a hurricane party, because you're at the end of the world's longest extension cord. You don't evacuate when you have 140 miles to drive between islands.* And Leslee—who seems able to inject humor and whimsy into almost anything—was good at talking about disaster in a way that was somehow realistic and not off-putting at the same time. *I'm not afraid of hurricanes,* she said unflinchingly. *But I'm really efficient when it comes to getting prepared for them.* In the Keys, Leslee had trained librarians and museum curators to make plans to protect their collections from storms. *If you're down to that seventy-two-hour window before a hurricane, you should know how many hours it takes you to do particular things,* like pack up antique furniture and pottery, board up windows, or move archives and rare books well above flood elevation.

In St. Augustine, Leslee worked her way through several roles at Flagler College, meanwhile also earning a Ph.D. in historic preservation from the University of Florida. She learned that the college's flagship building, another lavish antique hotel called Ponce de Leon Hall—also built by Henry Flagler and now serving as a student dormitory known affectionately as the Ponce—had sustained significant roof damage during Floyd.* Then in 2004, Hurricane Frances blew out a sliding glass wall under the red clay dome that capped the Ponce. A few years later the institution secured grant money to install hurricane-proof glass.

But even after this—and after successive storms like Hurricanes Ivan, Jeanne, Katrina, and Wilma, and Tropical Storm Fay visited Florida— St. Augustine wasn't prepared for a major hit from a tropical cyclone. Perhaps this was partly because the real estate industry and some property owners didn't want to own up to the danger, lest the admission damage property values. The Ponce and the Alcazar Hotel (that is, city hall) are arguably symbols of a Florida trait almost as defining and powerful as hurricanes—the pursuit of fantasy. Flagler College's namesake, the nineteenth-century entrepreneur Henry Flagler, is widely considered the father of modern Florida. In 1870, he cofounded Standard Oil (part of which later became Chevron) with John D. Rockefeller, then more

---

* Leslee Keys eventually wrote a book about the Ponce, published by the University Press of Florida in 2015. She calls the building "an imaginative and exuberant expression of the Gilded Age fascination with art and architecture" and "one of Florida's and the nation's most remarkable buildings."

than a decade later used the wealth he had amassed from oil refining to buy railroads, extend them around Florida, and put up a series of hotels across the state, including St. Augustine's most famous nineteenth-century edifices. The fantasy of Florida as a paradise for tourists and retirees has continued into the twenty-first century, and real estate developers, speculators, and property owners have sometimes been reluctant to recognize reality—that both pieces of Flagler's legacy, the oil industry and the coastal real estate boom, have set the state running toward a nightmarish scenario in which rising seas will collide with properties worth billions of dollars.

All of this denial was compounded by the city being full of new migrants—especially *snowbirds*, retirees who winter in Florida. Many of them head north during warmer months, effectively avoiding first-hand experience of part of hurricane season. Leslee and her husband had bought a 1928 house that stood on a small ridge (not big enough to be called a hill) in Fullerwood Park, two blocks from the Intracoastal Waterway, and she also noticed over the years how the streets around her filled with more and more water from tidal flooding, even on sunny days. But newer property owners wouldn't have perceived a worsening situation, just an ongoing problem for the city to fix.

How are we going to teach these people who have moved here in the last decade? Leslee thought.

The threat, not just to St. Augustine but to the entire nation's historic landmarks, became all too vivid in 2012, when Hurricane Sandy ransacked facilities and museum buildings on Liberty Island, where the Statue of Liberty lifts her lamp, and Ellis Island, the nation's historic immigration station. The reports of Sandy shocked Leslee, and she dedicated more and more of her time to helping historic preservationists plan for sea level rise and the amped-up storm cycles climate change was beginning to deliver.

She began collaborating with St. Augustine and with Jenny Wolfe on issues like resiliency planning—that is, proactively increasing a community's ability to withstand future disasters. In 2016, Leslee organized a workshop on building codes and flooding. It was scheduled for that November.

Then Matthew arrived.

In addition to trashing Jenny Wolfe's beloved garage apartment, the hurricane soaked Leslee's planned workshop venue—the room in St. Augustine's city hall where the city commission also meets. It also poured water onto the Flagler College campus and into the basement of the Ponce. She still held the workshop in a nearby nondrenched location. But her focus by then was also on helping the recovery efforts. Hurricane Matthew had howled a loud message at the city, far louder than anything she could have arranged herself. Matthew was like a test run for future flooding, laying St. Augustine's weaknesses bare. Major landmarks throughout the city fared relatively well. For instance, the Castillo de San Marcos looked as if the storm had barely touched it, while the National Park Service offices at the edge of the grounds got a thorough soaking. The city's nineteenth-century alligator farm weathered things fine: the alligators merely floated up in the surge and descended back down when the water retreated. But some private homes, including historic homes, had major damage.

Anyone who needed personal assistance from FEMA had to fill out paperwork and leap bureaucratic hurdles. *FEMA wasn't coming out to help us,* Jenny Wolfe observed angrily. *We had to go to each of our buildings in the city with our small staff and check a box* on a form. She and her colleagues enlisted volunteers. Leslee gathered a group of students to walk with her to homes in and around the downtown core and help people file for disaster recovery money.

Then, while still reeling from the damage to her own living space, Jenny began knocking on doors herself. She focused on the blocks south of city hall—the Lincolnville neighborhood. Signs throughout the neighborhood mark the St. Augustine Freedom Trail, houses and churches where Dr. Martin Luther King, Jr., gathered with other activists during the American civil rights movement. Jenny spoke with a mother and adult daughter here. *One of them has asthma. They had to relocate to an apartment off of US 1*, a busy highway. The mother's uncle had built the house himself and given it to them. But money was tight, and repairs seemed daunting. Jenny also met a Black grandmother sorting through her belongings inside a dark bungalow with a screened-in porch. The house was raised up about a foot above the ground on columns called piers, but it had still filled with water. Books, clothing, photographs were stacked on the floor, drying. *I just remember feeling her sense of bewilderment.*

In the months that followed, Jenny and her colleagues had to review dozens of applications seeking permission to either demolish or gut buildings in the historic district—erasing architectural details that form the city's story, little chapters and subplots vanishing. Dozens of homeowners filed requests with the city to raise their foundations on high piers; each had to be carefully reviewed so that the modifications wouldn't distort the aesthetic or legacy of St. Augustine. (One of the conundrums of historic preservation is how much to modify a building or property—even if in the interest of protecting it from disaster—before it loses the very qualities the make it historic.)

In May 2017, as the community was still taking stock of the damages, the state and NOAA released a second report on flooding and sea level rise in St. Augustine with more detailed recommendations. This report suggested infrastructure improvements such as upgrading the stormwater drainage system. It also offered up several warnings. "St. Augustine's historic districts are vulnerable, immovable, and irreplaceable," it read. Neither FEMA's policies on building elevations nor state guidelines for road and bridge designs had realistically taken rising seas into account, even after the city's experience with Matthew and "dire sea level rise predictions." Moreover, "many residents seem not to know what's coming."

Then in September 2017, a little less than a year after Matthew, Hurricane Irma arrived in the city. Irma was a Category 4 when it struck the Bahamas and Category 5 when it made landfall in Cuba, then weakened when it reached Florida. Like Matthew, the storm gave St. Augustine a serious dousing. The surge of water gushed into some of the same houses and buildings that had only just been repaired. Like Matthew, it turned people's belongings into debris and scattered them across the street. In the days afterward, hotels put mattresses out on the pavement to dry.

The city was more prepared this time. *We knew what to do to help us plan in advance*, Jenny said afterward. At the Ponce, after Leslee suggested and urged the idea, Flagler College put up door dams, barriers at the bases of doorways that help keep water out, and various property owners around town applied tape and caulk and other barriers to make their houses and buildings as watertight as possible.

Still, this second storm dealt a gut punch to some property owners who were just beginning to recover. And the picture it painted—of a life marked by flood after flood—looked disturbing.

In late October 2017, about seven weeks after Irma, Leslee Keys traveled to Maryland. Some of her colleagues elsewhere—preservationists in communities all along the coasts—had decided to expand the conversation about sea level rise, old cities, and saving history. In 2016, a group in Newport, Rhode Island, had organized an unusual conference called Keeping History Above Water. Lisa Craig, then still the Annapolis chief of historic preservation, had proposed turning it into a regular event and convened a second such gathering in 2017 in Annapolis. There they discussed solutions ranging from the conventional to the fanciful: from raising building foundations higher to designs for retrofitting historic buildings with "amphibious" foundations that would float on the water's surface when floods arrive.

Leslee's presentation, which she delivered on Halloween with colleagues from Stetson University and the University of Florida, was sobering. They called it "Preserving Paradise." Slide after slide of flooded streets, and an even more disturbing estimate of the damages: *The National Oceanic and Atmospheric Administration predicts sea levels will rise as much as three feet in Miami by 2060*, read one of the slides. *By the end of the century, according to projections by Zillow, some 934,000 existing Florida properties, worth more than $400 billion, are at risk of being submerged.* Her last slide was an image from *Planet of the Apes*: the Statue of Liberty sunk in sand.

Immediately thereafter, Leslee began organizing this same conference—to bring throngs of engineers, coastal scientists, historic preservationists, archaeologists, even the military to St. Augustine. Maybe a couple hundred minds could figure out how to protect this place.

When I called Leslee for the first time in the fall of 2018, she offered me a small lament. "I knew when I was twenty that I loved old buildings and I wanted to save them. And I wanted to use them to teach people about our history and appreciate them. So for basically forty years, I have been doing that. And I have saved statues and archaeological sites, and you know, it's a wonderful experience," she said. "I hate to think that everything—this sounds very conceited . . ." She paused; she was more used to understating her own work. "I hate to think that everything that I have done for the last forty years is now going to be gone because sea level rise has not been addressed. So it's very personal."

She invited me to visit St. Augustine the following year in May for the conference. She told me that it wouldn't be too hot yet or too crowded with tourists and that it would be mint julep season. She knew how to talk about disasters—and how to throw a party.

CHAPTER 4

THE FIRST HOME

In contemporary usage in Western society, the first definition of a home might be a weathertight place with a fixed address where you can sleep at night and go about your personal business by day. Even in this age of online placelessness, such a home is nearly an entry requirement to adult American society. Sure, in theory you can acquire a job, a voter registration card, and various forms of identification without one. Still, mainstream social norms dictate that most everyone should live under a roof and inside walls, and anyone else is either bending the rules a little or has no other options.

But at its core (or maybe its foundation), a home is an old invention, and its importance might be more about imagination than about structure. Many animals—from hummingbirds to great apes—make nests. Even a butterfly will look for a hiding place to rest. A number of animals are builders—termites, ants, beavers, weaverbirds. It is impossible to know fully how animals perceive or feel about these kinds of homes.

But at least for humans, home is far more than just engineering; it is also a combination of meaning, symbolism, and social function. If we are going to survive this mercurial moment in the planet's history, we may need to revisit the rules of what constitutes such a place.

What is the most essential nature of home? You could answer this question from an endless number of angles. One is to dig back to the root, to consult our ancestors, to sift through clues about the earliest human moments on this planet.

☙

In the late 1970s, a young anthropologist named Richard Potts embarked on an eighteen-month research trip to study what scientists thought were proto-homes, about 1.8 million years old, in the middle of the Serengeti Plain in Tanzania.

The Serengeti is, in most ways, one of the wildest, least human-engineered spots on Earth—an arid expanse of grass and acacia trees where more than a million shaggy wildebeests travel annually in a north-south circuit, pursued by everything that wishes to eat them—lions, cheetahs, hyenas, wild dogs. But one of the most human places stands in the center of all of this animal activity, Olduvai Gorge, probably the most famous fossil bed in the world, where a now-dried-up river carved through layers of sediment collected in a now-dried-up lake bed, exposing millions of years of bones and other remains. In the mid-twentieth century, Mary and Louis Leakey found some of the first known fossils of our hominid ancestors here, eventually demonstrating that human origins lie in Africa.

Potts quite literally followed in the Leakeys' footsteps—petitioning Mary for permission to reexamine the bones and stone tools that she had collected and that were kept at the National Museums of Kenya. He also spent a month with Mary at Olduvai examining the geology of the site. He was especially interested in the excavations at a place called Bed I, the bed of an ancient lake and streams where sediments are layered on top of countless bone fragments from zebra, hippopotamus, rhinoceros, antelope, pig, giraffe, elephant. And mixed with these were the bones of *Homo habilis*, the "handy man," the long-armed, strong-jawed human ancestor who made stone tools from flaked and sharpened rocks so he could butcher what he hunted and eat the meat and the marrow.

Then and now, anthropologists have debated what constitutes a human home. If you find a scattering of stone tools, does this evidence mean humans were at home here? Is a home a hut? Is it a hearth? Or is it just a spot where a person used to sleep at night?

Many researchers had assumed that the bones the Leakeys had found strewn about here, left behind by these handy men and women, were evidence of the earliest sort of home—or rather *home bases*, an anthropological term for an area to which humans returned again and again to eat and socialize, a sort of campsite. Mary also believed that a circular arrangement of rock she found at Bed I was the foundation of a hut.

But just the act of bringing food to a designated spot to share might be an indication that humans thought of this as a home. "What does it mean to be human?" Potts muses, four decades later. "We take it so much for granted that when we go out and we find food, we hardly ever just sit down in a supermarket, tear open the packaging, and just start eating," as a large predator like a lion might do with a recent kill. "We delay our eating of the food and bring it to some other place, often with an expectation of sharing with others," and the places where we eat together take on significance as part of home.

But as Potts studied the bones, he developed doubts about whether *Homo habilis* had really made a home out of these particular locations. On closer examination, the alleged hut Mary had excavated looked suspect, more like a pattern created by tree roots growing in a rocky substrate than by hands, without enough human artifacts within its perimeter to seem lived-in.

The bones revealed an even more complicated picture. Using a scanning electron microscope, which can distinguish details that are only a few microns wide, Potts came up with a technique for differentiating marks on the bones—those left behind by stone blades versus scrapes and punctures from animal teeth. On the bones from Bed I, he spotted "a heck of a lot of not only tool and butchery marks but also bite marks made by the teeth of hyenas and even larger-sized potential predators." These animals had gnawed on various prey and left teeth marks on a few *Homo habilis* bones as well. In other words, our hominid ancestors were most likely feasting in this spot sometimes, but so were large, powerful animals with sharp teeth. "And it struck me that these earliest supposed best examples of a home actually were pretty incredibly dangerous places to hang around."

To Potts's mind, you couldn't call something home if you couldn't safely bring your children, your elders, or the sick to gather there with you. At this stage in our evolutionary journey, perhaps we had learned to share but not yet to protect one another.

Maybe this wasn't the right place to seek refuge. Or maybe it was just too early. Maybe we hadn't found our way home yet.

ॐ

To follow the human fossil record through the next couple million years is like rummaging through the tattered belongings of long-gone, far-flung family members. When we manage to find any surviving scraps they left behind, we try to write a story about who they were, where we came from, and who we've become. Some of those scraps offer little clues about the habits of our ancestors and the ways they related to the planet. And some of them reveal as much or more about our modern presumptions about what makes a home.

Homo habilis had various descendants or maybe cousins (depending on how you draw and prune the human evolutionary tree). These people, our forebears, or perhaps our aunts and uncles, wandered across thousands of miles. Step one million years forward in the geologic record (and three decades forward in time from Potts's first study) and you arrive in what is now Israel. Here are the remains of what are probably the oldest-known hearths. In the mid-2000s, a team of archaeologists led by Naama Goren-Inbar found bits of flint, which makes a good fire starter because it sparks when struck. At the edge of a lake, the scientists discovered crumbs of scorched rock, wood, and charcoal and the skillfully severed bones of deer, butchered with stone tools—eight hundred thousand years old, based on carbon dating and other methods. Eventually the scientists sifted through and analyzed the flint bits for their *thermoluminescence*, how much light they emitted when heated and what that meant about whether they had burned in the past. The researchers made a compelling case that these were the oldest-known hearths, which may have been left behind by stocky, big-browed people with flat faces named *Homo heidelbergensis*. There were no huts found here or other signs of either simple or elaborate architecture, but this looked much more like a home than the place Potts had studied. There were hand axes, choppers, scrapers, awls made of stone—and what looked like the division of space into various tasks, toolmaking in one spot, fish-cleaning in another, nut-roasting elsewhere.* There was plenty of evidence of deliberate place-making and a set of activities that suggested the residents had learned to look after one another.

Step another half million years forward in the fossil record, and you arrive in the French Riviera, to a site excavated in the mid-1960s, about a decade before Potts's first fieldwork. About four hundred thousand years ago, *Homo heidelbergensis* made a camp on what is now a busy street in Nice. French archaeologist Henry de Lumley stumbled across the site almost by accident, when builders were clearing it for a new apartment complex and the bulldozers turned up "Paleolithic implements." De Lumley recounted the discovery in the popular magazine *Scientific American*. The place was "scarcely 300 yards from Nice's commercial harbor" and "near the corner of Boulevard Carnot and an alley romantically named Terra Amata (beloved land)." The builders paused, and then three hundred field researchers, including archaeology students and some excited volunteers, spent the next five months rummaging through the dirt.

Here de Lumley and his exuberant team found evidence of a series of huts, each in an elongated shape with curved walls made of branches (evident from fossilized postholes) and a hearth at the center. "One of the earliest of the huts is perfectly outlined by an oval of stones, some as much as a foot in diameter and some even stacked one on the other. The living floor within the oval consisted of a thick bed of organic matter and ash," he wrote. "A little wall, made by piling up cobbles or pebbles, stands at the northwest side of each hearth"—he interpreted these walls as windscreens to protect the flames.

---

* The human family tree is complex and difficult to chart in any straightforward way. In older interpretations, *Homo habilis* is the direct ancestor of *Homo erectus*. But more recent archaeological finds show that the two lived side by side and might have had a common ancestor. Other hominids (like Neanderthals) also shared the planet with *Homo erectus* and with *Homo sapiens* over various periods of time in the last two million years. We are all related, but scientists are still determining exactly how.

After de Lumley was done, the French built a block of apartments on top of the site, and ultimately placed a museum therein, the Musée de Préhistoire de Terra Amata, which remains there today.

But if home or home base are fundamentally human, it's hard to argue that these crude huts are the defining examples of them: the fossil record doesn't offer up anything quite like this for another 375,000 years or so. If walls and roofs define a human home, why didn't our ancestors build more of them over all of those millennia?

In the years since de Lumley's find, a few scholars have questioned whether the artifacts there really tell a clear story about the site's occupants, because over time, erosion, seawater, or animals may have intruded. Still for decades, Terra Amata has been heralded by many as the earliest example of human architecture, the oldest houses.

But the huts are not the only important milestones from this ancient era of human homemaking. Our ancestors had landed on some other elements that were perhaps more central to human lives, well-being, and survival: the sharing of food, cooperation, and the act of place-making—altering a space to keep yourself and the people you care about safer and more comfortable. Homemaking can be a simple and imaginative act. Maybe we would do well to heed what they knew.

ᛞ

Search through the many layers and decades of evolutionary science for the moment when our own species, *Homo sapiens*, began inhabiting the Earth, and you will find several origin stories colliding with one another. Such stories can be dazzling and fascinating, but under scrutiny, sometimes they reveal as much about the biases of researchers and of Western culture as they do about the habits of our species.

The earliest stories about *Homo sapiens* had a Eurocentric orientation—a reflection of the worldviews of the anthropologists and archaeologists who conducted the research. One version—which held sway for decades but is now considered incorrect by many scholars—goes like this: the first *Homo sapiens* (our species, the "thinking" or "wise people") who appeared in Africa had anatomy like ours but hadn't yet developed modern smarts. And for years, scientists dated the *Homo sapiens* birthday, the moment we first stepped on the Earth, at about a hundred thousand years before the present. Then about forty thousand years ago or so, *Homo sapiens* moved into Europe and Asia and perhaps triggered the demise, one way or another, of the Neanderthals. According to this story, it wasn't until around this moment that we also developed what anthropologists call "symbolic reasoning"—a technical name for what is fundamentally the human imagination, the ability to step back and engage in abstract thought, the tendency to solve difficult problems, tell inventive stories, and dream elaborate dreams. That capacity was on full display at places like Chauvet Cave in France, where the thinking people left behind graceful and intricate charcoal paintings of animals such as aurochs, ibex, reindeer, and panthers on a rock wall around thirty-five thousand years ago. Just before those paintings went up, the story goes, we were blessed with some sudden and magnificent change in our brains—a "neurological advance, perhaps promoting the fully modern capacity for rapidly articulated phonemic speech," wrote Stanford anthropologist Richard Klein in 1998. This was the so-called human revolution, some moment of biological enlightenment. Scientist and popular author Jared Diamond was even blunter: "That is the time when we finally ceased to be just another species of big mammal," he insisted in his book *The Third Chimpanzee*, originally published in 1991. (Diamond's statement is arguably anthropocentric and betrays a lot of cultural bias. Do we really have so much more sophistication than a humpback whale or an African elephant, species that develop their own cultures? Were we really not humans until we started to create the things that Western societies would recognize as "civilized"?)

These scientists contended that our newfound intelligence also led some twelve thousand years ago to the "Neolithic Revolution," the unfurling of human agriculture, the moment humans started planting grains and building stationary homes so they could tend crops. Many scholars argue that crop cultivation is the most impactful dietary, cultural, and societal decision humans have ever made.

And if this is your story about our species' arrival on Earth, you might say that the human sense of home arose from these two developments. We got smart. Then we started munching on grains. After we settled down in villages in places like the Levant in the Middle East, we transferred our "individual loyalty from the mobile social group to a particular place," argues Ian Tattersall, paleoanthropologist with the American Museum of Natural History, in a 2013 essay. Maybe only then did we feel at home, in a true modern sense.

If this is your story, the making of home might also be inseparable from our more destructive tendencies. "Adopting a fixed residence went hand-in-hand with cultivating fields and domesticating animals," Tattersall continued. The first human villagers in the Levant "locked themselves into a lifestyle, and to make the field continuously productive to feed their growing families, they had to modify their landscape. Today, we carry out such modifications on a huge scale, and nature occasionally bites back, sometimes with a vengeance." If this is your story, you might think humans could never be at home without dramatically rearranging our surroundings to suit our own needs and desires, potentially making a mess of the land around us—and, eventually, of the atmosphere.

But many other scholars have since contested these claims of sudden, epic transformations in the human mind. The view looks different from other continents. As scientists spent more time digging through the evidence from Africa, the "human revolution" concept didn't seem to resemble the reality emerging from the dirt. The origin of *Homo sapiens* moved backward in time as older finds appeared, including a pair of modern human skulls from Ethiopia that were originally discovered in 1967 by Richard Leakey (Mary Leakey's son) but weren't dated until 2005—when newer, more accurate techniques were available. Using radioactive argon (similar to carbon dating), scientists discovered the skulls were nearly 200,000 years old. And it seemed improbable to some experts that our species puttered along for at least 150,000 years and then had a sudden epiphany. In 2000, two scientists, including a colleague who worked with Potts in Kenya, Alison Brooks, wrote a hundred-page refutation of the "human revolution": the whole idea reeked of a "profound Eurocentric bias and a failure to appreciate the depth and breadth of the African archaeological record."

Over the next few years, some new findings effectively blew the lid off the old story. At one archaeological site, in 2007, an archaeologist named Curtis Marean announced that he and his team of researchers had found far earlier signs of symbolic reasoning. They had been nosing into caves on the coast of the Indian Ocean in South Africa for about a decade. In one cave in particular—where a local ostrich farmer built them a wooden staircase for easy access—they dug up fifty-seven bits of pigment, most of them ochre (rusty red-orange and made from clay that is high in iron), some ground into powder. These were left behind by members of our species who lived there more than a hundred thousand years ago. The only real usefulness of these materials to a literal caveman or woman would be for painting on rock or skin, an act that would be meaningless to a person who lacked imagination. There were also sharp and sophisticated pieces of stone blades mixed with layers of ash.* And there were piles of shells. The people who used this cave appeared to also be the first known seafood-eaters, feasting heavily on brown mussels and sea snails. But the tides here were fierce and the mussel-covered rocks treacherous, and Marean believed that the only way to successfully and safely harvest these shellfish would be to study and enumerate the lunar phases and movement of ocean currents. To live here, to feel at home in this place, you needed a sophisticated sense of time, an understanding of the past and future and the patterns of change on the landscape. You needed smarts and imagination.

Meanwhile in Kenya, Potts, Brooks, and a crew of about four dozen people had unearthed a new set of tools and other traces of very-long-departed people at a place about 120 miles northeast of Olduvai Gorge. Like Olduvai, this place was a dried-up lake bed sliced apart by erosion—in a flat, silty, open area with hills and ridges rising all around it. In the exposed sediment layers, the scientists found one set of artifacts that were about five hundred thousand to one million years old and a newer set from about three hundred thousand years ago. Potts began studying the older tools in the 1980s, but he and his collaborators didn't start investigating the newer ones until 2002. And in the more recent of the two eras, someone was making stone tools that looked a lot like the handiwork of *Homo sapiens*, acquiring pigments like ochre

---

* Marean would later argue they had been made via a complex method that involved heating them in a fire, more evidence of smarts.

and manganese (which is black), and collecting obsidian, a volcanic form of glass that can be made into especially sharp blades. None of these raw materials existed nearby; they were sourced across distances of up to sixty miles as the crow flies, but perhaps four times as far (more than two hundred miles) by foot through the ridged landscape. To procure such things across difficult terrain, you would probably need to trade with other people, the scientists reasoned. And to do that, you would require a mental map of these places, a sense of where you came from and whom you might be trading with, and a sort of early social network. You would need memory and imagination and learning.

The scientists dug and scanned and analyzed what they found in Kenya, and at the end, "we kind of looked at each other, my colleagues and I, and we said, 'This is us,'" Potts recalls. This was *Homo sapiens*, making our way through the East African Rift Valley long ago, with what seemed like a fully developed sense of human imagination.

In 2017, another skull—originally found in a Moroccan cave decades previously—made headlines after its age was reexamined with newer and more precise scientific dating techniques. It turned out to be three hundred thousand years old, much older than was previously believed, and probably one of ours, *Homo sapiens*.

We were an older species than scientists had previously believed, and we had been smart for a long time.

We had used these smarts and imagination to build deep relationships with the places where we started, to learn how to survive and adapt, and this was probably the true root of the human home.

Climate studies show that in the early existence of *Homo sapiens*, we survived an unpleasant and inhospitable ice age, when the planet turned cold and dry and threatened to evict us. Then around seventy thousand years ago, the human diaspora out of Africa began. From there we headed to five other continents—to places of ice and heat, forest and desert, coast and mountain. But there were no signs of enduring buildings for many more millennia. On the entire African continent, no scientist has yet turned up reliable evidence of any permanent dwelling structure made by early humans (though, of course, only a structure that is either exceptionally sturdy or well protected from the elements can survive that long, and even then, a scientist would need a certain amount of luck to find it). One of the oldest-known examples of early *Homo sapiens* architecture is a massive structure found in southwestern Russia and comprised of the bones of sixty mammoths, circa twenty-five thousand years ago. And even this doesn't look quite like home; in its enormity, it is more like an ancient cathedral or meeting hall. Maybe the things we built to live in before this date had eroded away. Or maybe we made ephemeral camps only.*

But in the newer retelling of the human story, if we were so smart and so inventive by the time we began to migrate around the world, it seems difficult to argue that we lacked a sense of home. Home feels so fundamental to the human psyche. Home is part of our "psychological machinery," Curtis Marean said, when I contacted him to ask about the origin of home. He puzzled over the question at first, because home is so often associated with a physical house. But then he acknowledged that there has to be something deep-rooted about our desires for home. "When I use those words, I'm talking about cognitive structures, and cognitive structures come from evolution." How could humans have wandered the planet for so many generations and had no sense of home?

Richard Potts argues that the quintessential characteristic of humanity isn't really a hut or an elaborate village. It is some combination of our adaptability and our imagination.

If we are looking for the human home, maybe it's not a thing but a story.

In some of the most ancient cultures on Earth, home is a place one belongs to, a place of safety and a gathering point for reestablishing social connection. "Home is a place or places on the landscape that you are somehow connected to," archaeologist Margaret Conkey, who has studied Paleolithic sites in France, said in an interview with Canadian science writer Jude Isabella in 2013. "It's also a conceptual and symbolic notion as to where people are from, where they relate to, and where

---

* The Indigenous San people of the Kalahari Desert, some of the few traditional nomadic hunter-gatherers left on Earth, move every several days, migrating along familiar routes.

certain important aspects of their lives take place. Home is a place where you reconnect with people or memories."

And in some of the most enduring cultures, home is a place you care for—as if it were your family. "We call the country mother," a man from the Warlpiri Aboriginal community in the Australian desert told anthropologist Michael Jackson. "The mother gives everything, like the land. . . . So when you think of where you were born you think of the country. . . . You have to look after it."

Humans have never before made a home under this kind of sky, within the sorts of climatic conditions that we have now created. In May 2021, the observatory at Mauna Loa on the Big Island of Hawai'i logged an average atmospheric carbon dioxide concentration of 419 parts per million. From measurements taken from glacial ice cores, the last time the Earth's reading was anywhere near that high was at least four million years ago, two million years before the handy people were chiseling tools at Olduvai. The climate has never changed so quickly or so violently in all of human existence. Because there is a time lag between the increase in atmospheric carbon and the shift in the planet's temperature, we have yet to witness how bad it may get. According to the IPCC, even with the most ambitious efforts to control carbon emissions, "it could take twenty to thirty years to see global temperatures stabilize." And with the worst case, the highest emissions, the planet could warm to such a catastrophic level that we would barely recognize this Earth at all.

I have inherited my human ancestors' imagination and capacity for storytelling. But I am not a Luddite or a primitivist. We will never go back to those first hearths. And I am not so naive as to think that a story can save us altogether.

But in the last several years, human societies have engaged in a project of unimagination, of ignoring or denying the signs of climate catastrophe, of distancing ourselves from the way the landscape is changing. We have built and engineered and expanded and created short-term gains for ourselves but neglected part of the equation of home—to know the place we are living on, to heed it, to take care of it.

And I think that we may not make it through this crisis if we forget that home isn't just a thing we build, but an awareness of and care for our surroundings and the capacity to imagine new ways of living in them.

We will need to learn to relocate, to re-create, to reconnect and collaborate with one another across vast distances and cultures, to find safety, to make shelter, to contemplate the maps and models of the world and grasp what they are telling us about where the planet is headed. In short, we will need to be the thinking, wise people. We will need a new set of stories about what it looks like to live on the Earth in a manner that doesn't destroy our future. We will need to figure out how to make a home in the greatest crisis we've ever known.

CHAPTER 5

THE THAW

home, n. *a refuge, a sanctuary; a place or region to which one naturally belongs or where one feels at ease.*

In 1993, when Lisa Tom was in the ninth grade, she finally persuaded her parents to let her move out of their place in Anchorage and set up permanent residence in her grandmother's house in rural Alaska.

Lisa was born in Anchorage. Her father, who is white, sometimes took her camping and fishing on the rocky, thick-treed Kenai Peninsula to the southwest. And every summer, her mother, who is Alaska Native, brought Lisa and her brothers (and sometimes Lisa's dad) to their grandparents' house—nearly five hundred miles west on the tundra in a Yup'ik village called Newtok. (Lisa's parents didn't marry until years later, when she was an adult, and her last name, Tom, belonged to her mother's family, one of several English Christian names originally given to Yup'ik men by missionaries, then later taken up by entire families.)

Then she and her siblings would return in time for school to the dim concrete city with its sky pierced by white mountains. Anchorage is ringed by natural beauty, but economic inequality has made parts of the city bleak. The early 1990s was a period of economic rebound as Kmart and other big retail stores sprang up from the pavement and increasing numbers of summer cruise passengers found their way into the downtown shops full of trinkets. But a large number of houseless people have long been stranded in this city by trauma or health troubles or poverty, including some Alaska Natives displaced from land hundreds of miles around.

By contrast, Newtok looks more ephemeral, at least on the surface, like a makeshift fishing settlement. But its residents have roots in the region reaching back at least centuries and likely millennia. Just twenty miles from the Bering Sea, which separates Alaska from Russia, the few hundred people who live here on an outstretched plain of grass and dwarf birch and willow are almost all Yup'ik, one of Alaska's eleven Indigenous cultural groups. Yup'ik is still the first language here, English the second. The village's Yup'ik name, Niugtaq, refers to the rustling of grass. To Lisa, this place of grass and water felt far more like home than the city.

Lisa's eyes change color—sometimes green or blue or hazel or gray—depending on the light reflected from her surroundings. A quiet person by nature, with a seeming tendency to contemplate each word before it passes her lips, she sometimes describes herself as stubborn (although that could also mean resolute, depending on the context). She begged her parents again and again to let her move to Newtok, and one day, they relented.

The 1980s and 1990s were a momentous time in Newtok's development. Decades later, people would recall how the land was higher and drier then, full of hillocks and golden grasses, and even the weather was calmer, infrequently troubled by storms. Winters were colder then, the snow arriving in September, the ice on the rivers hardening by October. In winter, a person could ice-skate along a seemingly endless chain of ponds and creeks that fed into one another. In summer and fall, you could wander the tundra uninhibited, pick fat, ripe cranberries, blueberries, cloudberries (locally called "salmonberries"), and crowberries that could stain your fingers red and deep purple. And you could mostly live off the land by hunting and fishing and foraging for wild vegetables.

Lisa's grandmother made fermented fish eggs in the refrigerator and fermented salmon heads (a dish called "stinkheads")* sometimes using coffee cans in the kitchen and sometimes via the traditional method, in a pit in the ground lined with grass. She cured seal oil the traditional way, in a sort of container sewn together from the animal's body.† She taught Lisa how to hang herring to dry from braided stems of grass.

At the time, much in the village was relatively newly built. ("Everything was clean and not broken," a resident would recall wistfully, many years later.) In the 1980s, Newtok had installed a new water treatment system—a big blue tank injected with chlorine and a brand-new roaring power plant (made up of a pair of oversized diesel generators). The residents had also laid down new wooden boardwalks that crisscrossed among the houses. Some people made use of materials provided by agencies that worked with the community to put up their own houses; some bought building supplies from Seattle, shipped by barge. Lisa's mother's cousin, Tom John, whom Lisa thought of as an uncle, built his house in 1994. Another uncle put up a new house a couple years later.

In light of all this, it was jarring when Lisa's grandparents returned from a community meeting one night and told her the entire village was going to move.

A threat stood just on the horizon. To the south, lay the Ningliq River (sometimes also spelled Ninglick)—a powerful, sinuous arm of water with its fingers touching the Bering Sea—and it had always been a ravenous creature. In 1984, engineering consultants hired by the village, with funding from the state, noted that the warmth of the water and the force of the river were thawing the ice and breaking up the permafrost that bound the soil together, shearing off blocks of riverbank, devouring ponds and clusters of willows and leaving broken mats of grass and

---

* Alice Rearden, a Yup'ik scholar and schoolteacher from Bethel, explains in detail: "Stinkheads are usually salmon heads fermented underground in a pit lined with grass and covered. The process is usually done in summer, and the delicacy is eaten right after it is taken out from underground. Salmon roe is also partially dried outside on racks and then fermented in containers usually in a cool place." Proper preparation is crucial, says Rearden. "Stinkheads have caused deaths over the years through botulism because of being improperly cured in plastic buckets."

† "These days, seal oil is rendered from seal blubber placed in buckets, but sometimes people will ferment it in a hollowed-out sealskin container," says Rearden. During my visits to Newtok, I witnessed the bucket method but not the older, more traditional one.

earth on the shore. Unchecked, the erosion would eventually pose an existential threat. "Relocating Newtok would likely be less expensive than trying to hold back the [Ningliq] River," the engineers wrote. By 1994, the Newtok council had started planning a relocation and examined six possible sites—the most likely of which lay nine miles away, across the river at a place called Mertarvik (in Yup'ik meaning "place to get water"), a campsite of generations past, with a few old graves, large swaths of berry bushes, and a meadow frequented by musk oxen. Mertarvik lies on Nelson Island, more than thirty-five times the size of Manhattan but home to just three other small Yup'ik settlements. The land, formed by a volcano, was underlain with bedrock. It would not fall apart or erode.

Next year, we're going to move to the new site, some of Lisa's friends kept saying at school. But they were too young to understand what such an undertaking meant. Charlene Carl, another of her schoolmates, used to wander the village, staring at all the houses, buildings, boardwalks, fish racks, and little huts that served as traditional Yup'ik steam baths,* and wonder how it would be possible to whisk all this up and transport it to another place. The kids couldn't know what an undertaking it would be to move an entire community.

Meanwhile, life in the village already felt more unsettled. In 1994, the village's only school caught fire and burned down, an act of arson. (A new one was built a few years later.) In 1996, during a flood, the massive Ningliq River ate away the bank of the smaller Kayalivik River, also called the Newtok River, turning it into a slough. The Kayalivik River began to silt up, and bit by bit, it became more difficult to move boats in and out of the village. And the village land was then exposed to the unabated fury of the Ningliq, which slammed against it every time winds came from the south.

But no one in Newtok imagined that the actual move would take up the next twenty-five years of their lives. In that time, a warming climate, the loss of winter, and the disappearance of ice would all make the prob-

---

* From Alice Rearden: "Traditionally, Yup'ik men would take 'fire baths' in the *qasgi*, the community house where men lived. There were fire pits in the center and no rocks on them. Later on, people started building the more modern steam baths that have an oil barrel made of metal. They put rocks on top that you splash water on in order to get the steam." Though the Yup'ik steam bath is similar in function to sweat lodges used in other North American Indigenous communities, it is also culturally distinct.

lem much worse, much more rapid than anyone imagined, and Newtok would become a harbinger of the dislocation thousands of others would experience, all across the North, all around the world.

ॐ

The first person I ever met from Newtok was Lisa's uncle, Tom John, who by 2015 was working as the administrator to the Newtok council.

I came to Alaska, like so many other observers, to talk to people who had firsthand experiences with the loss of permafrost. Nearly 85 percent of Alaska and about half of Canada and Russia sits atop a frozen layer of earth that is up to tens of thousands of years old and has, until recently, served as a kind of perennially frozen bedrock. Some parts of the Arctic are, according to a civil engineer at the University of Alaska Anchorage, like a "giant ice cube"—a bit of dirt suspended in a matrix of frozen water, as much as two-thirds ice by volume. But in an era of not-enough-winter and too-much-summer, such landscapes have been thawing and coming apart.

Scientists most often tell this story on the grand scale of the planet's atmosphere. How the Earth breathes over millennia, sucking carbon in, metabolizing and storing it in the body of the planet, letting it out. Since the last ice age, the planet has kept a great deal of carbon stored in the frozen tissue of permafrost—a self-reinforcing pattern called a feed-back loop. Locking carbon up this way keeps it out of the atmosphere, which keeps the Earth cooler, which allows the planet to form more ice and frozen ground.

But warm the planet up a little, and the feedback loop can spiral in the opposite direction. Heat permafrost up a little, and the bacteria within can awaken and begin their work, decomposing ancient carbon-based matter, metabolizing it, breathing back out hundreds of millions of tons of carbon in the form of methane and carbon dioxide. As carbon in the atmosphere raises the planet's temperature, more of the permafrost thaws and wakes, and releases even more carbon. Permafrost can amplify what is occurring at the scale of the entire Earth. Moreover, arctic regions are warming three times faster than the rest of the planet, possibly because sea ice, which has a cooling effect and reflects sunlight back to the sky, has dwindled. An excess of heat in the Arctic can desta-bilize the global weather system—weakening the jet stream and causing freak cold snaps in the winter or triggering warmer, drier summers in places like the Mediterranean and California.

In other words, what happens on the remote edge of the tundra has global implications.

But at the smaller scale of a normal human life—a small city, a vil-lage, a house, a road—the loss of ice and permafrost is part of the frac-turing of place and the loss of home. The land beneath everyone's feet and the places people have known for generations as home are now sinking and falling apart. When frozen, permafrost supports the land from beneath. In cold months, what is called "shorefast ice"—a thick frozen platform extending from the shoreline—armors the coasts of northern latitudes against the pummeling of storms and high waves. But when ice becomes thinner, scarcer, or more unstable and perma-frost thaws and warps, the land itself is vulnerable. Around the state of Alaska, chunks of the coastline have been collapsing into the sea during storm seasons. And the arctic and subarctic ground has been shifting and sinking. Buildings and houses slump and tilt at cockeyed angles. Buried pipes and sewer lines wrench and rupture.

Newtok has become part of an international emergency. The village is one of three Alaskan communities—along with Kivalina and Shish-maref on the coast of the Chukchi Sea—that have gained fame as home to America's "first climate refugees," among the first people who are being forced to relocate because of the impacts of climate change. As the North warms, a growing number of other communities find them-selves at the mercy of powerful erosion and facing the possibility that they will have to head for more stable ground.

My first attempt to reach Newtok was in late September 2015, the edge of rain and mud season in southwest Alaska, and the furi-ous northern weather had already foiled my plans twice. I had spent weeks setting up the trip and was supposed to meet Tom John in the village, along with an architect named Aaron Cooke. I had planned to watch a crew of Newtok builders assemble an unusual home: an energy-efficient demonstration house that would later be hauled to Mertarvik.

Designed by a Fairbanks-based organization called the Cold Climate Housing Research Center, which researches new ways of constructing in northern latitudes, the house was both quick to assemble and vastly more energy efficient than the drafty houses that populated most of Newtok—made of a series of trusses like a bridge, with thickly insulated walls. It would be mounted on top of a metal foundation that looked like a cross between a bed frame and a toboggan, on smooth metal skids, so that if the ground became unstable underneath it, it could be towed across the snow and ice to a new location. The construction was supposed to be an event—in the process, a group of Newtok residents would learn how to put together and maintain such a house. But a week and a half before my trip, Cooke had called to say that the barge carrying his housing materials still hadn't arrived in Newtok, and he wasn't sure if he would be able to meet me there.

I decided to visit the village alone. But when I arrived in Bethel, a town that functioned as the region's transportation hub, I learned nearly all of the propeller planes that served the region's remote communities, including Newtok, were grounded. Small, low-flying planes often can't use the same kinds of sophisticated instruments and guidance systems available to larger jets, and the cloud had gathered so densely, become so endless and impenetrable and watery, that the sky looked like an ocean. The airport terminal lobby, which was no larger than a ferry boat, seemed like it might actually be at sea. Beyond lay the Yukon-Kuskokwim Delta, or Y-K Delta, a mostly roadless expanse roughly the size of Oregon, with a population of twenty-seven thousand people* clustered into dozens of tiny rural communities like Newtok, surrounded by a tangle of ponds and river channels twisting like dense cords through a fabric of tundra grass, willows, and dwarf shrubs. Between spring and fall, hundreds of millions of migrating birds pass through this place. By contrast, human movement here is far more constrained by the terrain and the lack of major infrastructure.

The terminal filled with the squeals of restless children and the drone of two television sets, one set to sports, one to cartoons, and the buzz of

---

* I use 2020 census numbers here, rather than estimating the 2015 population, though the population grew some in the intervening five years.

conversation, most of it in Yup'ik. I waited more than five hours, and at the end of the day, as I was about to take shelter in a hotel, I heard a ticket agent call Tom John's name over the PA system. I pursued him through the airport and caught up with him—a wiry man of modest stature, wearing a sage-green cap low on his forehead and a gray winter jacket—as he was about to get into a taxi, just long enough to blurt out my name and offer an awkward handshake.

Still he recognized me again the next morning and offered a Styrofoam cup filled with coffee from the free pot that brewed at the back of the terminal. "It'll give you energy," he said gently.

Later, as we passed the time, I asked him about the move and the erosion. Parts of the Y-K Delta have always been inherently dynamic and erodible. But climate change has been worsening and speeding up the problem. The erosion was now serious, Tom John told me. "Probably less than a hundred feet until housing is lost," he said.

But when would they relocate? "People have been asking me how soon. If I had a crystal ball in front of me I would tell you," Tom John said, a ripple of irritation passing through his voice. "But I would say within a year."

This would later prove to be overly optimistic.

I sat with Tom, his son-in-law, and some of his neighbors in a row of hard, plastic seats for what felt like hours or weeks or some strange state of suspended time, but was actually two days, and caught a glimpse of what it meant to inhabit two worlds—stretched taut between the demands of mainstream America, a culture that relied on speed and planning and infrastructure and growth, and this part of Alaska. Still the United States, but also a place apart, defined by a different set of rhythms of land and weather. You might have plans in the Y-K Delta, but they would always have to bend to the demands of this intransigent landscape.

Tom John had been caught in this airport once before, he told me (though others had tales of being stuck here frequently, sometimes for a week or more). But he had once broken down in a snow machine on his way to Bethel and spent the night in a shelter made of grass. "When it got dark at evening time I got a little paranoia, when I see something, a dark shape, like a ghost," he recalled. "I was thinking of the Hairy Man," he said, and he and his friends, sitting nearby, chuckled. Tom John had a throaty laugh, like "heh heh heh."

"What is the Hairy Man?" I asked.

"Bigfoot," said his son-in-law gleefully.

A few hours later, two days after my arrival in Bethel, the ticket agent finally announced that passengers would depart for Newtok. With Tom John and his friends and neighbors, I clambered into a little bush plane, and the pilot launched us up toward the ceiling of cloud and over the speckled brown tundra.

We landed in Newtok in a light rain, and on arrival, everything wore a shade of gray or brown, the land, the river, the weather-washed houses, but with a faint golden hue escaping beneath the clouds at the horizon, as if we'd landed on a space between night and day, sky and land. It all took the color of my exhaustion and my unfamiliarity with the place and reflected it back to me. But I could imagine, if this were my home, how hard it might be to leave, knowing the return journey could always be this fraught.

ᦓ

Why would it take more than two decades to move four hundred or so people just nine miles? And what might this say about the kinds of fortitude any community may need to summon in order to prepare for a life on an unruly planet? The answers lay in the complexities of both the landscape and in the mismatch between it and the demands of twentieth- and twenty-first-century life.

For years, some of the changes around Newtok were like watching someone age, a gradual accretion of differences, the threads of gray creeping into a person's hair, a dimple furrowing deeper into the skin to become a well-worn line. The muddying of the land, the shrinking of winter, the worsening of rain.

The edges of the land frayed, as ragged as old fabric. One Newtok resident remembers, around 2000, walking down to the Ningliq erosion line with her friends and siblings. She was about six years old then. *There was higher land. I used to pick berries around there. There were rocks and sand. And the land was already breaking. Big chunks of land were breaking. We used to climb around them and hang out around there and play.* Between 1954 and 2003, the average rate of erosion in Newtok was 68 linear feet per year, or more than half a mile in five decades, although in some dramatically stormy years, as much as 113 linear feet of earth collapsed.

In 2003, the U.S. Congress passed a law that would allow Newtok to trade its twelve thousand acres of unraveling land for eleven thousand acres of much more solid ground owned by the federal government. Mertarvik, the new village site—and nearly all of the land surrounding Newtok—was part of the nineteen-million-acre Yukon Delta National Wildlife Refuge. But even after the land trade was sanctioned, it would not be easy to assemble a new community there. Mertarvik still had no roads or on-site power source or infrastructure of any kind.

Meanwhile worse storms had begun to arrive in Newtok, and the situation grew more urgent. In the fall of 2005, such a storm surrounded the village on all sides with rushing water and turned Newtok temporarily into an island. It also ripped apart the village's barge landing, making it difficult for the community to receive fuel and supplies from the outside world. The following spring another storm battered the village. Both tempests swamped the drinking water supply, spread raw sewage around the community, and destroyed food storage sheds. People worried that an even bigger deluge could inundate the whole place, and the only high ground would be the airplane landing strip, a wind-whipped shelterless gravel road, according to a report later compiled by the Army Corps of Engineers. The only place that could provide the "barest shelter" in an emergency was forty miles away by boat. The Corps estimated that the village could last just ten to fifteen years longer before erosion claimed the heart of the place—if it could survive the storms.

But Newtok's prospects began to improve after community leaders reached out to a planner named Sally Russell Cox, who worked for the state of Alaska. At the time, there was no agency at any level of government with authority over or experience with moving an entire community threatened by climate change. And certain bureaucratic requirements made it difficult to secure funding for a new settlement.

For instance, in Alaska, one must have children in residence to receive money to build a school. But given past experiences with boarding schools, parents with kids didn't want to move to Mertarvik until a school was actually available—creating a chicken-and-egg problem.

Cox helped Newtok find solutions to such challenges by gathering allies among state and federal bureaucrats and prominent Alaskan organizations. In 2006, the Newtok citizens and their allies formed an unusually tight collaboration called the Newtok Planning Group to try to help the village.

Around that same time, the Newtok council and administrators decided to put up three houses at the new site in the same manner that those in the village were built—with a few hands and some modest construction tools. Through a BIA grant, they acquired three housing kits—simple wooden frames and materials with assembly instructions. But when the barge carrying the kits tried to reach Newtok, a storm again swelled in the skies, and the ship dropped them instead in another village, where the materials sat in the rain.

By this time, Lisa Tom had fallen in love and left her grandmother's place for the home of her sweetheart—Jeff Charles, who also grew up in Newtok, a man who has sturdy hands and a terse manner of speaking but a near-constant smile. He had built his house more than ten years previously, saving every paycheck to buy the materials. *My honey was in his early twenties when he built it by himself,* she recalls. *He just used non-electric saws and stuff.* By the time they got together, Lisa had two small children from a previous relationship—Jimmy, named after her Yup'ik grandfather, and Ashley. And after she and Jeff moved in together, they built an addition to the house so they could all share it as a family—until it eventually became a cozy two-bedroom with an oil heater at the back and a common living room and kitchen. They also learned that they would add a third child to the household; Lisa was pregnant.

Still, in the summer of 2006, Jeff took a job, with eight other men, to assemble the then-dampened houses at the new village site across the river. During the construction season, he had to leave Lisa and the kids, camp in a tent at Mertarvik, subsist on instant noodles, and work ten- or twelve-hour shifts. By 2007, the houses were done. But few houses from the Lower 48 are designed to handle the conditions of the Alaskan tundra—the unstable, icy ground, the frost heave, the hundred-mile-per-hour winds, the bitter cold of winter, the damp springs and autumns. After prolonged exposure to the elements, the materials inside these first houses molded, the foundations eventually sagged, and no one moved into them.

The community was barely any closer to claiming its new home than before.

In the winter of 2006, Lisa and Jeff married, and she became Lisa Charles. They knew they would remain in Newtok—in their handmade blue and gray house with the fish rack and the steam bath beside it—for a while still.

☙

In the language that bears its name, the word *Yup'ik* means "real people."

The closest word to the English *home* listed in the *Yup'ik Eskimo Dictionary*, published by the University of Alaska Fairbanks, is *kinguneq*. Accordingly it also means *point of origin; area behind; time past*. In this definition, home is about both time and place.

But in Newtok, people often use a different phrase, *utercugtua nunamtenun cupegtua*: the first word means "I want to go home," and the entire phrase expresses both a desire to return and a longing for the land. It refers to a kind of homesickness that can be either mild or full-blown despair. There is another common phrase that describes people's feelings about the land, *Man'a nunakaput*, "this is our land." "It means that we came from the tundra," a former Newtok resident explained to me. "The land is like a person we have to respect. That's who we are, real people from the land."

On the day after my arrival in Newtok, Bernice John, Tom John's wife, offered to take me upriver where I could see the village's point of origin, its area behind, and its time past.

I had spent the night in the school, on a camp mattress on the floor of a stuffy, one-room library. There are no hotels in Newtok, and most village houses are too crowded to take in a stranger. Sunday morning had been eerily quiet as I wandered between the rain-washed houses. Even the dogs seemed to be hunkering indoors. But Bernice found me in the middle of the village on one of the boardwalks, as if she had known exactly when I would be there.

"Are you looking for somebody?" she shouted. Where Tom was watchful, Bernice was bright, with a chirrup in her voice like a sparrow's and arched eyebrows that sometimes gave her a quizzical look. She strode toward me in blue sweatpants tucked into tall purple muck boots, a grin on her face, her eyes encircled with delicate black wire-rimmed glasses, her hair in a long bob with bangs threaded with silver and brown.

"My husband is going to go check a blackfish trap," she said. "Wanna come and see?" It didn't entirely seem like a question. Quite obviously, I would follow her.

We traipsed along the boardwalks, the ground beneath so inundated that the wood made a sloshing sound with every step. I waited as we made a long stop at her house—a one-room dwelling occupied by the couple and several children and grandchildren (eleven residents altogether). From her kitchen, Bernice retrieved several pieces of her homemade fry bread in a Ziploc bag and gave them to me to take for the journey. Then the march back to the edge of the Kayalivik River, the smaller stream that ran along the east side of the village, where Tom had docked their aluminum skiff.

Tom ushered me onto the boat, and Bernice and their nine-year-old grandson stepped in after me. Our footsteps clomped like drumbeats against the aluminum hull. Bernice gestured for me to take shelter from the damp wind in a makeshift cabin built out of wood with a plastic tarp draped over it. I nestled in beside coats, buckets, a mug, a cracker box, a camp lantern, a metal coffeepot, and a frying pan. "We sleep in the boat when we travel out into the wilderness," Tom explained.

Gulls wailed overhead as Tom shoved a long pole into the mud and nudged the boat into the gray-blue water. Slowly, he said, "When we get to the old village site, I'll show you where these people came from." I understood that by "these people," Tom meant the Yup'ik founders of Newtok. Then he yanked the starter rope until the motor sputtered and then roared like a jackhammer.

Until a few decades ago, the people of the Y-K Delta maintained a seminomadic lifestyle, migrating between seasonal locations. Beginning each April, they were on the move, camping at the coast to hunt or journeying up the rivers in search of fish. Around freeze-up, they settled back into winter homes. Their feelings of home belonged not to any one of these places but to all of them. And they could relocate more easily as the seasons and the tundra required. Until as recently as the 1950s and 1960s, the architecture of the region was entirely local, nimble, functional, and well suited to this landscape of ice, water, and tundra. People lived in earthen dwellings, also known as sod houses, lined with driftwood and woven-grass mats. According to those who are old enough to remember them, sod houses were comfortable and efficient, kept warm with wood-burning stoves and bright with seal oil lamps. Many say the old sod houses were better and warmer than the flimsy wood-frame constructions people built in more recent decades. In generations past, with such simple dwellings, it was far easier for a community to pick up and leave when necessary.

But from the middle of the twentieth century, it was near impossible for the Yup'ik to continue their seminomadic traditions while also complying with government policies and participating in the larger American economy. And fixing to one place in such a dynamic region would require, in some cases, reengineering the landscape and building something more permanent than a sod dwelling or a camp.

Yup'ik people started to camp at Newtok in the late 1940s. The location was chosen haphazardly—it was the farthest point upriver that a barge was willing to travel. The law stipulated that Alaskan children must receive a formal education, and families didn't want their young kids shipped off to distant boarding schools. With access to supplies from commercial ships, the Newtok community could have a local school and keep its children close.

In the boat, the Johns and I zoomed up the river. Tall, tawny grasses and shrubby, waist-high yellow-leaved willows stretched from the banks to the horizon as far as the eye could see. The water was glassy and the river bloated from swallowing both rain and tundra soil. I crept out of the cabin and toward the stern to talk with Bernice, who pointed at a muddy bank ripped open by the river. "See where it's eroded?" she shouted above the motor. "It only stops when we're frozen."

As the frozen season has shortened and the thawed season length-
ened, the landscape has become ever more unsteady, and this instability
is threatening everything about the pulse of life here on the tundra.
Among other things, ice and snow are substitutes for infrastructure
in a region this cold. In a cold climate, frozen ground can be as hard
and sturdy as pavement. A river covered in thick ice can function as
a highway for snow machines. But as climate change raises the winter temperature even a few degrees at the edges of the season, frozen
ground turns to mud; ice cracks and thins and no longer supports as
much weight. When a person falls through broken ice into water, they
can quickly go into shock, develop hypothermia, and drown. In some
instances, Y-K residents have plunged through the ice to their deaths.

From the stern of the Johns' boat, we could see Nelson Island—
containing the new village site at Mertarvik—in a shroud of low clouds,
its rocky ridgeline faint and blue in the haze. But we headed the opposite
way, toward the place behind, where the people of Newtok came from—
the winter village they left only one or two generations ago. It bore the
same name as the river: Old Kayalivik. Bernice's family came from the
old village. "I never got to experience it myself," she said.

We arrived in about twenty minutes, and Tom slowed the motor,
which cut out and began to beep. "The grassy area," he said, gesturing
for me to look beyond the bank to our left.

Across the flatlands, a few tilted wooden crosses and fences marked
old graves—the influence of missionaries. The land rolled up around
them into a set of knolls. These were the remnants of sod houses, since
reclaimed by grass. They were so crumbled and overgrown that they
were barely visible from the boat, which was just offshore.

The old village site was some distance from the riverbank, and I was
wearing only calf-high rubber boots. Tom John was eager to reach his
fish traps and didn't seem to want us to go ashore. Instead, he jerked
the starter rope, and the motor rattled into action again. "If we were
living their lifestyle," Bernice shouted again, "we wouldn't be living in
this situation."

The river became smaller and smaller as we proceeded up the channel. When we arrived at the site of the blackfish traps, Bernice dropped a heavy metal anchor into the water, and all of us trundled out of the boat and onto the riverbank. The tundra here was grassy and springy under our feet, and a small creek meandered across it. Tom John wore long gray waders attached to his shoulders by suspenders, so he could wade in and check his traps, each of which was comprised of a wire cylinder about three feet long, with a wire cone mounted on the front, set into the creek channel to force the fish to swim in. Once they passed through the cone into the cylinder, the fish couldn't exit from the cone's funnel-like opening. Tom's traps had lured several fish, and he gathered them, slender and dark and wriggling, in a white plastic bucket. Tom and Bernice's grandson waved the bucket in front of me and exclaimed, "Fishy!" (Found only in Alaska and Siberia, blackfish are about eight inches long and often slow-moving but rugged. They can breathe air when necessary—surviving in the tundra mosses during a dry period—and live for months under ice. Like the people here, they are used to coping with whatever the elements dish out.)

To prepare the fish, you boil them, Bernice explained. "It's one of our delicacies." I could see the whole family's looks of satisfaction, the same I would feel if I had picked a bowl of ripe tomatoes from my garden, a sense of having tended the land and reaped something in return.

After the fish were secured, we boarded the boat again and turned back toward Newtok. The sky had blurred into gray once more, and drizzle dampened our clothes. Bernice sat with me in the back of the boat and mused about the experience of living in Newtok. She seemed to want me to understand.

"You won't find highways out here. Nothing paved," she said. "To me, it's a simple life." She paused. "I like simple, unlike city life."

꙳

Nearly everyone in Newtok grew up hearing a kind of saying or prediction, though the details varied depending on the teller. "The world is changing following its people," says the Yup'ik adage. According to Mary John, Tom John's mother, this meant weather would change when people ceased to follow nature's laws, and the snow would stop falling. She had heard it when she was young and remembered it still in her late seventies. Others believed the prediction was about people's moral behavior, their inability to get along, or their poor actions toward one another. Whatever the case, it was widely regarded as an ominous prediction, foreknowledge of climate change well before any scientists began talking about such a thing publicly.

And the implication seemed to be that a reconsideration was necessary—a willingness to reckon with the land and contemplate the values of the past.

Just after the first Mertarvik houses were complete, however imperfectly, the story of Newtok began to attract the attention of the rest of the world. In 2007, the *New York Times* named the people here "among the first climate refugees in the United States." Reuters also published a story describing how climate change was thawing and eroding coastal Alaska and endangering Newtok and other villages.

In the years that followed, a trickle and then a stream of curious reporters began to visit the remote village, and the community gained a kind of fame. The village was suddenly more than an isolated settlement on the tundra. By naming Newtok villagers as among the first, such stories seemed to hint that they were also not the last. Newtok was a symbol, and its residents were the harbingers of a much larger crisis, in which anyone might experience such dislocation, a prospect frightening even to those of us who imagined we were more mobile and detached from the world. "We're United States citizens. We're the taxpayers. We have military that serve in the Iraq war," Stanley Tom, then Newtok's administrator, told NPR poignantly in July 2008. "So we're part of United States. If we're being wiped out, you know, who's going to replace us?"

The attention helped the people of Newtok gather more support. But various hurdles still made it difficult to create a viable new village site at Mertarvik. For one thing, building on the remote Alaska tundra is quite different from doing the same task in, say, rural Indiana. Among other things, there are simply some aspects of mainstream society and infrastructure that you can no longer take for granted once you head into remote country in northern latitudes. In a temperate city, there are readily available sewer and water systems erected, maintained, and financed by a municipal or county government. Build a house, and you connect its plumbing to something that already exists. And there's the extensive power grid that the U.S. government rolled out over a period of about six decades until the 1950s, stringing power lines from urban areas to the farthest-flung rural farmsteads. Few people in the Lower 48 now have to think hard about the nearest power source when setting up a home, unless they choose deliberately to live off the grid, which is a tiny fraction of American households. Or consider even the most basic form of infrastructure—a road. None of these fundamentals come premade in a new village site in the far reaches of the Y-K Delta. And infrastructure has a longer gestation time than most people imagine. (A Los Angeles bureaucrat once told me that it takes at least twenty or thirty years to plan out and put together any major infrastructure, even with all the resources of a major city.)

In 2008, the community requested help from the Department of Defense through a program called Innovative Readiness Training, or IRT, in which military personnel practice skills similar to those required when building a base in hostile territory while "providing incidental benefit to communities"—a kind of service learning for soldiers. In 2009, military personnel swooped into Mertarvik in Black Hawk helicopters. The conditions were difficult, even for the army. Snow fell in July. Equipment sank into the tundra. Concrete blocks drowned in mud. Pieces of the road, made of a sturdy mat-like material called Dura-Base, washed out in the high tide. A barge ran aground. But over multiple construction seasons, the IRT crew assembled a working set of roads, blasted a rock quarry into the bluff, and built a barge landing.

In 2011, the community and the Newtok Planning Group outlined a formal vision for Mertarvik. The plan was full of hopes, both simple and ambitious. It envisioned a greenhouse, a big school, a new store, and wind turbines that would spin out clean energy when the wind rushed through. The plan also harked back to the traditions of the past. It referred to four phases of relocation: *uplluteng*, or getting ready; *upagluteng*, or pioneering, a word that also evoked the process of moving with the seasons; *nass'paluteng*, meaning "transition"; and *piciurlluni*, meaning "we made it."

The community had already spent a long time getting ready. So in 2011 and 2012, Newtok made a second attempt to build at Mertarvik, this time with a new set of houses funded by the Association of Village Council Presidents. Once again, Lisa's husband, Jeff Charles, was asked to help as a crew supervisor. The houses would have a slightly better design this time: the walls would be made of something called "structural insulated panels," preassembled in a factory with insulation sandwiched between the wood. The crews sited the houses away from the water, upslope, with a wide view of the river. But the work was much the same as before, hand tools and portable generators, dinners of instant ramen, long days under the Alaskan midnight sun.

Two of the houses were nearly done at the end of the first season, and in 2012, Lisa and the kids joined Jeff for part of the summer, camping out in a mostly finished house, from which they could watch seals swimming across the water on a clear day. At the time, she worked as a cook at the Newtok school and had the summer off. She had just given birth to her sixth child, Jeffery Jr., and she could only travel back and forth if the water was calm. *I wouldn't want to bring him on a bumpy ride, because you know like shaken baby syndrome*, she remembered years later. There was no electricity. The kids slept together on the floor of the living room, because they were too scared to be separate from their parents, even in an adjacent room. Lisa's mother came to visit. *She would babysit, and I'd get to go fishing*, Lisa recalled, pole fishing for salmon in a small stream nearby. And there were ample varieties of berries to pick. And it felt like the move was *finally, really happening*, Lisa thought. She was eager to start a new life in Mertarvik, without the threat of erosion. From the window of the house where they stayed, *you could see everywhere. We've seen musk oxen before from the kitchen window of that house. You could see them going up the hill and around.* But the family couldn't stay. Lisa had to return to her job on the other side of the water. *It would be expensive living off portable generators with family and bills to pay, loans, that kind of stuff.* They needed to go back to Newtok.

A group of elders agreed to travel seasonally back and forth to take care of the new houses. The community called them "the pioneers." But still, no one lived there year-round.

By this time, tensions and stress levels had risen in the community. Stanley Tom had achieved a national reputation. (One article in the *Guardian* called him "the most powerful man in Newtok.") But some members of the village grew impatient with him and the others who had presided over the council for many years. They began to clamor for new elections.

The community rift ballooned into a legal dispute. After a new council (which called itself the Newtok Village Council) won repeated elections in 2012, the old council (called the Newtok Traditional Council) refused to yield their offices, their records, or their bank accounts. The state began auditing the old council's spending and found discrepancies and other evidence of problems. Some village residents refused to acknowledge the new council; others felt the old council hadn't been managing the relocation properly.

Work stalled on the houses and on a new Mertarvik Evacuation Center, intended both to serve as a gathering space and to house the community in an emergency. The latter remained less than half finished, a skeleton frame with no walls, and the former had incomplete wiring. And underneath this newer set of houses, the land settled, and the foundations sagged.

In early 2013, Lisa Charles found herself in an unexpected position. She had served on the old village council years before (between 2006 and 2008) but was never involved in anything related to the council dispute. A levelheaded, mild-mannered person, she was a good choice for community leadership. So, although she was shouldering the demands of raising six kids and working, she was also asked to be part of the new village council, which would eventually include Lisa's father-in-law, Paul Charles, and Bernice John.

Meanwhile, storms continued to snarl and gnash at the edge of

Newtok. A particularly monstrous one arrived in November 2013. *Some roofing material flew off some houses. There was one home that built a wooden cover on the top of their roof that got blown off. It was pretty heavy,* Lisa remembered. *I think we lost a lot of land that day.*

❧

In May 2014, a legal ruling from the U.S. Department of the Interior settled the village council dispute, affirming the results of the 2012 votes. But by the time I visited Newtok in 2015, the rift between the two parties remained. The old council refused to speak with me, even though I walked to the houses of some of its members and knocked on their doors.

But Lisa Charles heard that I was visiting Newtok—the lone reporter who had been stranded in a storm. She dropped in on me in the evening in the school library as I rummaged through the belongings and instant camp meals I had strewn about the room, and asked if I needed anything. She had a generous spirit, and perhaps she took some kind of pity on me, because the day after my boat trip, she invited me to lunch at her house, a couple minutes' walk north of the Johns' house on the boardwalks.

The house was blue and rectangular, with a satellite dish jutting out one side and a roughly hewn entryway on the other, stacked with boots. Someone, presumably the kids, had taken the liberty of labeling the front door with blue and purple markers in an uneven scrawl: "Welcome to Lisa's House. Lisa's House Real." Then to the left of these notices, in more marker ink, a list of all the household members, including kids, parents, and dogs.

Inside, the house was full of lively, comfortable clutter. Family photos hung all over the walls and were papered across the refrigerator door. Cooking implements sprawled across the counters. A shiny stand mixer sat on top of the oil stove that heated the house. Two teakettles stood on the kitchen range, and a crowded spice rack hung on the wall above. Clotheslines were strung across the ceiling, from which hung hats, coats, and gloves, some store-bought, some handmade of furs from animals hunted locally.

It was a house striving for modernity in some of the most inconvenient circumstances. By the door stood an enormous plastic trash bin that was used to hold not garbage but water, hauled from the school.

No one in Newtok had plumbing or running water. Lisa's son had just left to retrieve more water a few minutes before I arrived. Meanwhile, her house was full of appliances made for a space with working pipes. The sink had a spout but no water came from it; Jeff had drilled a drain so that they could run the dishwater out of the back of the house. Their new washing machine was jammed into a tight space behind the wood-stove. It was automatic and, when connected to pipes, designed to sense how much detergent and water was required by each load, but Lisa said they had to dump some in by hand from the tank whenever they heard the device trying to start up. "I heard we're going to have running water in Mertarvik, and I hope it's true," she said wistfully.

The television was hooked to satellite service. But internet and other telecommunications were glacially slow, and cell reception, which had come to the village less than ten years previously, was spotty. (Lisa had bought the first cell phone in the village in 2007.) But her two smallest children, Jodi, then age five, and Jeffery Jr., three, were mostly ignoring the cartoons running in the background and amusing themselves quietly. Jodi had her hands in a Play-Doh jar. Jeffery seemed to be laughing at his own private jokes.

Lisa and her cousin, Brittany Tom, a nineteen-year-old who was babysitting for her, were seated at a small oak table with white chairs. I had come toward the end of their meal, and the two had mostly finished eating. But Lisa had laid out various plastic bags full of fish she and Jeff had caught and dried themselves, and she offered them to me. "These are dried halibut," she said, gesturing to some pieces of white meat that looked like jerky. "There's dried salmon or herring." Skeptically, I took a piece of dark pink salmon and bit into the chewy meat. But the flavor was rich and deliciously fishy and bittersweet with smoke. Beside the fish was a jar of seal oil, and Lisa encouraged me to dip the meat in the oil. But this was a stranger and more pungent flavor.

I took a sniff and suddenly realized what I had been smelling all over the village, in the houses and the school gym and the stores and drifting into the air. It was seal-hunting season. Hunters were going out every day to catch the animals and returning to cut and cure and process the meat. And the odor of rendered seal oil hung over the entire place.

The hunting of seals in this part of the world has often been misunderstood by outsiders. "The White Seal," one of the stories in Rudyard Kipling's *The Jungle Book*, is set on the island of St. Paul, three hundred miles southwest of Newtok, and portrays seal hunters greedy for fur pelts. An animated version of the story, filmed in 1975, aired often on television through the 1980s and 1990s, depicting lumpish hunters with dark, clownish faces wearing pale parkas and carrying clubs wielded on seal pups. The story and the movie left in my childhood imagination the idea of cruel people persecuting intelligent, sad-eyed animals. But the people of St. Paul, the Unangan, were forced to carry out commercial hunts by Russians and later by the U.S. government. The Yup'ik seal hunt has always been a subsistence practice—undertaken with deep reverence for the land and sea. When someone successfully hunts their first seal, they often give the meat away to an elder, and the family sometimes organizes a celebration. Newtok also hosts annual dances and invites villages from the region; at these events, families often celebrate hunting and fishing accomplishments. In some traditional stories, such as those recorded by Alaskan anthropologist Ann Fienup-Riordan, seals have their own societies and villages, as humans do, and the hunt is a reciprocal relationship, in which seals offer themselves only to those who treat both the natural and the human world with respect.

I took a taste of the seal oil and inadvertently made a face, and Lisa giggled. "It's something you have to grow up with to like. When I was a kid I was crazy about seal oil." I turned back to the fish and the stacks of big crackers also known as "pilot bread," popular in rural Alaska. "When you can't get fresh bread, those are a big help," Lisa explained.

Lisa held up a bag of wild tundra root vegetables to show me. Known as "mouse food," people in the village harvest them from the caches of mice and voles. Each root was like a tiny parsnip but the width of a fine paintbrush handle. Brittany exclaimed, "Those are good in seal soup."

Brittany Tom now lived with an uncle and several other relatives in Lisa's old house, their late grandmother's house, which had since become the closest dwelling in the village to the erosion line of the Ningliq River. "Do you know when they're gonna move grandma's house?" Brittany asked Lisa. "Maybe spring or winter. We gotta wait for the FEMA update," Lisa said. FEMA was assessing the condition of Newtok's housing stock. Lisa's grandma's house was old and spare and covered with chipped red paint. But the family was still waiting to hear if FEMA would allow them to save that house or the one we were sitting in and haul them across the river.

After a half hour of watching me eat her fish, Lisa left for work, and Brittany and I lingered at the table. Brittany was a petite person, casually wearing multicolored leggings and one red and one blue sock tucked inside sandals, with a soft, soprano voice and an obvious enthusiasm for storytelling. "I graduated in 2014, last year, and I was supposed to go to college. But then my grandma wanted me to help to take care of her. And she ended up passing away in April," she explained. She planned to apply for the spring semester. "I can't wait," she said with enthusiasm.

But she still worried for her community. "Did you see how close the water is toward the houses over there? When I was younger, I remember the land being able to stretch way far out."

She pulled out her cell phone and began showing me photographs as the television murmured in the background and Jodi banged the bottom of the Play-Doh container loudly on the kitchen table. There were images of traditional dancing, with dance fans made of feathers, fur, wood, and grass. Then another of someone harvesting a seal.

Then an image of Brittany's daughter, who had been born when Brittany was sixteen. She had let the father's parents adopt her. "Even though it was hard for me, I didn't want to take her away from what made her happy," she mused. The daughter lived in another village, called Chevak. But Brittany had a hard time imagining her own life anywhere other than in Newtok. "I tried to move away for a bit, but I didn't last. I ended up coming up back here," she said.

And yet, her whole life, she had known Newtok would move. "I thought it was going to happen right away, but here I am graduated and everything, and they've only built a few houses."

Growing up, she had heard elders talk about how the climate was changing and how this would create hardship for people. "It makes me sad," she said. "Like the kids who wouldn't be able to grow up where we grew up, and wondering how they're going to end up, like how their lives are going to be changed."

$\mathscr{F}$

On the day that I left Newtok, the weather was beginning to turn. Frost settled on the surface of the boardwalks, rendering them extra slippery. I spotted a little boy making a game of running toward the boardwalks and sliding across the icy surface. My own boot slipped off the boardwalk and splashed into the muck beside it. But the morning was otherwise auspicious. Early sunlight crept into the horizon and tinged all the buildings rose gold. The sky turned a fresh, pale blue, and I walked the village, photographing everything. The shack of a post office, not much larger than the average backyard storage shed, tilted at a strange angle by the settling ground beneath it. The ghost of the old school that had burned in Lisa's youth—looking like a rotted-out corpse, with partially boarded-up windows and pocked and frayed walls. I could document the damage, and also the beauty. The boats moored at the edge of the river. The golden grass glinting with puddles of light. The clotheslines hung with laundry and drying racks loaded with cuts of wild fish. Birds piped overhead, and I could see that it was quietly lovely here. I could see how, if you knew the place, if you were a real person of this land, you would worry for it and how it was withering away, the way you would for the body of an ailing friend.

Lisa Charles gave me a parting gift of a tiny hand-sewn ornament, boot-shaped, soft, sewn from beaver fur and calfskin—along with some coffee poured into a Styrofoam cup. Tom John, whom I found in a small building beside the school that served as a community office, simply shook my hand. "See you in the future," he said, as if the future were a destination on the landscape that we might both arrive in.

One of the last photos I took, before departing the village on another tiny propeller plane, was of an empty building pad made of Dura-Base, the same material that had been used to forge Mertarvik's first roads. Neither the barge—the one with materials for the new energy-efficient house—nor the Fairbanks architect who was supposed to meet me here originally had arrived.

Weeks after I left Newtok, I learned the barge had first caught fire and then later been blocked by storms and sailed on to Nome, three hundred miles north along the coast. As usual, the tundra and the Alaskan weather were not going to cooperate with human designs.

𓆉

Later that fall, again with help from Sally Russell Cox and other Alaskan officials, the community entered a design competition sponsored by the U.S. Department of Housing and Urban Development. The application envisioned green design—with energy-efficient houses specifically suited for the tundra. The $60 million grant would have allowed most of the community to move quickly. But in January the following year, their request was rejected.

In the summer of 2016, Aaron Cooke, the Fairbanks architect who never met up with me in Newtok, made it to Mertarvik, along with the materials for the new house. He brought his dog and a skeleton crew, camped out in an unheated shed built by the military, and hired workers from the community to assemble the house over the next seven weeks. (The crew also had to dig their own outhouse.) In the end, it was a good house: so efficiently insulated that it would use less than half of the heating fuel required by an average rural Alaskan home. It was wired to accommodate solar panels or wind power, should those ever arrive in Mertarvik. This house, unlike its predecessors, would neither sink nor grow mold. Still it stood empty; it would be hard for anyone to live here full-time without the basic trappings of community and working infrastructure.

Three more years would pass before I would return. As I followed Newtok's story remotely, I began to wonder if the community could survive. Storms reached farther into the land, often smacking against the houses. Sometimes the wind was so strong people could feel it roaring beneath their floors and watch it warp the windows, so that it looked like the glass was breathing. And afterward, each time, more land had vanished.

In late 2016, the Newtok council asked President Obama to officially declare that the village was suffering from a major federal disaster, "due to a combination of periodic flooding, persistent erosion, and permafrost degradation"—in other words, the disaster of climate change. The declaration would have helped make more money available for Newtok to move. Romy Cadiente, who had been hired by the village council a few years previously to work full-time on the relocation effort, spoke to a public radio reporter: "We just need to get out of there," he said. But the policy was usually for discrete events—like hurricanes—and had never yet been applied to a slow-moving, multiyear, inexorable crisis. In January 2017, in the last days of the Obama administration, the White House denied the request.

But the worst tragedy for the community was not any rejection by outsiders. On a Sunday morning in late March 2017, Tom John set out from Newtok on an Arctic Cat snow machine. The sky was clear. The sun would have reflected off the glaring white tundra, and in black pants and a black coat, he would have cut a dark image on the land, like a silhouette. He left to hunt seals, as many generations of Yup'ik men have done every season. He was the last hunter in the village who still pursued the animals by kayak, the traditional way, as his father had taught him. This time, he pulled a brand-new red-and-yellow fiberglass kayak (though he owned and had always previously used a handmade wood and canvas one, the kind you might see in a museum).

But Tom John didn't return that evening. The next day, the village summoned the Coast Guard and asked for help. By Tuesday, he had still not appeared. The search team found his snow machine and some of his gear at the edge of the ice, eighteen miles west of the village.

For weeks, a group of fifteen volunteers from regional search-and-rescue teams looked for him, towing hooks and dipping poles into the bottom of the Bering Sea and scanning the depths with sonar.

No one could determine exactly what happened. (While some hazards, like unstable ice, become worse as the seasons grow unsteady, the Y-K Delta wilderness has always held inherent dangers even for the most seasoned backcountry traveler—volatile weather, difficult terrain, vast spaces where help is unreachable.)

Tom John was never found.

This was not, however, the end of Newtok's story, simply a grievous reminder of the precariousness of the place and the fragility of the lives held there on the broken land.

CHAPTER 6

THE EXPLOSION

home, n. *the place where one has been born or reared.*
home, v. *to go or return to one's home.*

When Doria Robinson was a little girl in the 1970s and 1980s, her bedroom window in Richmond, California, framed a view of two distinct landmarks. In the foreground loomed an oil refinery—a colossus of steel and concrete with a tangle of pipes that resembled a monumental Rube Goldberg creation. In the far distance, a blue and hazy hump of mountain shimmered across the San Francisco Bay. She never knew the mountain's name until she was an adult: Mount Tamalpais. A playground of tourists and wealthy San Franciscans, it seemed to lie in a landscape of privilege that was not her own. But she learned early what the refinery was. Every day, she would watch it fume like a roiling thunderhead.

The city of Richmond grew up around a refinery, built in 1902—by the very company that Henry Flagler had cofounded, Standard Oil—along a dusty road next to a railroad settlement. By the 1940s, the place was ripe with promise, the "Wonder City," the port and shipyard humming with World War II manufacturing. There were ample jobs building navy vessels in the vast shipyards run by Kaiser, and when there were not enough white men to fill these positions, the company recruited a significant number of women and people of color. Doria's great-grandparents, on her father's side, arrived during this period from rural Louisiana and Arkansas, part of the Great Migration, the decades-long exodus of Black Americans from the South, seeking to escape the violence of Jim Crow segregation.

But here the refining and manufacturing boom didn't yield the sort of wealthy fantasy-life that it had offered some white Floridians. The Robinsons found many of the same troubles in California as in the South. At that time, Richmond enforced segregated sports and recreational facilities, movie screenings, and Boy Scout and Girl Scout troops, and the police could arrest and jail Black people if they couldn't prove they were employed in the war effort. There was not nearly enough housing for the multitudes arriving to support the new industrial economy, and the workers also rented "hot beds" (for sleeping in shifts) and slept in movie theaters and even chicken sheds. Racial covenants that restricted homeownership to whites were codified into some deeds and rental agreements. New housing was explicitly segregated: local authorities barred Black residents from living in the sturdier, better-constructed properties—and forced them to flimsier, more temporary housing along the railroad tracks and near the factories and the refinery in an area sometimes nicknamed the Black Crescent. Still, there were jobs, so some hope of making a life here. And the labor of these migrants became the economic engine that transformed the San Francisco Bay Area into a modern economy. "They settled in the foothills of west Oakland and Richmond, far from the wealthy white cliffside mansions and nearer to the shipyards. They planted their collards and turnip greens, and let chickens forage out the back," writes author Isabel Wilkerson of the southern migrants who settled in California. Richmond's population quadrupled between the late 1930s and the mid-1940s. Its neighborhoods were full of greenery—kitchen gardens and flower nurseries—and farm animals. And its downtown restaurants and movie theaters were thronged, like "carnival night every night every hour," according to a 1945 account in *Fortune* magazine. That year, the housing authority also tried to evict thousands of tenants from wartime housing projects, and Black residents staged a series of demonstrations and a no-rent strike, which eventually forced some local officials to resign, a small victory.

Some chroniclers of Richmond insist the city's slide into economic hardship began when the war ended and the shipyards and the Ford plant closed. (Ford had plenty of demand for its cars but moved its operations to a cheaper rural location.) Others say that "white flight" ultimately brought the place to its knees. After the war, real estate agents and bankers designated parts of the downtown (including the neighborhood where Doria's family lived) as zones of "blight." The term was originally associated with agriculture and the withering of crop plants, but in housing policy, "urban blight" was often a code word for the places where people of color lived. The Federal Housing Authority generally wouldn't offer mortgage insurance to homebuyers in such areas,* keeping ownership out of reach for many Black residents, while many whites left for "nonblighted" parts of the city.

Meanwhile, after the wartime industries closed up, the most significant economic muscle remaining in Richmond was the refinery—which became Chevron in 1977. By the latter half of the twentieth century, as environmental awareness dawned across the United States, people in Richmond also noticed how the refinery could turn the city air so foul that it was difficult to breathe and how many young kids suffered from asthma. The refinery would often send out flares, balls of flame that were supposed to burn off excess gas. The flares could reverberate like an earthquake, rattling houses. Even the company's seemingly benign gestures sometimes felt tainted: in 1976, after a shopping mall opened in a renovated oil tank farm that Chevron sold to a developer, the competition shut down many downtown retail businesses.

Still, the first few years of Doria's life were in some ways a halcyon time. They were also marked by all the strains of growing up in a single-parent household with limited means. The old green bungalow where her mom, Kathy, raised Doria and her brother—mostly alone, after their father left when Doria was two years old—stood in the Iron Triangle, a neighborhood that jabs across the map like a sharp arrow and is named for the

---

* This kind of discrimination occurred nationwide. According to Nikole Hannah-Jones, writing in the *New York Times Magazine,* "As part of the New Deal programs, the federal government created redlining maps, marking neighborhoods where Black people lived in red ink to denote that they were uninsurable. As a result, 98 percent of the loans the Federal Housing Administration insured from 1934 to 1962 went to white Americans, locking nearly all Black Americans out of the government program credited with building the modern (white) middle class."

railroad tracks that run along three of its boundaries. A half block away and around the corner was a local history museum, formerly the city's first public library, an ornate red Classical Revival building with imposing white columns—built in the early twentieth century with funding from steel magnate Andrew Carnegie, like a temple to American industry.

Doria could spin through the neighborhood on her bike with the kids on the block and walk or ride to the Apostolic church where her paternal grandfather served as minister. Her grandfather remained close with her mother even after the divorce. *He was a huge-personality man who gathered a congregation of people from all over,* she said, years later. Most had been *farm folks* before arriving in Richmond, and they knew how to grow something from nothing, literally and figuratively. The church bought a 320-acre ranch about thirty miles outside the city where the community raised beef cattle and set up a small mixed farm with fruits and vegetables, horses, chickens, pigs, and rabbits. They also bought rental properties around the city. They used the money from these ventures to pay for young Richmonders to attend trade schools and learn skills like carpentry and electrical work.

The church itself was an act of barn raising. Doria's family and the parishioners helped tear down the old building and put up the new one by hand, the kids pulling nails with their stubby fingers, her uncles running the plumbing and the electrical. The church community would also organize to help one another in times of difficulty. *Somebody's house burned down. They'd build them another one. Things that should have been devastating tragedies for people became a way that people rallied and strengthened the church,* Doria remembered. *So growing up in that community made a huge impact on me and what I believe. My default for what is possible is totally different than for many people.*

On weekdays, the parishioners ran after-school programs for Doria and the other children in the neighborhood. Afterward, the kids could pick up sweets for a dime or a quarter from a neighbor who ran a candy store across the street. Doria knew most of the people living in the blocks around the church—aunties, uncles, cousins, Mr. Thomas, Sister Easely, Sister Brown, Sister Clark. They left their doors unlocked, and she could wander in anytime and visit. Her great-grandfather lived about a mile away in old war-era housing and sold homemade mini-pies you could hold in your hand—loaded with pecans or peaches.

But in the mid-1980s, the crack cocaine epidemic spread across the country, seizing the younger generation, especially Black youth. It ripped through Richmond with stunning swiftness. *People got strung out*, Doria said. *People started shooting each other and being crazy.* Locks and iron security bars appeared on the doors. There were murders. Drive-by shootings. The police, instead of helping, would often harass neighbors—hollering insults at the very people who had tried to report the crimes in the first place, dragging people violently out of their houses. *We had no police protection*, Kathy Robinson recalled. *We just decided as a neighborhood we were going to look after each other.*

As before, the Robinson family took an undaunted, do-it-yourself approach to this terrifying set of circumstances. Always a whirlwind of energy, Kathy tried a thousand ways to pull her family and her neighbors out of danger and poverty—she ran her own craft business, making clothing and jewelry, launched after-school programs for hundreds of kids at the local YMCA, taught herself how to fix her own car so she wouldn't have to pay the mechanic, took a few law classes, and sat in on the traffic court hearings of white people so she could figure out how to represent herself when she had her own complaints. (She would eventually sue the Y for discrimination after realizing they had paid her less than her less-industrious white coworkers.)

But this was not enough of a countervailing force. In his teen years, Doria's brother was increasingly pressured and threatened by gangs in the Triangle. There were shoot-outs on the streets, and the house fronts, fence posts, and mailboxes became pocked and scarred with bullet holes. Eventually Kathy decided they had to leave. In a less-troubled area near Nicholl Park a couple miles away, she scoped out a vacant house—a fixer, a raggedy place cluttered with the unclaimed belongings of an elderly woman who had recently died. But the house had good bones, a lemon tree in the back, and a lush yard wrapped around the property, even if overgrown and weedy, so Kathy called a real estate agent and bought it in 1987. The family moved the year Doria celebrated her thirteenth birthday. In the years thereafter, six kids who lived in a house next door to their old place in the Iron Triangle—the children she used to ride bikes and play tag with—were shot and killed, not all at once but in a series of moments in the wave of senseless violence that passed through the city.

All of this is to say that by the time she entered her teens, Doria was already a kind of refugee: she had survived an economic disaster and a public health crisis that hardly anyone else in America seemed to be paying attention to. And for a while, the family just cleaned house. In the new place they cleared out cobwebs, dust, worn-out furniture, and an attic full of the previous owners' papers and unanswered letters. Kids from a foster home for boys where Kathy worked at the time helped rip out the carpet. And the house became an oasis from the urban crisis beyond it.

In any good and righteous story of survival and tenacity, you would think this would be enough. You would imagine that a family so tenacious could reasonably rebuild their lives from here and prosper. But it is harder than this to escape the penumbra of a place like Richmond.

You couldn't see the oil refinery from Doria's new bedroom, but no one ever forgot it—the way it simmered on the horizon. Then, on an April day in 1989, the Robinsons were sucked back into another spiral of catastrophe. The Richmond Chevron Refinery exploded. *It sounded like a bomb*, Kathy recalled. *And everything got black.* The flames rose a hundred feet. A curtain of dark smoke unfurled above the city and much of the eastern San Francisco Bay. At times, the plume of pollution eclipsed the sun. *It reminded me of the pictures you see of the atomic bomb explosions.*

The Robinsons hunkered in their new house as instructed by authorities. Ash rained on the city, and the smoke didn't dissipate fully for another six days. Afterward, Kathy inspected the two cars sitting in their driveway—a newer Volvo and a red sedan she'd bought new only to have it break down soon after. Whatever was in the air and the ash had eaten away bits of the paint from both of them.

Kathy avenged the situation a little by demanding money. In the months that followed, when Chevron was issuing compensation checks of a few hundred dollars to the people of Richmond to cover damages, she met with their lawyers and insisted they pay her $5,000 for her cars.

If you don't want me to go to the newspaper and tell them what you're doing to poor people down here, you're going to have to give me more money, she insisted coolly.

But it was so far from being enough, and Doria figured she was done battling this place of fire and decay. *Oh my God, what a lost cause*, she thought. When she was in high school, she realized she wanted to get out of here as soon as possible.

She couldn't then imagine how Richmond would call her back. How something here had rooted into her, a strange and tenacious bit of hope and belonging, and she couldn't so easily transplant herself. She would eventually try to grow something new here, something to make this unruly town into a place that might survive the twenty-first century. She would eventually realize that the disaster of Richmond was also the catastrophe of fossil fuels and of climate change. The refinery that loomed over the town was also one of the largest producers of greenhouse gases in California—and thus it had an outsize role in disturbing the planet's climate. Both Richmond's nightmares and its dreams were entangled with thousands of other calamities and communities around the world.

※

My first brief trip to Richmond, California, in my midtwenties, was not as a journalist; it was a social outing. But the image I formed of the place remained resonant later. My boyfriend—a jovial man with dark, curly mop-hair who later became my husband—invited me on an adventure to the San Francisco Bay Area to meet his friends, including Myk, who ran an unlicensed punk recording studio and performance space out of an old commercial building and who seemed like an inspired and degenerate character you might find in a Kerouac novel. The space, called Burnt Ramen, was well-known and beloved by punk fans. (One reviewer on Yelp called it "the most revered DIY venue in the Bay Area punk/crust/hc [hardcore] scene.") And I would later learn Doria Robinson was also a punk fan and had a few times been a concertgoer at Ramen's.

We rode there one night with a group of friends, who parked their cars in the street outside. Then we toured a set of rooms painted in bright and intricate art and graffiti. We watched Myk, a sinewy white guy with a goatee, and his girlfriend, a tall Black woman with a reedy voice, howl through a duet in his recording studio. Myk introduced us to his pet boa constrictor—who was lithe, several feet long, and fed on live rats purchased from the pet store—and his housemates, one of whom was too broke to pay full rent and lived in a kind of crawl space.

It was all a display of weird, inventive, and necessary creativity—the kind you might need to survive in a precarious place. When we left that evening, Myk's girlfriend called after us, "Oh, you guys parked on the street!? Ooh, if you only knew! The other day someone came through here and smashed every single car window."

But I didn't think much about Richmond until a few years later, when I traveled back to California to research a series of articles on resilience, which was by then becoming a popular buzzword to capture everything from addiction recovery to preparing for the apocalypse. I arranged a meeting with Mateo Nube, cofounder of the environmental and climate justice organization Movement Generation Justice and Ecology Project. Mateo directed me to the address via BART, the San Francisco–area regional rail system, about a half-mile walk from a station.

When I got there, I was surprised to find myself not at an office building but in front of a terra-cotta-colored house with blue window trim. On the porch a woman tended a potted plant, her hair pulled into a long white braid with delicate tendrils escaping into her face. This turned out to be Mateo's mother, who smiled and told me to head around the back, where I was immediately awestruck by a cluster of spiky five-foot-tall plants so robust that I nearly took them to be cactuses, until I realized they were homegrown artichokes. Mateo's family home doubled as his organization's office. One of his colleagues lived next door, and they had pulled down the fence between them and established a shared garden, with a chicken coop, a set of rain barrels, and a composting toilet. Mateo met me in the yard. Tall and lean, with a pointed chin and a closely trimmed beard, he was dressed casually in gray cargo pants and an athletic jacket. He gestured for me to sit at a glass patio table across from him, and, with a cacophony of bird-chirruping and rooster-crowing in the background, he told me about the first time he'd had a revelation about climate change—around 2005, the year that Hurricane Katrina ravaged New Orleans. Mateo was descended from a family of German Jews who had fled to South America to escape the Holocaust. He had come to the United States from Bolivia as a college student, and perhaps partly because of his family background, it wasn't hard for him to connect the dots between the science of climate change and the human chaos it might cause—and from there to an experience of profound grief. "When I first started reading about climate change, I got really depressed. I freaked out," he told me in a gentle tenor voice, clasping his hands, clad in black fingerless knit gloves, in a namaste prayer formation. "My first child had been born. She was a year old. I was like, wow, what has she been born into?"

Founded in the early 2000s, Movement Generation had brought together a loose network of activist groups that fought for policies related to equity, such as fair wages, tenants' rights, police reform, and cleaning up the kinds of pollution that plagued communities like Richmond or Bayview Hunters Point in San Francisco—a neighborhood adjacent to a toxic waste dump and, until 2006, a polluting, coal-fired power plant. Collectively, the activists would brainstorm and trade strategies and ideas about the most effective means of creating lasting cultural and political change. Climate change was not originally part of these groups' purview—at the time, it felt too distant, too overwhelming, and frankly, Mateo admitted, too much associated with white environmentalists who rarely welcomed the concerns of people of color.*

But Hurricane Katrina offered a prelude to the horrors the world might face in the climate crisis. And it was plainly not an equal-opportunity disaster. When the levees that were supposed to protect the

---

* Despite often feeling excluded from the mainstream environmental community, in study after study, Black Americans and Latinx respondents consistently are more concerned about climate change than whites. In a 2009 poll, 57 percent of voters of color felt global warming was very or extremely serious, compared to just 39 percent of whites. The pattern has held over time. In a survey more than a decade later, less than half of whites said they're alarmed or concerned about climate change, compared to more than two-thirds of Latinx respondents and 57 percent of Black Americans. People of color also expressed more willingness to participate in environmental activism.

city broke, they released floodwaters into a majority Black neighborhood called the Lower Ninth Ward, and the news coverage of the storm and its aftermath filled with images of suffering, including stories of Black New Orleanians in often desperate situations, standing thigh-deep in flood-waters or trapped on rooftops. Thousands of people were stranded inside the Louisiana Superdome, the sports arena turned shelter-of-last-resort that quickly became a kind of hell, with people penned behind metal barricades, with no sanitation, patrolled by armed National Guardsmen who did little to stop horrific crimes from occurring inside. Afterward, it became clear that the same sorts of discriminatory housing practices that placed the Robinsons in the path of refinery pollution also confined Black Louisianan homeowners to low-lying, flood-prone land. Black homeowners in New Orleans were three times likelier to suffer flood damage than white homeowners. But the federal government dished out recovery money based on the appraised value of houses—simply recapitulating the long and deadly history of inequality. After the hurricane, all of the problems of racial injustice worsened in that city. Child poverty rates soared; the chasm between rich and poor yawned wider.*

Mateo and his colleagues observed these patterns from afar and could see how they would play out in cities everywhere on a hotter planet. Already, people of color and anyone living in poverty are far more likely to reside near polluting facilities. And as the planet heats up, higher temperatures worsen many kinds of air pollution. Global predictions also suggest food and water scarcity could be on the horizon, problems that would strike the most vulnerable first, especially those in poverty who can't afford to pay premium prices for necessities. A 2009 study found that in Los Angeles, Black Americans were twice as likely to die from a heat wave, in part because they live more often in neighborhoods that are physically hotter, with more concrete, fewer trees, and less shade. Already, Black American communities face greater

---

* "The income disparity between rich and poor is so great that [in 2015] Bloomberg declared New Orleans the country's most 'unequal' city," writes Gary Rivlin, author of *Katrina: After the Flood.* "And it's hardly just the poor who are suffering. The median Black household income in New Orleans in 2013 was $30,000—$5,000 less than it was in 2000, adjusted for inflation. By contrast, median household income in the white community increased by 40 percent over that same period and now stands at more than $60,000."

risks from flooding and hurricanes, especially in the South along the Atlantic coast. In the Bay Area, historically, most communities of color have occupied the "flatlands," forged out of the San Francisco Bay from fill, dredged sediment, quarried rocks, and construction debris. In sea level rise projections, these landscapes are quickly dampened by the tide in a matter of decades.

Everything about climate change could make life worse for the communities these activists were fighting to protect. "To put it bluntly," Mateo said, "if this set of issues isn't taken on, all the other super-urgent issues we're working on kind of become irrelevant in fifty years."

So in 2007, Mateo and his Movement Generation colleagues began organizing a series of annual retreats in rural Sonoma County at an eco-farm and center for permaculture, a form of sustainable agriculture. There, away from the city, surrounded by fields of heirloom vegetables, a group of about thirty peers and collaborators from the Bay Area activist community—eventually including Alicia Garza, who would go on to cofound the Black Lives Matter movement—talked earnestly and seriously about climate change and what it meant for their work. And though it was a global crisis, they felt their approaches to it should start at home, informed by the perspectives of the vulnerable communities they lived in and worked with. One strategy, for instance, would be to advocate for policies that could prevent the kinds of horrors experienced during Katrina from recurring—fight against laws and planning decisions that might further divide cities into elite, gentrified, eco-friendly enclaves and poor, polluted ghettos. Simultaneously, they would campaign for policies that would be both sustainable and economically fair: for instance, infrastructure to make more public transit accessible to many rather than just to allow green cars to be purchased by the wealthy.

A second approach was to master a kind of urban self-reliance through, among other things, permaculture, partly so that when the crisis hit, poor communities could take care of themselves and collectively hatch their own solutions. (Growing food is also an act of renewal and sustenance, especially for communities that have often been denied the opportunities for beauty and solace that exist in wealthier places.)

The activists had also begun collaborating with other national and international social justice advocates in Boston, Miami, and Los Angeles, and the New York–based Right to the City Alliance—a group founded on the ideals of twentieth-century French philosopher Henri Lefebvre, who believed that profound societal change could come from the sharing and reclaiming of urban spaces.

All of this was markedly different from the approach of some mainstream, mostly white environmental groups, who were spending much of their time pushing for solutions that would reallocate large sums of money—through policies such as cap-and-trade and offsets, both of which put a price on the privilege of emitting carbon—with the idea that you could, in this way, overhaul the old economy and conjure a new, green one.

At first glance, Movement Generation's approach sounded a little rogue but full of unusual creativity born out of crisis. "Some people say we're survivalists," Matéo reflected, but the word sounded too individualistic to him. "It's quite the opposite. We're really trying to build collective solutions."

Mateo was also devotedly anticapitalist and insisted that you would have to pull the old economy up by the roots to fix climate change, most especially yanking out the fossil fuel economy—which has always been by far the biggest source of climate pollution, about two-thirds of global carbon emissions.

For the moment, he was also focused on the tangible and immediate things he and his collaborators could do to rely less on that economy and more on what they could produce for themselves. Permaculture projects would help provide a model of another way to live, he explained, part of a cultural shift, what he called a "transformative narrative."

Permaculture encourages its adherents to think of human communities as part of nature. In an urban setting, it reconceptualizes a city as an ecosystem, often a damaged one that you might want to restore. In doing this, you might try to root out the things that were polluting the urban environment and grow new ones in their places. Cities are a large component of climate change—home to 55 percent of the global population and responsible for simultaneously 80 percent of global gross domestic product and more than 60 percent of carbon emissions. Starting in a few small neighborhoods, in communities that carried the most risk and the greatest burdens of pollution and poverty, you could transform the urban landscape, and maybe the rest of the world would follow. This was radicalism in its truest sense—change from the root.

To deepen their collective knowledge of permaculture, Movement Generation had helped organize a series of hands-on workshops to teach skills such as composting or installing systems for collecting rainwater and gray water (recycled from activities like laundry). By 2010, the organization and its collaborators had built a number of urban gardens—including one in San Francisco in raised beds on the roof of what was then Alicia Garza's office building, one in Oakland at the headquarters of a Latina immigrant rights organization, and the one behind the patio table where Mateo and I sat.

But one of the best examples of their strategy and thinking predated the Sonoma gatherings—a set of gardens and farms in Richmond, California, an initiative led by Doria Robinson.

There it was again, unusual creativity—a rain barrel and a composting toilet, an artichoke plant the size of a Christmas tree, a garden full of kale and squash as the tiniest first steps to begin reimagining the world, perhaps steering it away from a future of hurricanes and fossil fuel pollution and injustice.

To make better sense of all of this, I decided I would need to go back to Richmond.

A day later I rode BART to the edge of the Iron Triangle, Doria's neighborhood, along the old Union Pacific railroad lines.

ॐ

In the language of a twenty-first-century environmental activist, Richmond, California, is a *frontline community*, a place that has always faced the immediate consequences of fossil fuel extraction, long before anyone else woke up to the more global consequences, the problem of climate change. This is battle language—as in, the front lines of a war—and also a language of power. The dictionary definition of *front line* includes "a position at the forefront of new developments or ideas, or at the center of a social, political, or ideological debate; the vanguard."

But in the 1990s and early 2000s, many people thought of Richmond and places like it—in the inner city, along the Gulf of Mexico, in the shadow of gargantuan industrial facilities, next to coal mines, sliced open by massive highways—as disposal sites, throwaway communities, landscapes of waste.

In 1991, poet Adrienne Rich wrote of this tendency to waste human resources as if it were a tragic (but not inevitable) part of the American national character: "Waste. Waste. The watcher's eye put out, hands of the / builder severed, brain of the maker starved / those who could bind, join, reweave, cohere, replenish / now at risk in this segregate republic." And Richmond was like a testament to this idea, full of both toxicity and "human wreckage," writes labor organizer Steve Early. Leaking battery cases lay buried beneath a shoreline park. Carcinogenic polychlorinated biphenyls (PCBs) from old utility equipment leached into local groundwater. Land owned by old pesticide companies lay overgrown and full of dangerous chemicals. Radioactive material lurked in the landfill. Crime continued to soar in Richmond; 1991 saw the city's peak number of homicides, sixty-two deaths. "Meanwhile, Richmond's African American community was decimated from within by the highest rate of AIDS transmission in the Bay Area," writes Early. "In search of personal security, better housing and schools, or improved job prospects, middle-class Blacks joined the trajectory of Richmond's earlier white flight." If such a place was your hometown, you learned a particular story, that to succeed in life, your first job was to escape. The story said you had to get out of poverty—as if poverty were a place and not a condition.

In her teen years, Doria launched a furious campaign to extract herself from her hometown. First, she applied and was admitted on a scholarship to a private college-prep school in Berkeley for her last two years of high school. To cover her extra expenses and tuition, she worked multiple jobs including evening shifts at a Berkeley pizza parlor. After midnight—when BART had closed—she rode her bike eight miles back to Richmond. The pizza shop owner took an interest in Doria's educational ambitions and gave her generously large payroll advances so that she could afford to join the school's international field trips to places like Egypt and Israel. She tossed dough and served pizzas for a year after high school in order to pay the remainder of her tuition. Then in 1993, she left for the opposite side of the country—Hampshire College, an experimental liberal arts school in Massachusetts. In 1995, she fled farther—to the opposite side of the world, a year abroad in India and Nepal, through what was supposed to be a formal program in Tibetan studies but turned into a free-for-all. The institution that was hosting her and her schoolmates stopped offering classes a couple months in, and her peers, whose families gave them ample travel money, abandoned their schoolwork and left for backpacking excursions through Asia. Doria had no cash to spare. So she bartered for a room in a monastery by offering English lessons to the monks. There, she immersed herself in ad hoc spiritual study with a series of meditation teachers who lived nearby.

In the retelling, the experience was almost mythic. Doria knew that a person from her background, from a beat-up town full of pollution, wasn't often given such free access to the world. *I got all transformed. I didn't talk the same*, she thought. *I should be doing something important. I wanted to lead a spiritual life.* She thought about staying in India and becoming a Buddhist nun. Survivor guilt—her successful, improbable escape into another life while other kids from her street never made it to adulthood—haunted her meditations. She wept often.

The person I was working with was a meditator in the mountains. He literally lived in a rock house, an hourlong hike in the foothills above Dharamshala. One day she had a meeting with him there to ask him questions about her meditation practice and her thoughts of becoming a spiritual devotee, removed from the traumas of her past. His response was abrupt. *He was like, "You need to go home."* She was crestfallen. *At the time, it felt like he was saying, "You're useless." In retrospect, he meant, "You don't make any sense here. There's work for you to do in Richmond."*

She didn't want to heed the advice. *I was like, I'm not going back there.* She moved to San Francisco, at first working the graveyard shift for an organic produce distributor, all night, unloading trucks, hauling pallets into a cooler. She found it meaningful but exhausting and left for advertising work in the early dot-com sector—higher paid but, to her, vapid.

She started taking art classes and reconsidering what she might do next with her career. Then she became pregnant with twins, a boy and a girl. She had met their father at an art festival, but he was a wanderer, with a tourism job that sent him on stints through Canada, California, and Mexico. And Doria suddenly wanted to be tethered to a place, to something real, some solid piece of earth where she could raise her young children.

She also wanted to get her hands dirty again, literally craved her own plot of vegetables, but San Francisco community garden spots had seemingly endless waiting lists. *I felt like I had all these skills, and I'd learned so much and everything. I need to go where I can actually make a difference.* She finally found a job based in Richmond doing watershed restoration—tending plants, greening urban lots—and her paternal grandparents (the minister and his wife) asked if Doria wanted to manage her great-aunt's house, *a little, little house with a very big yard*, three bedrooms, stucco on the outside, and a wide porch slung across the front. It was about a half mile from where Doria spent her early years and about two blocks from her grandfather's church in the Iron Triangle. The area was thoroughly urban, and yet in her childhood, her uncle had kept a horse and a cow in the backyard.

She went to see the house. *It was a heap. It had been destroyed. Random renters and different church members had come in with huge families. It was the kind of thing where you open the door, and there's that smell that hits you.* The walls were painted drab gray with paint and plaster falling off, full of holes, with mildew and mold growing in strange places. She found grease stains behind the stove and gunk caked in the refrigerator. The yard and the shed were full of piles of old junk.

But she remembered her great-aunt, how the house used to be clean and the yard full of roses. And Doria felt she had to revive the place. She had entered a plot twist in the story she had previously told herself. Poverty no longer felt like a place you ran from; it was more like an old weedy lot, a bit of broken earth, a crumbling fence, or a run-down house that needed repairs. She was tired of wasting things and people. She moved in when the twins were about two years old and began clearing the space, creating room for bicycles and toys and garden implements. About six years later, she painted the house marmalade orange and midnight blue, then a few years later repainted it to sunflower yellow. The family adopted a feral kitten, a gray-and-black tabby that the kids fished out of a trash can and named Tobias (after author Tobias Wolff). Doria planted a peach, an apricot, and a pomegranate tree in the yard. When the trees started bearing fruit, elderly women would stroll past the yard on Sundays, on the way to church, and ask to pick the peaches for making pie. Giving away peaches became one of Doria's fondest, simplest pleasures.

Circumstances had improved only a little in Richmond since she was a teen. Chevron* had not cleaned up its act much; between 2001 and 2003, the EPA noted nearly three hundred pollution spills from the refinery. "These are highly toxic, often cancerous, chemicals spilling directly into residential communities of families, children, the elderly, and the sick," writes investigative journalist Antonia Juhasz in her book *The Tyranny of Oil.*

Nor did the town feel safe. Violent crime had been decreasing for the past couple of decades, but the city still had a homicide rate about five times higher than the California average.

But there has always been a forceful undercurrent of grassroots activism running through Richmond and numerous efforts to raise up the community—from Black Panthers offering free breakfast to schoolchildren in the 1960s to antipollution activism aimed at Chevron beginning in the 1980s. Communities for a Better Environment—a feisty California-based environmental justice organization with a track record of bringing legal and political challenges against the fossil fuel industry—opened an office in the community in that decade. Together, they and a homegrown Richmond environmental group, the West County Toxics Coalition, founded in 1986, brought a successful class action lawsuit against a Richmond-based chemical facility over a fire

---

* Regulatory violations have been almost routine across the refining sector, and the industry often fights back when regulators try to hold it accountable. According to an extensive 2011 investigation by the Center for Public Integrity, "Regulators have little sway over refineries. Between 2000 and mid-2010, refinery owners contested 53 percent of all violations cited by state or federal safety inspectors, allowing companies to put off improvements and save money. On average, some twenty months pass before a contested case is closed. No other industry with more than a thousand violations appeals such a large proportion of findings."

that released noxious fumes. That win in the 1990s provided settlement money for, among other things, a warning system, so that residents would at least be alerted to the next industrial emergency, including a refinery fire. (Once a month, Richmond tests the sirens for less than three minutes, a reminder of the potential for disaster.) The Asian Pacific Environmental Network began organizing the Laotian community in Richmond around the same time, and pushed the city to set up a multilingual emergency warning system after a major refinery fire in 1999.

By the 2000s, there were additional political efforts to reclaim the city that didn't exist during Doria's youth. A group of local activists had also organized a series of grassroots campaigns to run against the old guard of Chevron supporters on the city council. By comparison to the well-moneyed efforts of their opponents, the progressive election campaigns of the 2000s were scrawnily funded. But in 2004, Gayle McLaughlin, a former schoolteacher, a democratic socialist, and member of the Green Party, improbably grabbed a seat. In 2006, she won the mayoral election, making Richmond the largest city in the United States with a Green mayor. Doria volunteered for her campaign, knocking on doors, walking precincts, and making phone calls. *It was the fact that she was saying, "I'm not going to take money from corporations, and I'm going to shift the conversation in Richmond to quality of life and to bringing the kind of industry and the kind of jobs that we really need." What she was trying to do is what people wanted, at least people who vote.*

That same autumn, locals set up a sit-in and tent encampment in Nevin Park, less than a block from Doria's childhood house in the Iron Triangle, to protest a wave of deadly gang shootings that had occurred earlier that year. In a convenience-store parking lot across the street, people assembled a shrine to someone who had died in a recent shooting. (Their efforts would help inspire McLaughlin to push her colleagues in the city to set up an office of neighborhood safety.)

Such things felt like seeds, and Doria looked for more ways to dig in. She heard about a new community gardening organization called Urban Tilth, formed by a teacher named Park Guthrie, who lived just outside of Richmond, in an unincorporated part of the county. Park had proposed an ambitious vision for the old Santa Fe railroad line corridor, a place that seemed to symbolize all that was wrong and all that was possible.

In Doria's childhood, the railroad right-of-way had been the awkward shortcut to her grandparents' house, the makeshift path everyone used as a pedestrian thoroughfare. You could walk the rocky track ballast and the crooked wooden ties—in some places, the tracks had been removed, and in others, the metal still jutted up. It was an interstitial landscape, hairy with coarse grass and tall stems of aggressive fennel gone wild, streaked with graffiti and littered with broken glass and used syringes and condoms.

For decades, community activists had clamored for this moldering two-mile strip of land to be converted into a paved bike path and green corridor, and, in the mid-1990s, while Doria was away, the city had pulled out the old tracks and regraded the ground so it was level. It wasn't until 2007 that the bike path finally opened. In the beginning, few people used it. Doria tried to jog the path, and when she would approach someone from behind, they would sometimes startle at the sound of her footfalls and turn toward her with their fists clenched. *They thought I was gonna jump them.*

But Park wanted to turn the Richmond Greenway into a real gathering place, and a description of his ideas circulated on a community email list. *Berryland, with all different types of berries that kids could just come up on their bikes and eat. It sounded so ridiculous then. But it was beautiful, like really engaging and inspiring, and I called him up, and I was like, "Let's make it happen."* She assembled a group of neighbors in her backyard one Saturday to discuss, and a week later, they held a volunteer workday and staked out where the beds would be.

The word *tilth* refers to cultivated land and soil. One of its oldest meanings is also "honest labor," and another definition refers to the cultivation of knowledge, humanity, and decency. Urban Tilth wanted to bring all these things forth from the old battered ground of the Iron Triangle. But the organization had little money. Everything was donated and most of the labor came from volunteers. They had to haul clean dirt to the site to make sure it was uncontaminated by whatever chemicals might lie buried in the ground of the railyard, especially arsenic. But within a year, there were eighteen kinds of berries planted along a stretch of the Greenway, and the old railroad tracks were transformed from a shadowy shortcut no one was supposed to take to a small oasis. When the berries ripened, you could find people wandering here, their hands stained with red and blue fruit.

Doria began to see vacant land wherever she looked in the city—and instead of the old empty lots seeming like eyesores, they appeared to have potential. A year after volunteering with Urban Tilth, Doria became an employee, and she, Park, and her fellow local gardeners looked for places to create gardens wherever they could find them. At the high school, at a decommissioned and largely abandoned middle school, in church lots, and in private front yards, they would grow food.

In 2007, a thunderous bang, a ball of golden flame, and a cloud of ash announced that another fire had broken out at the Chevron refinery after a corroded pipe failed. Across the bay, Antonia Juhasz witnessed the incident (and described it later in *The Tyranny of Oil*): "In January 2007, most residents of the area (I among them) thought the refinery had exploded. . . . The five-alarm fire burned for nine hours, and the 100-foot flames could be seen with the naked eye in San Francisco," roughly ten miles away.

As the sirens wailed, Doria was loading plants into her car. *I remember thinking to myself, do I go inside to shelter in place? Or do I try to pack the car fast? Then I was ranting to myself as I was driving to work, thinking, why do they get to do this? Why don't I have a say in this? Who owns the sky?*

In the end, wind blew much of the smoke away from Richmond and other population centers, but Doria's rage would not dissipate so quickly. But she wouldn't try to run from Richmond this time. She would stand her ground. She would nourish this polluted place.

꙳

For an out-of-towner, BART is clean but noticeably loud. The train screams and shrieks along the tracks. A day after meeting with Mateo, I watched through a train window as the landscape transitioned from the leafy streets of Berkeley into miles of strip malls, empty lots, junkyards, and abandoned warehouses with rusting corrugated roofs.

Doria Robinson met me in the Richmond BART parking lot, wearing black sweatpants with a racing stripe down the side and a purple shirt printed with a drawing of a bicycle and the words BEAU-TIFUL MACHINE, her hair in long, tight dreadlocks. She greeted me by announcing that her car had broken down, and we'd have to wait for a colleague to meet us. (Years later, she would confess that she'd been out of gas money and subsequently run her tank empty but was too embarrassed to tell me.) We stood awkwardly on the pavement for a few minutes, and Doria filled the time by offering me a summary of her organization's philosophies—and talking a mile a minute. "One of our principles is that there's plenty of resources. There's just not enough all in the right place," she told me. For instance, there was land abandoned by industries, businesses, and institutions that had packed up and left decades ago—it stood vacant, gaping, wasted. And everywhere there was need and hunger—for hope, for health, for community, for opportunity. The trick was to understand how the first problem could actually be an answer to the second. "We get all kinds of donated stuff from the city or from people—mulch and all kinds of straw and logs. We'll build our raised beds out of old eucalyptus trees that have been cut down and whatnot. So we're kind of scavengers. We don't have any gardens that we operate on private lands. We do everything on public land."

Several minutes into this discussion, a hardy, brown-haired twenty-two-year-old named Adam Boisvert drove up in a pickup truck, and he and Doria urged me to climb in after them. Thus began our whirlwind tour of the city, zipping through a patchwork of empty lots, old warehouses wreathed with barbed wire, churches, and rows of bungalows with small, sun-dried brown grass lawns and metal grates over the windows and doors.

In the four years since Doria helped plant the first berries, Urban Tilth's efforts had grown into a rangy, gangly, scrappy effort. She had collected staff with the same philosophy that guided her other efforts—many were people whose worth and labor and know-how would otherwise be wasted. Broadly speaking, most of them were still kids, some just out of high school. Many had come up through an apprentice program that the organization's staff had first launched with kids from Richmond

High School. (Adam was an exception, a graduate from the University of California, Berkeley, who had met Doria serendipitously on a bicycle trail and was chasing his passion for organic farming. "I always joke that Adam is the one token non-Richmonder," she said.) Eleven staff worked at eleven gardens, mostly part-time and seasonally, with a donated office space and gardening supplies, subsisting on an almost impossibly meager budget. (She would later admit to me that they were often running on fumes, so to speak, in those years.) To fill the gaps, they sometimes took on paid private work through a program called Farm Your Lawn, in which an individual property owner might hire the kids to transform a yard into a food garden. Above all, the organization ran on conviction and faith. Doria's goal was for the city of Richmond to grow at least 5 percent of its own food. In 2009, Urban Tilth had harvested six thousand pounds of produce. (This was simultaneously a lot of food and a drop in the bucket relative to what an entire city would eat. But it was also a visionary achievement—proof that you could coax something nourishing from an otherwise neglected urban landscape.)

The truck arrived at a linear strip of land that ran perpendicular to the street with a paved bicycle trail threading through the center. The land stretched out on either side of the pavement—behind a school, some old warehouses and a corrugated metal building that served as an auto body shop, and alongside some residential buildings. The three of us disembarked. A wooden sign announced the location: BERRYLAND, on a bright yellow and black background splotched with images of ripe raspberries. Tawny bark mulch covered the ground—so that walking felt like a spongy, springy activity—and the garden stretched across it, various rectangular raised beds edged with wooden boards and logs. The breeze carried the scent of eucalyptus. WELCOME TO THE LINCOLN SCHOOL FARM. *BIENVENIDOS A LA GRANJA*, read another sign with rainbow-colored lettering. Berryland had expanded to include a vegetable garden farmed by a nearby group of elementary schoolchildren. "They come out and maintain it and harvest and learn about photosynthesis and whatnot," Doria explained. "And people from the community can come and harvest whatever they want."

Behind us was a raised bed lush with strawberry plants, and plump fruits dangled from the edges. There were other beds with boysenberries, raspberries, blueberries, pineapple guava, gooseberries. "The berries don't usually stay for very long once they come up. People just eat them. And that's the point. And you can come at all hours, all times, and people are harvesting something."

At the edge of the garden, a mural along the warehouse walls depicted a series of bright puzzle pieces—each painted by kids from local foster homes with a different image, mostly animals, people, bugs, trees, butterflies.

"A student from San Francisco Academy of Art came up with the idea," Doria said. "And the kids connected the puzzle pieces all together—that's kind of how community works. Like it all kind of connects, even though people can be so different."

Other beds were full of fruits, herbs, and vegetables—tomatoes, green beans, onions ("because people really, really eat the onions"), mint (beloved by members of a local mosque, who used it to make mint tea), peas, melons. Some of the vegetables were specific to local tastes and histories: next to a bolted parsley plant stood a tall brassica called purple tree collard, a varietal that had reputedly traveled from Africa to Black communities in the South and eventually to Richmond. These stood beside more murals with images of flowers and mushrooms.

A petite woman wearing gardening gloves wandered into the scene, and Doria introduced her affectionately as "Ms. Tania," Tania Pulido. "This girl has her hands in everything."

The twenty-year-old now managed Berryland, Doria's former job, and oversaw a group of teenage apprentices, including a pair of cousins who lived in a house around the corner. Doria had hired them after they had wandered onto the Greenway more than once and started offering to help move mulch and logs around.

"You know, they were kind of hanging out, playing Nintendo, trying not to get into major, major trouble. They'd be like, 'You want some help with that?' And finally, we're like, 'Do you want to work with us?'" Once they signed up, they began to organize extra volunteer crews made up of their friends.

On the second Saturday of every month, the crew would host a barbecue with community members. At the most recent, Tania's family had made chicken mole.

It was a subtle but crucial shift not just in the landscape of Richmond but also in its culture and self-image. (To borrow the words of Adrienne Rich, it was the "meticulous delicate work of reaching the heart of the desperate / woman, the desperate man /—never-to-be-finished, still unbegun work of repair.") "People really bunker down," Doria explained to me. "People go inside of their barred houses and hide from the world after they get off work and don't go out in the street. Kids don't play outside. The gardens are kind of central to giving people an opportunity to live a little, come out of their bunkers."

❧

There is a strangeness to our collective relationship with oil and fossil fuels. Many of us can't fully imagine what our lives would look like without them—what it would mean not to fill up at the pump, not to turn on the gas furnace—and yet our continued reliance is poisoning both present and future. For precisely this reason, we have called fossil fuels an addiction—a dependency so toxic, especially because of its role in climate change, that it could wreck the very foundations of our society. But unlike with drugs, not every petroleum addict suffers the immediate consequences. Communities at the margins and in the shadows of mainstream society have to bear the worst impacts: high lead levels, increased childhood asthma, low birthweights, respiratory disease, lung cancer. The hideousness of pollution and toxicity is unavoidable in Richmond. By contrast, the hideous impacts of climate change—caused primarily by our collective reliance on fossil fuels—have only become obvious to the majority of people in the last decade or so.

In the face of all this, it is a radical thing to imagine you can grow food in an oil town.

A little after Berryland began on the Greenway, another Urban Tilth project launched at Richmond High School when a teacher and friend of Park Guthrie wanted to fix up an old garden that had been neglected for more than a decade. Behind the rust-colored trailers that served as extra classrooms, he, Park, and a group of students revived a series of garden beds, planting them first with fruits and flowers and later vegetables, and spruced up an old greenhouse. Park and the teacher also organized an urban agriculture club, and then a class as a partnership between Urban Tilth and the high school. Shortly thereafter, Urban Tilth hired a Richmond High student named Jessie Alberto, first to work on the Greenway. Then, after he graduated, he also assisted with the class and with another garden at a second high school, called Kennedy. The mascot of Richmond High School is an oil well, and the team is called the Oilers. But the Urban Tilth class curriculum was either subversive or simply honest, depending on your perspective—about ecology, fairness, pollution, oil, nutrition, food deserts, and why some people get left out of the economy.

Park eventually stepped back from the organization to take a teaching job himself, and Doria took the helm. Then she and the other Tilthers decided to expand the gardens at Richmond High. The high school had ample grounds where they could imagine cultivating plants, but it was a difficult time for the community. In the fall of 2009, a teenage girl was beaten and gang-raped outside the Richmond High homecoming dance, one of the worst acts of violence the school had ever known. It would take years for this trauma to even begin to heal (and for the eventual criminal trials to resolve).

The school and the community still needed a reminder that it was possible to grow promising things here. On one winter weekend in 2010, a massive volunteer effort of high school students, teachers, and others from the community built six more long raised beds behind the Richmond High football field—nearly eight hundred square feet of cultivation space where they could grow more fruits and vegetables—and another thirteen crop rows and 2,600 square feet at Kennedy.

The kids of Urban Tilth usually had a different vision of the city than many of its older residents. Tania Pulido came to the gardens after *barely graduating high school by a hair*, she said later. In school, she cut class often. Her family, originally from Mexico, nearly lost their house after her dad was fired from his job during the economic slump of the late 2000s. She went through a dark moment, questioning, she said, her very existence, and began dressing in all black. A lot of kids she knew were excited to receive a check from Chevron when they turned eighteen—settlement money from a previous refinery incident. But Tania formed doubts. She loved documentary films and watched a series of movies about oil, eventually deciding that she would campaign against the company. She began attending protests. She then met Jessie Alberto and started volunteering in the garden at the same high school she had previously avoided attending. On Martin Luther King, Jr. Day in early 2011, she helped with an Urban Tilth festival at the Greenway. Three hundred volunteers turned out to clean up, tend the garden beds, and feast on vegetables harvested there, and the organizers provided a stage where youth performed Mexican dancing, spoken word, and a comedy routine.

Alfonso Leon first became aware of Urban Tilth during his sophomore, junior, and senior years of high school, when he was working on a local nonprofit project that sent him on a quest through the city to map creeks and rivers. Then he enrolled in the Richmond High urban agriculture class. He vividly recalled how it had felt when he was just six years old, in 1999, and the Chevron refinery had caught fire once again, sending 1,200 people to local hospitals. He remembered how he had sheltered first at school and then in his family's house as if a storm were passing through. *Even as a kid, I just knew what the refinery was. This is like the freaking pits of Mordor. This place is pure evil.* But he didn't think oil would always be the city's destiny. When Alfonso looked at Richmond, he didn't just see crumbling industry or barbed wire or vacant lots. He saw nature.

Just after Alfonso graduated in 2011, at the age of eighteen, he was hired as an Urban Tilth employee, supervised by Jessie Alberto. One of his primary responsibilities was to cultivate food on the grounds of his former middle school, which had since been shut down because of earthquake safety concerns and city shortfalls, then abandoned, and at one point, set on fire by arsonists. At some of its first school sites Urban Tilth had secured keys from staff and teachers and persuaded groundskeepers to switch on the water, then later asked the administration for formal permission. Doria, Park, and their burgeoning army of youth often took an act-now-apologize-later strategy to their endeavors. But by the time serious cultivation began at the middle school site, the school district had negotiated a formal land-use agreement with the organization. On breezy, grassy land beside the school, Alfonso and Jessie Alberto planted a one-acre fruit and vegetable farm, large enough to supply weekly boxes of produce to a couple dozen Richmonders through the long California growing season.

It was part of a larger push-pull happening in Richmond more generally, a feeling that the community was trying to change its identity but could only do this by struggling with its past. The mayor and her progressive allies sought out green businesses and actively tried to entice them to move to Richmond. McLaughlin also spearheaded efforts to create more renewable-energy training options within a local jobs program. In 2009, the old Ford assembly plant by the waterfront reopened partly as a headquarters for SunPower, a company that assembles rooftop solar panels, partly as an event space where dancers and derby skaters and even hundreds of meditators at a mindfulness conference would gather in the light of the tall glass windows. It also housed a company that made countertops from recycled glass. The plant's former oil house became a visitors center for the National Park Service.

Meanwhile, a Safeway closed in Richmond, leaving the city with only three full-size grocery stores. Meanwhile, a new study said childhood obesity rates in the city were rising. Meanwhile, a graffiti artist nicknamed Nacho left his scrawl all over the Iron Triangle while eluding the befuddled police. A small Richmond company sold vegetable seedlings and also produced lettuce fertilized via the by-products of aquaculture—that is, fish manure. A world-famous submarine designer worked on building a minisub in Point Richmond that could explore the world's deepest deep-sea trench, near Guam. A congregation in the Iron Triangle was forcibly evicted from an Apostolic church after it fell behind on mortgage payments on a loan some felt was predatory; police handcuffed the minister and his wife and eighty-year-old mother and led them away.

As ever, Richmond stood at the edge of promise and peril.

※

In 2012, a group of Urban Tilthers—Tania, Adam, and a rotating cast of other coworkers and friends—were sharing a three-bedroom house the color of old dishwater in the Richmond Annex, at the south edge of Richmond and the neighboring city, El Cerrito, about four miles from the refinery and two from where Doria and her kids lived in the Iron Triangle. For the Tilthers, it was a moment of little money but much camaraderie. *A lot of us were super broke. It was a friend's grandmother's house. It was very inexpensive to live there*, Tania recalled later. It was also a moment of transformation for Tania. After struggling academically in high school, she had attended a community college, then been accepted to a program in peace and conflict studies at the University of California, Berkeley. *I have to make my parents proud, and I'm gonna prove to myself that I can do this*, she thought. She tended the Richmond Greenway garden part-time, transitioning to full-time in the summer, while Adam worked at the Richmond High farm.

The housemates set up a chicken coop in the backyard. In the evening, they often cooked big meals together, usually frittatas or stir-fries with eggs and garden vegetables in a combined kitchen-dining room with an odd assortment of red and black cabinets. *We would have these feasts. We would laugh sometimes. Yeah, we're poor, but we eat good*, Tania said. One window faced north to El Cerrito. The other looked toward Richmond and the tail of refinery smoke that always curled up on the horizon. Because it was so well lit they called it the sunroom and would linger there often, as late-day, orange-ember California sunsets angled through the western window.

A multiyear drought spread across California, the worst to hit the Central Valley and the California coast in 450 years. It would eventually cause an unprecedented die-off of trees in the Sierra Nevadas, decimate ranches and farms, and set the stage for a series of tragic megafires, including the 2018 Camp Fire. But Urban Tilth wouldn't feel the worst of this. The organization held its biggest summer apprenticeship program to date—forty-seven kids working the soil, turning out fruit and veggies.

That summer was also the 110th anniversary of the Chevron Richmond refinery. The Richmond Museum of History, in the historic library building around the corner from Doria's childhood house, announced that in the second week of August an exhibit would open to mark the occasion with music and refreshments.

On that Monday, August 6, the kids in Urban Tilth's apprenticeship program gathered at the abandoned middle school's one-acre farm and at the Greenway and the Richmond High garden for one of their last work days of the season; the beds were loaded with crops such as squash, zucchini, corn, basil, and tomatoes.

And at about 6:30 that evening, a fire started in the Richmond refinery.

It began when a pipe ruptured, enveloping nineteen refinery workers in a vapor cloud. Eighteen of them fled, and one lingered in firefighting gear, which kept him alive as leaked fuel ignited and burst into a fireball inside one of the engines. Within minutes a cloud of white smoke enveloped the facility; then a more ominous black plume, opaque as fabric, like heavy black robes, lifted up into the sky, then orange torrents of flame. The smoke rose to four thousand feet, higher than the summit of Mount Tamalpais across the water.

The sirens awoke and screamed a warning across the city.

Tania and Adam were in the backyard when they heard the noise, and Tania knew instantly this wasn't a drill. Something was wrong. They saw a large plume of smoke and halo of distant flame over the fence line, and everyone at the Tilther house headed indoors to the sunroom and watched, mesmerized, through the window as a black cloud spread over the city. One housemate took panoramic photos, and they sat for hours, as each neighborhood disappeared beneath the smoke. Tania's mother, who lived nearer to the refinery, called her and said the smoke had blotted out the sun. But there was nothing to do but wait it out. *I was thinking about all the people who had to deal with their asthma, and all the animals who live outside,* Tania remembered. *I was thinking about my community, my friends, my family, my loved ones. And then, of course, I was thinking about the garden. And it was just a deep sadness—and I was very enraged.*

Closer to the center of the city, Doria Robinson had just arrived at her house when the smoke appeared in the sky. *It's all happening again,* she thought. But it was bigger than anything she remembered. The flames were visible over the tops of the houses. *I was like, What the hell is going on?* She was alone that day; her kids had gone to visit their dad, and she was grateful that they didn't have to take smoke into their young lungs. She walked into her house and called her mother to make sure she was safe. Then her staff began calling: she told them to make sure all the Tilthers and kids went home to shelter in place.

Only after all of this did she have time to consider how the soot was falling on the gardens, possibly rendering the produce inedible, untouchable. All the food, all the labor, all the hope, what if it was ruined in an instant? And for the first time in years, she would ask herself if she should really be doing this work. Was it safe to grow anything in this beleaguered city?

That evening, she posted an image on Facebook of the tall and sickening conflagration rising above Chevron. It looked like a bomb had been dropped on the city. She wrote, *This is what it means to live on the front lines! WHO OWNS THE SKY???!!!*

☙

A few weeks after the smoke cleared, I would sit with Doria in her yellow house and drink tea, and she would tell me what it all meant to her—this ugly, shocking, and powerful moment of revelation.

But in the moment, everything was dark. In the moment, there was rage, and debris to deal with.

PART TWO

CHAPTER 7

THE HOME FIRES BURNING

home, v. *to proceed or direct attention toward an objective.*
(Example: We're homing *in on a solution.)*

The story that takes place after an acute disaster such as a fire or a flood is generally unglamorous, especially when compared with the adrenaline-charged heroics of fighting fires or rescuing people who are trapped or injured.

But sifting through all of the wreckage, putting things back in order where possible, salvaging what still has value—these tasks are no less important, and in some ways require even greater mettle.

Immediately after the Carlton Complex Fire burned through the town in the summer of 2014, the people of Pateros, Washington, began en masse cleaning up trash and debris.

There were yards full of ash and rubble to dig up; truckloads of melted and warped scrap metal to haul away; concrete foundations to be excavated or buried on-site; burned trees and brush to remove. It was like an archaeological dig in reverse. Could you take a major catastrophe and hide it, bury it, haul it away, so that people could move on with their lives?

Meanwhile, a stream of donations from around the country started piling up in the Pateros fire hall, the city hall, and the school, and someone had to decide what to do with all of it. Some things were useful—water, food, clothing in good condition. But many were not—broken appliances, old bird cages, tattered swimsuits, a rusted push mower. There were enough items to fill multiple warehouses. They required perpetual sorting and reorganizing.

Carlene Anders, the firefighter, threw all her energy into helping manage this messy recovery process. You could call it a karmic adjustment or a return of generosity. Twelve years previously, she'd given birth to a premature baby boy, twenty-four weeks' gestation, one pound and ten ounces, and Carlene had never forgotten how many people from Pateros had stepped in—retired teachers who had run her daycare business while she spent 128 days in a hospital in Seattle, people who donated money and gas cards and phone cards. Her son had since grown into a healthy adolescent, and she felt she owed a debt to this community.

Disaster recovery is its own professional field—a mix of science and social work with a generous dollop of bureaucracy thrown in. Carlene had known nothing about it, then felt as if she was cramming an entire university course of study in the subject into just a few months—every day scouring websites, making phone calls to navigate the convoluted processes of applying for government aid and philanthropy, getting access to heavy equipment, dealing with cleanup of wastes both hazardous and benign, and addressing miscellaneous government requirements. It's useful to have what's called a "long-term recovery group," a committee of people who know what's going on, who can take in resources and donations and distribute them to the right people. Carlene helped set up the Pateros-Brewster Long Term Recovery Organization and worked without pay until she couldn't any longer. In September, the group's board cobbled together a salary for her, and she took the helm as its executive director. Just a few months later, she took charge of the organization leading the entire countywide effort, the Carlton Complex Long Term Recovery Group (funded initially by an anonymous donation from a local apple business).

The U.S. Federal Emergency Management Agency eventually supplied money to rebuild public buildings and infrastructure in communities damaged by the Carlton Complex Fire, including the Pateros water towers, but refused any aid to private property owners, a decision that frustrated many in the valley, including Carlene. FEMA did, however, assist in another way: by calling in the legions of disaster volunteers. Many of the major Christian churches have disaster response wings, some of them vast and well organized. Such volunteers are not supposed to preach, only help. "Disaster chaplains," clergy and some laypeople trained to support survivors of disasters, usually commit to a code of ethics that includes this fundamental rule: "Do not proselytize." A FEMA employee began calling Carlene to ask if she wanted to invite various relief groups to town. By the fall of 2014, teams of volunteers from Christian Aid Ministries, Western Anabaptist Mission Services, and Mennonite Disaster Service had arrived.

Carlene's new organization housed some of the volunteers at Alta Lake—a tiny resort area two miles south of downtown Pateros that had been ravaged by the fires. They stayed in a motel that had survived the disaster, but the keys, in a now-torched outbuilding that had served as a clubhouse, had all melted. *So we had to crawl through windows and open the doors.* Then the volunteers were dispatched all over the area—to help with cleaning up, clearing debris, and providing emotional support.

In Pateros and Brewster, disaster chaplains helped people sift through ash so they could try to recover lost belongings, valuable jewelry, ceramics. A group of retired veterans and firefighters made house calls and cleared debris. *They were fast and furious*, Carlene recalled. With Carlene's mom, they buried the rubble from the family home, though she insisted they leave the foundation exposed.

It was a massive endeavor. Some locals set up entire new business ventures based on the cleanup effort. For instance, a Pateros mom of a teenager started a scrapping business at the behest of her son, to help pay to rebuild her own home.

Meanwhile, there was the question of where to house people who'd been displaced. A third of the firefighters in the Pateros fire department had lost their own homes while they were out trying to contain the Carlton Complex and save the homes of others. Some children in every grade of the school district were suddenly houseless. People camped around the city in tents and trailers. The Pateros mayor stepped down shortly after the fire because her house had burned down, along with her mother's and uncle's homes—and she needed time to support her family.

Carlene let two wildfire survivors stay at her late grandparents' place in Brewster. Meanwhile, the Recovery Group brought in dozens of trailers, many donated, some acquired on Facebook, and asked people with vacation homes to house the displaced. In the fallow months of winter, a local orchard let wildfire survivors move into its farmworker housing.

The first person to rebuild was a retired teacher named Sue. She was one of the lucky few to make a successful insurance claim, and she hired contractors to put up a new house where her old one had been, near the golf course at Alta Lake. The volunteers still helped clear her yard. One photo of the community's work on this house would become iconic: Carlene and other community members and volunteers raising Sue's first wall with their hands. In the end, it was a modest house, tan and brown with an ample garage—done by Christmas. The symbol of things to come, Carlene hoped.

As the piles of ash and debris shrank and disappeared from the landscape, she and her collaborators—including a large team of philanthropists she'd assembled from around the area—decided they would build more houses for those who had little to no means to recover on their own. And there were many in this category. The people of Okanogan County—wherein sit Pateros, Twisp, Winthrop, parts of the Colville Reservation, and a scattering of other small communities in a landscape roughly the size of Connecticut—have one-third smaller household incomes on average than other people in Washington state as a whole and are nearly 70 percent more likely to live in poverty. People who had already been bearing this kind of strain had suddenly also lost residences, material possessions, and life savings in the fire.

First the Recovery Group tried buying manufactured homes for people—and in one case, experimented with a yurt. But eventually they decided they wanted to give people something better than that, real houses that might outlast the next disaster—with fire-resistant siding and metal roofs, which are not generally combustible and are unlikely to trap embers that could ignite other parts of the house. The Recovery Group borrowed house blueprints from Mennonite Disaster Service, but the designs were originally intended for developing world countries, mostly in warmer latitudes. They had to make revisions so the roofs could handle snow and the walls could contain a bigger load of insulation.

Anyone who got a house would have to meet certain criteria. They needed to own the land and agree to live there for five years (barring extraordinary circumstances). They had to learn fire-readiness, a series of strategies for preventing a house from catching fire and making it easier for firefighters to access the property if flames did arrive.

By the spring, the Recovery Group had chosen eleven households and begun raising the millions of dollars required. The first home would go to a Latinx family who had been in Pateros for twenty years. The husband worked at the school district, and the kids attended the high school. In April 2015, troops of volunteers laid the first four foundations. They started drywalling in the summer.

At this point, Carlene believed she was charting a path out of the previous summer's devastation. It had been a once-in-a-lifetime disaster, she thought, but her community would survive and rebuild, even if it took years.

But the weather of 2015 was as strange as the previous year. The winter brought normal precipitation but too-warm temperatures—causing a "snow drought" in the mountains that starved the streams of meltwater in the spring. A heat wave hit the Pacific Northwest in June, and a few places clocked record, over-100-degree temperatures that month. Another dry summer followed.

In mid-August, thirteen months after the wildfire that assailed downtown Pateros, Carlene was attending an emergency response and recovery class led by FEMA on the west side of the Cascade Mountains, twenty miles outside Seattle. There were four dozen emergency managers in one room, and suddenly the air filled with the chirruping and buzzing of cell phones and pagers. A group of fires had lit and were spreading around Omak, a town about thirty miles north of Pateros. Another called the North Star had ignited on the Colville Reservation, and four blazes were burning in Chelan County, to the southwest of Okanogan County. The evacuations had already begun.

The class came to an abrupt halt, and its attendees hit the road.

Carlene returned home that afternoon. Over the next few days, she helped patrol the area in a small fire engine called a rescue rig, equipped with Jaws of Life for prying people out of cars. She was ready for any emergency that might come to the area. Pateros was spared this time, but up the valley, west of Twisp, a tree branch was tossed against a sagging power line by the wind, igniting another wildfire, called the Twisp River Fire.

Carlene had the emergency radio on, and in the afternoon, she heard a caller describing a dire situation in Twisp: a crew was entrapped in the fire, and someone needed to be helicoptered out. She panicked. Her daughter had been out in a fire engine in that part of the county. *I couldn't get ahold of my daughter. And that was her region, that area, Twisp River Road. And there were only two of them. And I thought, "Oh my God, that's them!"* Carlene drove out of town, still suited up in her firefighting gear but in her own SUV. Heading northwest, she passed a long line of cars—a parade of evacuees moving in the opposite direction. They were all going to Pateros. She called the disaster chaplains from the road. *Please call everybody you can,* she begged. *We need help right now in Pateros!*

Just before the tiny town of Carlton, she spotted the car of a local newspaper reporter, flagged her down, and asked her what had actually happened. A fire engine—not her daughter's—had lost visibility in the blackness of the smoke and teetered off the road. Three firefighters perished, and Carlene knew one of the dead. He was twenty years old. She had taught him to ski when he was a little boy and had fought fires with his father.

The people of the Methow Valley would grieve this loss for years, but for now, there was an immediate crisis to manage. Five days after the Twisp River accident, news reports said the Okanogan Complex topped the size of the 2014 megafire and called it the largest single fire in the state's history. (Officially, not all the blazes merged. One, called the Tunk Block Fire on the Colville Reservation, remained separate, which disqualified the complex of fires from breaking a new size record.) In the end, 120 more homes were lost.

The survivors and the emergency responders turned to Carlene and her group. *I remember everybody looking at us and going, well, you're going to take this on, right? And there was a point where I literally got physically sick and thought, can I do this? Can I live through this? And I thought, well, who else could actually do this? We have to take this on. So then I stopped panicking a little bit, and said, okay, how do we strategize to make this work?* She and her community had learned a great deal over the past year. They'd figured out how to raise their town up from the wreckage, bit by bit. They had learned much about how to recover from disaster—and this was knowledge many others would need in this precarious era. They could teach them.

You could scratch a story of hope out of the ash. It existed in what could grow back and what could be learned—on the land and among the people.

⚶

The 2014 wildfire season shook Susan Prichard, the forest ecologist, to her core. *It's kind of unfair to compare it to a war-torn country, because none of us knows what that feels like. But there's a similar trauma in living through these wildfires*, she reflected later.

Still, she felt that there was a lesson in all of this fire—a way to live more safely that might be found on the landscape itself. Fire was part of this place, and in the twenty-first century, it would have stronger presence and forcefulness. To look after home here, Susan believed people would also have to take care of forests and sage lands—they would need to accept and attend to fire.

After the Carlton Complex struck, she kept her eyes on the new burn scars. One September day, a month after those fires had been fully contained, Susan led a tour with a group of fire scientists, some of her research collaborators, to Loup Loup Pass, up a highway that was blackened on both sides, down a Forest Service road, and over a ridgetop. The aftereffects of the fire there had been mixed. There were some places where trees were scorched and dead. Other parts of the forest had only a charred underbelly—especially where the underbrush and trees had been cleared either with heavy machinery or with prescribed burning. Her colleagues were a loquacious group—pointing and cracking the occasional joke. But when they rounded a corner, they caught a view over the edge of a canyon, and everyone went silent. *We got out of the cars. It kind of startled me how quiet everyone became for so long, a good five minutes. And what you see is just ponderosa pine completely obliterated by fire. You look over this vast area, thousands and thousands of acres, really steep drainage. You're seeing into this whole watershed, and there's not a single tree living in it.*

With its heavy, corky, vanilla-scented bark, ponderosa pine is more fire tolerant than many trees. Arguably, the tree deliberately courts a certain kind of fire, a strategy that it has evolved to live in dry places; it drops its needles in layers to create a place where flames can run. But the Carlton Complex had been especially hot and fast here, and the trees hadn't survived. On top of this, the rains that had fallen after the area burned had also scoured out all of the topsoil. The landscape was almost nude. It would be a lifetime before this place regrew, and what would it become in a changing climate? Perhaps not another pine forest anytime soon—there were no trees, no cones, no source of new pine seeds nearby. It might reemerge as something else: some future state and perhaps some different ecosystem than what had been here before.

What was the lesson here? Would this be the fate of the pine forests of the inland Northwest? Or could the damage be lessened and tempered if people changed their relationship with fire?

To answer this, Susan and her colleagues had decided to develop a series of computer simulations of the 2006 Tripod Fire, her first up-close megafire. Because the Tripod had been confined mostly to forest and wilderness, it made for a simpler experiment than the Carlton Complex. They would create a series of "what-ifs," as Susan called them—building alternative worlds in which fire was treated differently. Based on detailed maps of terrain, wind, climate conditions, and habitat, how would a fire like the Tripod have burned if people had allowed other flames into that ecosystem in the past seventy years, as Indigenous communities once had generations before them? What if people had allowed every lightning strike and natural fire start to take its own course? Alternatively, how would the Tripod have burned if firefighters had successfully put out every previous blaze over the same period of time? It would take a few years to develop this model, but Susan and her collaborators expected to see two patterns emerge in the simulations. One would be a collage forest in which the Tripod Fire never got enough fuel to damage the entire landscape but burned in little patches, some quieter and some more intense. The other would be a "boom and bust" pattern, in which a forest dense with trees and growth and thick understory fueled a fire so hot that it ripped across nearly the entire area.

This was not just hypothetical. You could also find small examples, like test cases.

When the 2015 fires burned—they lay on the other side of the county from Susan's home—she hadn't incorporated them into her research.

But there were other people across the mountains thinking similar thoughts about fire and how to live with it.

🜚

When the Okanogan Complex Fire first swept through the area, another scientist and practitioner watched a sort of small ad hoc experiment unfold next to his old house. Dale Swedberg worked thirty miles from Winthrop as the crow flies, or about ninety miles if driving over and beyond the mountain pass full of scorched ponderosas—at a place called the Sinlahekin Wildlife Area, a fourteen-thousand-acre strip of pine forest and lake and sagebrush.

Dale hadn't thought much about fire when he first took a job overseeing the Sinlahekin nearly twenty years previously, in 1997. Fire had been a neglected subject when he got both of his degrees in wildlife biology. So when he moved with his family into a drafty farmhouse at the center of the wildlife area that served as the supervisor's residence, he didn't consider how much fire would be part of both his workplace and his home place.

Then in 2000, he drove out to see the aftermath of the Rocky Hull Fire, which had burned down more than thirty houses and scorched over nine thousand acres in an area about twenty-four miles northeast of the Sinlahekin. He hiked through part of the burn scar that ran through federal land, and at first glance, it seemed like just a scene of destruction. He noticed how the needles had been toasted in a horizontal direction by flame and wind. He came across the grisly sight of a buck deer that had burned to death and a bighorn sheep with badly scorched feet, and he could hear hordes of insects chewing on the dead trees. But he kept returning, watching what happened and what regrew. He observed how quickly the land greened back up. Even plants like bitterbrush—a favorite food of deer and, according to some common wisdom, a species that didn't like fire—sprouted right back. To Dale, all of this was an eye-opener, and it led him to question the role of fire in the wildlife refuge he was then overseeing.

He began reading books about fire—smokejumping, firefighting, fire history. In one volume, he ran across an anecdote from an anthropologist who had driven through the Methow Valley with Indigenous elders from the region, the Methow people (who became one of the twelve bands of the Colville Reservation after their traditional lands were taken in the nineteenth century). "When we had gone through about half the valley, a woman started to cry," the scholar recounted. "I thought it was because she was homesick, but, after a time, she sobbed, 'When my people lived here, we took good care of all this land. We burned it over every fall to make it like a park. Now it's a jungle.'"

I had come to the realization that I was managing a fire-dependent ecosystem, Dale reflected later. The longer he worked at the wildlife area, the more evidence he gathered, eventually collaborating with a local fire ecologist who knew how to find evidence of past fires by taking samples from the centers of living trees and from dead stumps and logs. This researcher studied the rings of wood in the Sinlahekin's trees and the fire scars within them, and over the next several years, he and his team were able to reconstruct four centuries of fire history in the Sinlahekin Valley forests. This tree-ring exploration revealed that past fires were frequent here; any one spot might have burned roughly every five to fifteen years. The evidence corroborated Dale's intuitive understanding. After years of gathering data, he and his collaborators inferred that humans set many of the fires, since lightning would probably not have come often enough to explain them.

At the same time that this research was unfolding, Dale decided he would start conducting prescribed burns inside the wildlife area. In 2005, he undertook the first. He had to leap over various bureaucratic and regulatory hurdles, especially those imposed by the state. *It was a battle royale*, he recalled later. But that fall the state provided a fire crew to burn an area near a lake. It was the first prescribed fire of many that were done under Dale's watch.

Over the years that followed, prescribed burning in the Sinlahekin was an ongoing process of trial and error—some fires too smoky, some too hot. In 2010, Dale hired a "burn boss," a longtime fire manager from the Methow Valley named Tom Leuschen, who knew fire like an art form.

Prescribed fire is sometimes called controlled fire, and a skilled burn boss knows how to both tame and manage flames and smoke. First, the burn crew draws a line around where the fire will run in order to confine the flames—either by wetting the ground or by digging down to mineral soil, often by hand with a tool called a Rhino, which resembles a curved hoe. Then they use another tool called a drip torch—which looks a bit like a gasoline can but with drops of pre-lit fuel emerging from its tip—to dribble bits of fire in a line across the landscape. The flames are lit in small strips, usually opposite from the direction that the fire would want to travel. In other words, if a crew is burning on a slope, they would terrace the fire down the hill, since fire likes to travel up. They would burn against and not with the wind. An expert burn boss would study the terrain, the fuel, and the weather, and be able to turn the flame lengths and the heat up and down by adjusting the size and orientation of the strips. The flames can even be directed, to some degree, to protect certain parts of the land, avoid an old standing tree snag that woodpeckers like, clear away parasitic mistletoe, and kill off weeds.

As Dale and Tom invited more fire back into this place, the land responded in miraculous ways. After fire, buckbrush—another favorite of deer, with glossy green leaves like wintergreen and puffy white flowers—sprouted in profusion. (Its seeds, Dale learned, could live underground for two hundred years and then resprout after a fire.) Elderberry trees, gangly and shrubby with clusters of bright oval leaves and dark purple berries, regrew with gusto. *They get burned down to just nothing but a skeleton. And I recall one that was burned, I believe in April; I took a picture of it in May when there were just green sprouts around the base where the original tree was. And by October it was over ten feet tall!* This alongside flowering currant bushes, cottonwoods, willow, a shrub with cloudlike flowers called ocean spray—all leafing out profusely after being scorched. Basin wild rye, a plant that germinates more heartily when exposed to woodsmoke, spread out, grew tall, and turned golden in the fall. Ponderosa pines so enjoyed a good fire that they developed *stretch marks*, as Dale called them—new growth splitting apart the old scorch marks and healthy bark appearing in between.

He called these *fire effects*, the stunning ecological responses that you could get only by allowing flames onto the land. Dale began to think of the land as thirsty for fire—a provocative if paradoxical metaphor. If fire were a kind of sustenance, then the land craved it, almost as much as it craved water in a drought. He thought of an uncontrolled fire—especially a wildfire that was so severe that it couldn't be tamed by firefighters—as a *feral fire*, a fire gone rogue. By failing to set prescribed fires, people had allowed feral fires to take control across the West.

By the time of the Carlton Complex Fire, Dale and his wife had moved out of the wildlife refuge headquarters and into a house about twenty miles away in a little town called Tonasket, along the Okanogan River. He had been promoted to a new role; he still managed the Sinlahekin, and he added three other refuges to his docket: one near Winthrop called the Methow Wildlife Area; another just south of Sinlahekin, called Scotch Creek, where he was also running prescribed burns; and a third, Sherman Creek, just above the Columbia River. He had planned a prescribed burn for the Methow Wildlife Area also, but the Carlton Complex roasted the place before he could organize it.

Afterward, Dale visited the site and also drove the stretch of highway over Loup Loup Pass, near the place Susan had brought the scientists. He was pretty amazed by how hot the fire had burned, how it had killed a lot of the big trees. It was the first time that the impact of climate really *came to roost* with Dale. *Climate was starting to drive the fires to make a megafire, and it was in the backyard,* not just in the distance. But this only reinforced what Dale already felt. *We need more prescribed fire.*

The true test of his prescribed-burning efforts came the next year when part of the Okanogan Complex burned into the Sinlahekin.

He had driven out of town on the day when the Lime Belt Fire, one of the five that would merge into the complex, began. By the time he came back, fire was traveling across a nearby mountain, over the crest and back toward the Sinlahekin from the south.

He spent the next few days driving back and forth between the Scotch Creek and Sinlahekin Wildlife Areas, talking with firefighting crews and sometimes joining them in battling fire. Around Blue Lake, a foot-shaped body of water at the center of the refuge, he watched the fire make runs, embers spotting and tumbling down the dry, grassy slopes into the draws and woods, then lighting flames that ran back up in a sort of zigzag pattern. He watched an osprey glide smoothly through the smoke, seemingly unfazed.

He and Tom Leuschen had done a series of prescribed burns all around the bottom edges of this lake, and here was a sort of proving ground. The firefighting crews were trying to light what were called "back burns," setting deliberate fires in the path of the advancing feral fire, so that the two would meet in the middle, eat up all the fuel, and hopefully peter out. *When the fire came through it was really pushing hard. It was coming up over the slope, and they were trying to burn out the area that we had thinned and prescribed burned. And they couldn't get a fire going in there for beans.* Neither the feral fire nor the back burn had enough fuel to sustain themselves here. The old prescribed fires had already protected the forest around the lake.

Farther north, the crews made a fire line at the top of one ridge, digging down to the mineral soil with a bulldozer to create a spot where they could safely try to defend the area. They back-burned in strips from the top of the ridge down. A back burn is different from a prescribed fire in that it is sometimes less artful, less focused on ecological goals, and more concerned with the immediate aim of quelling an existing wildfire. In the end, they burned from the base of the ridge upward in one hot, fast whoosh of fire. Afterward, Dale was exasperated: they had worked so quickly that they had killed many of the trees.

Up the road was yet another area that had seen prescribed fire twice in the past decade—in 2005 and 2014. Dale knew where the edge of that burn was, a straight line the crew had dug in the ground by hand. And the Lime Belt Fire seemed also to know that a barrier lay here. It would burn up to the line and stop. Or it would cross the line, but then *kind of meandered, skunked around, and really didn't do anything.* The side that had never seen prescribed fire turned *black, completely, utterly black, burned really hot.* And the other side looked as if it had barely been touched. Dale watched and took videos. *I was getting pretty cocky at that point. I started calling it the I-Told-You-So Fire.*

August 18, 2015, was Dale and his wife's thirty-ninth anniversary, but he fought fire that day and stayed the night at a bunkhouse in the Scotch Creek Wildlife Area. The following day, Tonasket, the town where Dale and his wife were living, was evacuated and its residents sent to Brewster. But Dale stayed on and was *up the next morning for more firefighting and eating smoke.* In places, the flames were as tall as thirty feet. On August 19, Tom Leuschen joined Dale at the refuge to help, but he left in the afternoon after he heard that his wife had to evacuate their home because of threat of the Twisp River Fire. (Tom had also known one of the firefighters killed there since he was a baby, the same kid whom Carlene had taught how to ski. He had known the parents before they were married. *That really had him shook up,* Dale remembered.)

Eventually, a Boeing 737 "water bomber" plane, flown by the state's firefighting division, passed overhead and dropped flame retardant on the refuge.

North of Blue Lake, Dale, four other people, and two fire engines beat back the fire and stopped it from advancing farther north. It had burned about seven thousand acres, nearly half of the Sinlahekin.

In a report issued after Dale's 2016 retirement, the Washington Department of Fish and Wildlife would describe this fire as mixed, "with extreme fire behavior in some locations and slow creeping in others." But the area recovered quickly. The land grew back green, and the grasses and wildflowers flourished, just as they had before. Prescribed burning had "aided in slowing the fire and in some cases stopping the fire completely," the report went on.

But Dale and the crew's efforts were, of course, not the end of the Okanogan Complex.

Outside the wildlife refuge, the fire was not an experiment but a terror.

People need to accept the fact that no fire is not an option. Period, Dale said afterward. But fear and loss can be hard to reason with.

⁂

The acceptance of disastrous fires and other such crises is hard, I think, for a society like ours that has such trouble relinquishing control.

The denial of climate change has always been partly fed by an unwillingness to let go: if you acknowledge that the atmosphere has limits, then you must also place limits on human desires. Similarly, if you acknowledge that fire cannot always be quenched, then you also have to accept there will be losses. Sometimes the things you hold dear may even burn down. Sometimes you will have to reconstruct who you are, what you want, how you imagine the future will play out. And however much you might want to believe that individualism and pluckiness will save a person from disaster, any recovery will always be a collective process. You can't just resurrect a single house stranded in the center of a burned, drowned, or ruined town. You have to rebuild the community, or you will never have enough to sustain people in the long term.

On a couple of days in late October 2019 I got a small glimpse of what it means to recover when Carlene Anders invited me to a "donation for donation sale" that she and her colleagues had organized in Omak—a town of about five thousand people at the center of Okanogan County, south of Tonasket, east of Loup Loup Pass. Here in 2015, the Okanogan Complex (144,000 acres) and the Tunk Block (about 166,000 acres)—had surrounded the community, burning through both forest and sagebrush. Here Carlene's group—which was renamed the Okanogan County Long Term Recovery Group to encompass the entire five-thousand-square-mile area—had jumped in again and helped people get back on their feet, especially those who had lost their homes in this second round of disasters. Four years later, the Recovery Group was offering up many of the unclaimed items that had been donated and asking for small monetary contributions in return, in any amount. It was like a combination pop-up thrift store, fundraiser, and giveaway. And the event gave me an excuse to explore the area.

Omak's most famous landmark is the rodeo, an arena called the Stampede. Its grounds stand at the western edge of the Colville Reservation, a stretch of tribal land twice as large as the state of Rhode Island, made of mountain and range and pine forest.

And in every direction around Omak, the dry hills rise up. Forest alternates with a scrubby, shrubby landscape that some would call high or "cold" desert. But more precisely it is sagebrush-steppe—covered with tough but fragrant sage, along with bitterbrush, sumac that turns a deep red in the fall, and dry golden grass. Along the roadsides you can spot clusters of aspens with pendulous leaves that turn golden in autumn and, along the creek beds, lush elderberry, cottonwoods, alder, birch, and willow.

Omak was vividly sunny that weekend, as it often is, with a brisk wind huffing over the hills and through the streets. I wandered much of the place on foot, traipsing through the tiny downtown, past a movie theater with an old triangular marquee announcing FRESH POPCORN TO GO, a furniture gallery, a store that sold sewing machines, a tavern, an indoor nursery full of tropical plants, and a combined Mexican and Chinese restaurant. On Main Street, I noted that the natural foods store bore an image of the town's founder on one pink stucco wall—a surveyor originally from Illinois, photographed in a cowboy hat and suspenders with his two children. It was almost Halloween, and when I walked the pathway around the rodeo grounds, a group of teenagers, outfitted as zombies with a combination of white and blood-colored makeup on their faces, asked amiably if they could eat my brain.

The donation sale was in an unassuming green shed on a street corner in a residential neighborhood. Inside were tables stacked with old pots and pans, fuzzy hand-crocheted blankets, piles of bakeware, decades-old cookbooks full of casserole recipes. Carlene was not in attendance, but two of her staff were there: Jessica and Renae, who were officially "disaster case managers" but seemed to do anything required—sorting boxes of donated goods, managing accounts, overseeing volunteers, lending a sympathetic ear, and more generally, helping those who had lost their homes to a wildfire get what they needed to reassemble their lives. Jessica told me that whenever she met a wildfire survivor, she would say, "I'm really sorry for your loss," acknowledgment that to lose roots and a home and all of the artifacts of your life was not so different from losing a loved one. All grief could be disorienting and burdensome and tormenting in the same ways.

Meanwhile, Renae enthusiastically sorted piles of donated flatware and arrayed them across some of the blankets. She told me she had nearly lost her own place in 2014. She and her partner were then in their thirties but had lived off the grid in a solar-powered house in a neighborhood full of retirees. The woodpiles around her house had caught fire, and the chicken coop had burned down (though the chickens had escaped and managed to survive). But she and her partner had defended the place by wetting the perimeter with a sprinkler system—and with luck.

As locals trickled in, the two women introduced me to some of them. Many of those who had lost their homes lived in the hills above the town, the hills that had burned. One woman had lost her house of forty-some years and only just moved back to the property a year ago, after the Recovery Group helped her rebuild. A man told me he had seen the fire coming toward his house, smoke rolling in dark and thick, and had to flee far down a dirt road to escape. When he and his wife came back, nearly everything his family owned was gone—the house, their trucks, the trees that had shaded them.

But what I sensed from these conversations, even more than grief, was gratitude. "It was kind of hard to believe," Rob Stafford said of the new house that the Recovery Group had built for him and his two sons.

He lived on the Colville Reservation, which had suffered damage in the 2015 Tunk Block Fire. A dozen houses burned on tribal land, several along a single road that sloped down from the highway. Rob's was one.

Rob said he would be honored if I would visit his new place and hear his story. So the day after the donation sale, I drove out of Omak, past the rodeo grounds, and up a highway onto tribal land to visit Rob's rebuilt house.

This new house was not on the same road as the old one, but on a turnoff into a neighborhood called Bigfoot, down a crooked drive that felt more like a walking path than a residential street, woodsy, tall-treed, unscorched by the last round of fires. After I found the place, Rob invited me to sit on his back porch overlooking the pine-filled woods and drink tea. He leaned back in the sun with his arms folded and gestured down the hill.

"I just put a sweat lodge down here. You can just see the ribs of it right there." Under the trees below us, a series of cut red-osier dogwood branches were bent into a dome shape and tied together.

Rob was not a tribal member but a "descendant," he told me—connected to one of the Colville bands on his mother's and grandmother's side. "I grew up here, and then I spent eighteen years over on the Spokane reservation." His son and the son's mom belonged to Colville, and the land we sat on was officially part of the tribe's "trust land," managed by Colville and held by the federal government.

"I was an outlaw for several years," he volunteered. I sensed he had a longer story than the one I would get to hear on this visit. He mentioned that he had struggled with addiction but now was clean and worked at a local mental health treatment center. He had navigated more than one kind of recovery in his life.

"I was diagnosed with PTSD because of the loss" from the fire, he explained. "And it wasn't the loss of material things. But it was the loss of seeing what my kids had to go through." His sons, the two who lived with him, had been twenty-one and eight years old at the time of the Tunk Block Fire. "We had birthdays; we had Thanksgivings. And then my older daughter would come over. We had Christmases. We had sweat-houses. And we had peyote meetings. And we had rock and roll jam sessions there, because I had a big studio, which I lost everything out of, too.

"And so for a long time, I had nightmares of the fire creeping over the hill." The day before the fire burned his house down, Rob was staying in Omak with his fiancée, "and I was calling different people up here on how far the fire was going." He drove up in a little white sedan to evacuate his pets and salvage a few belongings. But he hadn't realized how close or fast the fire was approaching. "What I grabbed out of my freezer was two or three gallons of huckleberries. And then I grabbed a blanket, and I grabbed some other things that were sacred to me. I grabbed a bunch of beadwork and things like that. I thought, well, I'm just going to get this stuff, just in case. And then we start driving back down the hill. Once we got to the bottom, the flames were coming like fifty feet at us. It was basically a firestorm." Rob had to drive more than three hours out of his way on tangled, twisting mountain roads to escape the path of the fire. Later that evening, one of his relatives sent a message on Facebook to say the old house was gone.

"What really laid heavy on me was, how do we tell Aydan?" The younger son had been away visiting his mother in Western Washington during the fire. Rob knew he had to tell the kid in person, and he wanted to organize a ceremony, based on many he'd been involved in over the years as a singer of Indigenous music. "His mother came over here, and we took a drive up. Everything was still smoldering." His son saw that other houses along the road were gone. "And he would say, 'Well, what about our house?' And we still didn't say anything. And when we're coming down to our house, everything is visible. And the look on his face when he saw it was all black and gone was—it was something that I could never explain. It was amazingly sad." But he hugged his son, and they all cried and sang a traditional song.

It was community that made it possible for Rob and his family to put their lives back together. Rob was astonished when the Recovery Group told him he would get a new house. "I kept coming up. And it was kind of hard to believe that it was even going on. And I would come up here and meet everyone that was doing the foundation. And then I'd come up and see the people that were doing the drywall."

In mid-2019, it was done, and there was an official house dedication. A small group of Mennonite volunteers sang a hymn in robust harmony, and a Colville tribal member led a blessing with an eagle feather and cedar-smudging. Their new life in this house was a paring down of things. "Everything is wiped clean like that. There's a gratitude. And we do the gratitude song because of the gratitude that you feel in your heart. You know even though you've lost everything you've still got family."

I had never been through such a loss. I couldn't imagine how I would cope. And yet I knew that many more people, myself included, would need to find this kind of gratitude, to live in a world that might wipe you clean, erase what was familiar to you, and move forward still.

Every house the Recovery Group had built also had to be ready for the next fire. I noticed there was a buffer of clear space around Rob's house—where the sunlight broke through the trees and where a fire engine could maneuver if a burn came through here.

You could never keep fire out of a place like this, only hope to keep it in check. "The Indigenous worldview emphasizes the dual nature, creative and destructive, of all forces," write Indigenous scientists Robin Wall Kimmerer and Frank Kanawha Lake. "Fire can be a force for good as it warms homes and stimulates grasses, but it can also be immensely destructive. The role of humans is not to control nature, but to maintain a balance between these opposing forces."

Among the Colville bands, there were other long-standing traditions about fire and the land. Later when I spoke to Colville's natural resources director, Cody Desautel, he felt that, even in the severest of fires, the reservation had not taken as terrible a hit as it could have—because of its careful land management practices. Like other western Indigenous communities, Colville's knowledge of fire has deep roots through generations and centuries, and today the reservation has one of the most significant prescribed fire programs in the Northwest. His gaze was focused not just on the homes that had been lost but those that had been saved. "I think we saw the potential for a Paradise-type scenario," he said, referring to the 2018 Camp Fire that had destroyed Paradise, California. But he feels prescribed fire helped insulate the community from worse losses. The reservation has treated nearly two hundred thousand acres of forest with some combination of prescribed fire, thinning, and forest restoration. The "old-timers," Desautel said, always tell him to burn even more. More fire, more fire, to clean the land, to renew it. A remembrance that the wild land demands things of us and sometimes takes things away from us as well.

In the early spring of 2020, just after the pandemic began, Susan Prichard and her son, Travis, burned small patches around the ponderosa pines in their own yard. It was the moment when a ring of snow first melts around the base of the trees and then expands into small islands of thaw, a time when there's little danger of the fire escaping. The pines had laid down a thick bed of needles, like kindling. *What I could see from the ponderosa pine's perspective*, Susan said, because she sometimes tried to think like a tree, *is that these trees hadn't seen fire before*. Still, the trees shook off the flames. Ponderosas have bark that can separate into flat, corky chunks, like puzzle pieces, and when flames run up the trunks, the tree drops bark in little curls and flakes. Susan and Travis watched the pines fling pieces of fire. It was a show—like a fire dancer. By the end, the flames had pruned the lower buds and branches, which would soon fall to the ground. A sort of fire-cleansing.

Several months later, on the west side of the mountains in midsummer, I hiked an area called Norse Peak, east of Mount Rainier and above a ski resort. It had burned furiously three years previously and was a graveyard of firs and hemlocks, blackened and bony but still upright, looking weirdly like coatracks or old broomsticks. But at the ground level was the most stunning, almost absurdly vivid and garish display of wildflowers I'd ever seen—a spilled paint box of purple lupines, nodding heads of red and yellow columbine, explosive spiky blooms of red paintbrush, fuchsia-petaled asters. These were the "fire effects" Dale Swedberg had talked about. And the scene left me dazzled with a sense of possibility, the promise of rebirth.

By then, most of the 2020 Pacific Northwest fire season (which used to run from June to September but now can reach into May and October) had passed without major incidents. In late August, the Palmer Fire scorched nearly eighteen thousand acres in Okanogan County, not technically a megafire but exacting its own harsh tolls. That fire threatened more than eighty homes and forced several rural communities to evacuate.

Still, as summer came to a close, I thought maybe the Northwest would make it through the season, and months of pandemic and upheaval, without also bearing the tragedy of fire on a grand scale. And that would be some bit of grace. The *Seattle Times* alluded to the same. "But so far this year, most of Oregon's and Washington's 2,611 wildland blazes have stayed small," wrote reporter Hal Bernton on August 21. Washington had even sent fifteen fire engines and personnel from more than a dozen agencies to help California, where fires raged north and south of the San Francisco Bay Area.

I had long before made plans to drive south down the coast in early September (and meet with Doria Robinson) and was keeping an eye on the California fires, hoping some of them might be under control by the time I was supposed to leave. As Labor Day approached, a local television news website published a cheerful assessment: Puget Sound, the area around Seattle, would see "plenty of summer sunshine hanging around." But "along with the heat comes some warning too," a Red Flag Warning signaling serious fire risk. I felt a knot of anxiety in my stomach. Even on the west side (on a hike I had just done near Mount Rainier), I had noticed that the ground was so dry that it crackled underfoot.

But when Labor Day arrived, it was the wind that set the whole chain of events in motion. In many parts of the world, there are stories of "persistent malevolent winds," as California writer Joan Didion described them, the Santa Anas in Southern California, the mistral in southern France, the khamsin in Israel. But no one in the Pacific Northwest had ever seen winds as wicked as those of Labor Day 2020. Vicious, dragon-breath, gale-force winds, they seemed to take nearly every little fire start, every candle-size flame on the landscape and transmute it into a monster that could potentially dash across thousands of miles.

In Okanogan County, a fire called the Cold Springs started up on Sunday night on the Colville Reservation just outside of Omak, not far from the rodeo grounds. By Monday, it had mushroomed to 150,000 acres. In fierce winds, embers can travel far, and the fire spotted across the Columbia River. Here it was renamed the Pearl Hill Fire. By Tuesday, the two fires together had burned more than 330,000 acres. A young couple who had been camping abandoned their truck near the Columbia River and tried to run from the flames with their one-year-old son in tow. The parents were ultimately rescued, but the child didn't survive.

More than a hundred miles to the southeast, a 15,000-acre fire ate up and mostly destroyed a century-old railroad town called Malden and a neighboring burg named Pine City.

Simultaneously, the normally wet west side of the region also burned. On Tuesday, a fire in the Bonney Lake area, east of of Tacoma, destroyed four houses. The next day, another fire started beside a Target store in the same community, prompting evacuations. "I've never seen anything like this," a local fire chief told a public radio news crew. By the evening, nearly 300,000 acres had burned in Washington state in a single twenty-four-hour period.

In Oregon, "firefighters fought at least thirty-five large blazes . . . with a collective footprint nearly twice the size of New York City," reported Reuters. "Absolutely no area in the state is free from fire," said the fire protection chief for the Oregon Forestry Department. A fire ignited along the Clackamas River in Oregon and grew from a small thing to 138,000 acres overnight, forcing residents of the Portland suburbs to evacuate.

On Monday night, the smoke also descended on Seattle. I knew it was coming and closed all the windows, though it was stuffy and warm indoors, and switched on a tiny air purifier that was never designed to clean my entire house. On the evening of Labor Day, I dreamed that I was trapped in a campground with fire on all sides. My husband ran into the grass to find a way out but never returned. I tried to drive out, but there was nowhere to go. I hunkered in a car, as flames rolled over it. Even in a dream state, I could feel heat. And then I woke, gasping.

The next day the sky was the color of cigarette stains, faintly luminous like a dying fluorescent bulb. I could taste it a little: the flavor of tobacco, dirty water from an old pipe, rancid crackers. It was poisonous: the air pollution gauges registered astronomical levels in Seattle and many other parts of the region. I sheltered indoors and ran a box fan with a furnace filter taped to it, a makeshift means of cleaning the air. It helped but didn't entirely remove the feeling of slow asphyxiation, the stinging in my eyes and throat. My head was clogged with thoughts of burned towns, burned houses, burned lives. I watched the news. At night, as I tried to cook dinner, I burst into tears. All I could think about were the tens of thousands of people who were losing everything—and how onerous and terrifying it was.

By the middle of that week, half a million acres in Washington and more than 800,000 in Oregon were torched. The governor of Oregon reported that five towns had been "substantially destroyed." At least a couple of the fire starts—including the small brushfire that launched the gigantic Almeda Fire, which burned down more than 2,600 homes in Oregon, and the Cold Springs Fire—were investigated as possible arson attempts. Rumors surged on social media that the fires were the work of radicals—either the Proud Boys or antifa, depending on which seemed like the more disturbing bogeymen to a particular audience. And 911 lines and sheriffs' offices, which were already overwhelmed with the responses to actual disasters, became further clogged with reports based on such gossip. Though most of the fire starts were human and not lightning, neither conspiracy tale was true.

There were fires along the major highways—the interstates and the Pacific Coast Highway, which wends south along what is usually a quiet, drizzly terrain of coastal cliffs and dunes and seaside towns.

In California, the August Complex Fire—formed by the merging of more than thirty separate fires in the Coast Range Mountains, grew to become the largest in that state's history, more than one million acres by the time it was over. Three other fires burning simultaneously would be among the state's biggest on record. National Guard helicopters airlifted two hundred campers and hikers trapped in the center of another fire in the Sierra Nevada.

The fires burned through many square miles of forest and sagebrush-steppe (an ecosystem that, like forests, has been starved of regular fire), across rangeland and grassland, through towns, jumping roads and rivers. In some places, the flames were aided by cheatgrass, an invader grass from Eurasia that grows thick and burns hot. They were driven by dry winds and heat. They were driven by climate change. They couldn't be suppressed. You could have called them the West Coast Complex—as if one epic conflagration had raged up the edge of the continent from south to north, stretching from mountains to coast. As a combined event, this was unprecedented in the recorded history of the American West.

The smoke rose into the atmosphere and drifted a thousand miles to the west, where it entangled itself in a Pacific cyclone, spiraling across satellite images.

It spread east to Manhattan and then to Northern Europe.

The whole world was breathing the residues of our fires, the ashes of burned towns and scorched land.

The stories that sprang up online after the 2020 wildfires seemed to dwell mainly in devastation, and, certainly, Susan Prichard told me, the events had been "jaw-dropping," even to fire scientists and climate scientists. The future foreseen in climate models—the era of megafires— was arriving even faster than anyone had imagined. (A few years previously, two scientists had run numbers and estimated that climate change alone had doubled the amount of forest burned in wildfires in roughly the past three decades.)

But everyone I met who dealt with wildfire on a regular basis seemed to know despair yet be able to live in a practical strain of optimism. Not the same as hope—not anchored to expectations about the future. But the kind where you size up a catastrophic situation, decide what is available to you, and get to work, by whatever means are available.

There were two kinds of work. One was the work of renewing the world, the fire-dependent ecosystems that people lived beside. "People have lived with fire for millennia," Susan said, her views on the subject undimmed. "If we stop being afraid of fire and allow fire to do its work around our communities and actually engage with fire, the next fire that comes will be much more benign." She sent me images of her simulations of the Tripod Fire. They depicted a series of hypothetical scenarios, including landscapes that had known many past burns. The images of these looked a bit like pointillistic paintings, with patches and spots representing many different kinds of forest habitat. These were more resilient, she and her colleagues had written, more likely to support habitat for animals like lynx, even after a hot fire. They represented a better world—in which forest could be preserved by people who were willing to let some fire burn.

Then there was the work of recovery and rebuilding. The fires of 2020 had been frightening, Carlene Anders* told me. A few hundred evacuees, including farmworkers from the town of Bridgeport, had fled to Pateros and Brewster. She had driven through both downtowns and surveyed the scene just afterward, families sleeping in parked cars everywhere with dogs leashed to their side mirrors, clothing wedged in their windows to approximate privacy curtains. "I thought, how do we do this? How do we do it safely in a pandemic?" She had spent six years by then developing a strategy for getting food and shelter and help to people after a fire. "Now you've added another layer that makes it incredibly difficult." Community leaders couldn't make announcements at social gatherings, for instance. It was harder to temporarily house people and give them separate air to breathe. But quickly a plan came together. In the years since the Carlton and Okanogan Complex Fires, Carlene had become a sought-after expert on disaster. She had offered guidance to survivors of the fire in Paradise, California, and, after the 2020 fires, she began advising the town of Malden on how to set up a recovery group like the one she had helped put together and run.

There are stories of grief and horror from the 2020 fires that will never really vanish. But Cody Desautel from the Colville Reservation sent me photographs of two stands of trees, both full of ponderosas, both visited by a severe blaze that season. One had seen good fire in seasons past—Desautel had burned it himself two decades previously—and another had not. The untreated patch of forest was entirely blackened and all the pines dead. But the patch that had experienced prescribed fire twenty years ago looked almost unscathed by the 2020 blaze, except for a charred snag on the ground.

---

* In the November 2021 mayoral election, Carlene Anders lost to another candidate. It was, however, not a bitter campaign, and her opponent praised Carlene for working "long and hard towards the recovery efforts after the fires" and dedicating "countless hours helping those who lost their homes try to find housing or rebuild."

Eventually, I went alone that fall to visit the blackened shores of Omak Lake, an inland saltwater body on the Colville Reservation, full of cutthroat trout—down a road that wound out of the town of Omak, through a canyon and onto a ridge. Most of this landscape was sagebrush-steppe and a bit of farmland and orchard. Both sides of the road and much of the eighteen-mile lakeshore were black as tar. In some places, nearly every bush, every branch, nearly every pine was burned. Charred stems of bitterbrush and sage prickled across the hills like oversized thorns. I saw a mother doe and two fauns dash across the roadside, searching for something edible, finding it in a still-green irrigated field. The scene shocked me. But I tried to see it differently, knowing that some things would regrow.

If we are going to make it through this unruly era intact, I think we will need to keep remembering how to renew the world.

We will also need to change our relationship with other things that burn—the ancient forests that now exist as coal and petroleum, the fuels that are driving up the planet's temperature. But that is a story for another chapter.

CHAPTER 8

FINDING HOME GROUND

In her childhood in Kentucky, Black essayist and scholar bell hooks experienced two kinds of homes. One, her grandmother's, in farm country, was full of belongings in both senses of the word, beautiful and personal objects and also a profound connection to history, meaning, and identity. "Her house is a place where I am learning to look at things, where I am learning how to belong in space," she writes in the book *Belonging: A Culture of Place*. "In rooms full of objects, crowded with things, I am learning to recognize myself." She recalls bittersweetly how fields of tobacco surrounded the house—a crop that offered rural Black families a measure of economic self-reliance, even though it would later be linked to serious health conditions—and that her grandparents harvested it themselves, "the leaves braided like hair, dried and hung." She remembers the practice of stringing "red peppers fiery hot" to cure from a thread hung in the window. But her parents' urban home, the one she grew up in, was "an ugly house"—far less emotionally compelling than her grandparents', more just a piece of property. It contained "a great engulfing emptiness," she writes. She laments how capitalism and real-estate markets reshaped her community's notion of home—turning it from a place where you could belong into a thing that you should possess. In her parents' house, "space was not to be created but owned. . . .

Consumerism began to take the place of that predicament of heart that called us to yearn for beauty."

In modern American parlance, the meaning of "home" is entangled in real estate. Homeownership—especially the possession of a single-family home—is supposed to confer legitimacy. In young adulthood, I learned that to be a "grown-up" was to buy your own house, even if it left you with a crippling mortgage that would devour most of your earnings. The idea lies at the core of the American Dream, that postwar fantasy that exerts a stubborn influence on our culture no matter how outdated it becomes or how many will never attain it—or have sometimes been deliberately excluded from it. Few things drive our society as forcefully as the real estate market, which can buttress or unravel our economy entirely, as in the Great Recession of the late 2000s.

Meanwhile, this second interpretation of home—home as meaning and identity, as emotional space, as a relationship with the place around you—has been sidelined in America. The lack of property or walled dwelling makes you *homeless*, a word that neglects the intangible kind of home. (Some advocates and community organizers have urged the adoption of the terms *houseless* or *unhoused* in recognition that someone can have a meaningful home even if they lack a physical dwelling.) To have an attachment to the *local* means you are also perhaps *parochial, provincial, limited, confined, insular*. Dig further into word associations and synonyms, and you arrive at *small-town* and *small-minded*. *Localism* is a *preference for a particular place or region, especially that in which one lives* and also the *limitation of ideas, sympathies, and interests resulting from this*. People who are too attached to a place—especially certain less-desirable places—are allegedly backward or narrow. Stereotypes often hang over rural towns, places of urban poverty, and regions of little economic opportunity. Culturally and economically, we expect people to escape such locales and reach for what is global, cosmopolitan, sophisticated—but placeless.

As a student at Stanford University, bell hooks worked on shedding her accent: "It was a way to avoid being subjugated by the geographical hierachies around me which deemed my native place country backwards." She was simultaneously homesick, longing for the Appalachian countryside. "In my mind and imagination I was always returning to the Kentucky hills, to find there a way to ground my being." But so many factors had conspired to estrange her from the land—segregation, racism, traumas both personal and collective—that for decades, hooks felt it was impossible to return.*

She tried on other places: Wisconsin, Connecticut, Ohio, New York. She achieved career success—which gave her the opportunity to own real estate—but she felt ungrounded, incomplete, melancholy. Then, "I plundered the depths of my being to see when and where did I feel a sense of belonging, when and where did I feel at home in the universe." The answer was to return to her origins. Thirty years after she had left, bell hooks moved back to Kentucky, to settle in the college town of Berea, a community with a long history of progressivism where she could make a home full of meaning, like her grandmother's house. She defined for herself an ethic of place: "A culture of belonging rooted in the earth."

Reading these words, I feel a visceral longing for the kind of rootedness that hooks found. (I have uprooted myself many times, and my search for a home full of meaning is long and still ongoing.)

But I have wondered if such rootedness is even still possible.

Is it risky in the twenty-first century to love home as much as hooks did? Might you feel too much solastalgia—the homesickness of climate change—to be able to cope with the future?

Is it backward or unreasonable or stubborn to attach too much to a place, when we must all be prepared to uproot ourselves?

What sort of home should we seek out now?

ॐ

There's a story that says it's suspect to be too fiercely attached to or defensive of your home turf, even in service of a good cause. And that logic allegedly extends to environmental problems—our relationship

---

* Even before climate change, the uprooting of Black Americans and other communities of color has been a frequent pattern in America. In the rural South, for instance, discrimination in property law, lending, disaster assistance, and real estate practices ultimately forced 95 percent of rural Black farmers off their land over the last century.

to land, ecosystem, and planet. The acronym *NIMBY* probably first appeared around 1980, short for "not in my backyard," describing a person or group who resists some kind of change or development at home—and usually meant as an insult.

The original NIMBY was someone who opposed nuclear power or nuclear waste in their "backyard." Sources differ on who coined the term, but the man who carried NIMBYism from obscurity into our common vocabulary was a British politician named Nicholas Ridley, or Baron Ridley of Liddesdale.

With a family fortune derived from shipbuilding, coal, and steel, Ridley became a Conservative member of Parliament in 1959. He was widely known as a pit bull for the free market, an antagonist of socialism and unions, and a curmudgeon who overindulged in cigarette smoking. When Margaret Thatcher was elected prime minister in 1979, Ridley became a close ally. He helped her usher in her own *ism*, Thatcherism, one flavor of the larger ideology neoliberalism, which has transformed the British and European political landscape, upended American politics, and shaken up the world. Initially, neoliberalism was partly a reaction against communism, a pendulum swing in the opposite direction: in the simplest rendering, the ideology aims to starve government and the public sector to the barest of skeletons and the most minimal functions, privatize public services when possible, shrink the power of organized labor, and let the corporate economy run unfettered. Neoliberalism also "redefines citizens as consumers, whose democratic choices are best exercised by buying and selling," writes British columnist George Monbiot. This notion reduces our individual power to act in the world via democratic or collective process and makes us tiny actors in the global market. And in this schema, home is not a thing you belong to.* It's a thing you buy. Not the home of meaning, but the home of possession.

In the backdrop of Thatcherism, in the late 1970s and 1980s, the British government was pushing for new nuclear power development.

---

* I am simplifying and narrowing this discussion of neoliberalism to how it affects concepts of place and home. But many volumes have been written about why this ideology is dangerous and antithetical to addressing climate change, protecting democracy, and preserving much of anything for the public good. Naomi Klein's works, especially *This Changes Everything* and *The Shock Doctrine*, describe the flaws of neoliberalism in great detail.

In 1981, the U.K. set up an entity called Nirex, funded by the nuclear industry and regulated by the Department of Environment, to search for places to dispose of radioactive waste. But much of the public was understandably apprehensive: even so-called intermediate nuclear waste has to be isolated for several centuries before it is considered safe. Meanwhile, Pennsylvania's Three Mile Island nuclear power station had melted down in 1979, and in 1986, the Chernobyl catastrophe in what was then the Soviet Union sent shudders—and radiation—around the world. In response, environmental groups and citizen activists rose up to protest and fight proposals for nuclear waste dumps. The same year as Chernobyl, Thatcher appointed Nicholas Ridley to the post of secretary of state for the environment.

Ridley used the term NIMBY with frequency and enthusiasm to suggest that the communities that stood in industry's way were parochial, narrow, too concerned with the local, backward, limited, and insular. They were selfishly advocating for their own local concerns and impeding the global march of progress. They were ignorant Luddites. (Around the same time in the United States, Wendell Berry observed a public meeting held by industry representatives and government regulators who were proposing a new nuclear power plant in rural Indiana, and noted how the conveners applied the same sort of disrespectful tone to the residents in attendance. "They had come to mislead us, to bewilder us with the jargon of their expertise, to imply that our fears were ignorant and selfish," he wrote. In Berry's estimation, these industry experts placed "no value at all" on the meaning of home and community.)

In some cases, the nuclear industry and its champions won their arguments, but in others, the activists couldn't easily be deterred. In 1995, Friends of the Earth revealed that Nirex had interfered with the British government's efforts to publish reports questioning the safety of nuclear waste disposal. In 1997, the group leaked a memo from a Nirex advisor cautioning that poorly designed storage might contaminate groundwater with radiation. After this revelation, the industry's handling of nuclear waste in Britain would be the subject of controversy and investigations for decades.

Meanwhile, the shorthand NIMBY stuck around. Its meaning has shifted—often the NIMBY is less of a localist and more of an elitist, someone who might object to, say, the construction of a halfway house out of concern that it would dent their property values. (Ridley himself was later labeled a NIMBY, after he opposed a housing development that would have obstructed the view from his home in the Cotswolds.)

In 2011, *Time* named NIMBYism number five in its "Top 10 Green Trends." Journalist Bryan Walsh credited NIMBYism for one of the climate justice movement's most defining victories, "pushing President Obama to postpone the Keystone XL pipeline that would have brought crude from Canadian oil-sands development across the Midwest." Nebraska ranchers and Great Plains tribes were especially influential in this battle, galvanized in part by the potential for the pipeline to break or spill and foul the Ogallala Aquifer, a major water source that underlies eight states. They joined national environmental groups in a groundswell of protests in Washington, DC, and around the country—also raising the alarm about the climate consequences of fracking vast reservoirs of Canadian tar sands oil, exporting them via pipeline, and burning them. But Walsh insisted NIMBYism has a "dark side," also holding NIMBYs responsible for stopping some renewables development. "Environmentalists may welcome NIMBYism now—but it could bite them in the future."

Cape Wind, a proposed offshore wind power development off the coast of Cape Cod, famously languished for years and eventually failed in 2017 after locals, especially wealthy and influential residents such as Robert F. Kennedy, Jr., fought bitterly against it. And in many interpretations of this story, the NIMBYs were especially villainous actors— rich people whose personal aesthetics and property values were more important to them than whether or not the planet roasts. Kennedy insisted his primary concern was that the wind farm would damage a sensitive location and likened it to building in Yosemite National Park (though some evidence from the U.K. and Belgium suggests offshore turbines actually help create better habitats for fish). In response, a group of prominent environmentalists pleaded with him to change his position, arguing that "nothing threatens the Earth's most special places more than global warming." He did not.

But when interrogated, the whole idea of NIMBY starts to fray at the edges. For one thing, there is no clear definition of who NIMBYs are, and the term is used sloppily to describe anyone who opposes the building of anything even slightly close to home—from a chemical-weapons incinerator or an oil pipeline to a food pantry—no matter the reason. After conducting a series of surveys, two University of California, Santa Barbara, researchers concluded that many opponents of wind turbines simply distrust either the industry or government authority or dislike the whole notion of wind power—and then find it especially objectionable when a wind farm is supposed to go up near their place of residence. In practice, it can sometimes be difficult to impossible to locate real people who fit any definition of NIMBY. After the world's first major tidal energy generator—a form of renewable power that uses wave energy to produce electricity—was installed on the bay side of the elbow-shaped Ards Peninsula in Northern Ireland in 2008, a University of Exeter researcher traveled to two nearby villages in search of NIMBYs. But he couldn't find them. Instead, he noticed that the more attached someone was to life on the peninsula, the more they loved their place, the more they supported the new energy project.

But none of these studies seems to reckon with the simple understanding that bell hooks developed so early in her life. There is more than one reason to attach to a home and to resist its alteration. You can cling to home as property—fight for yourself and your own financial gain. Or you can love a home and belong to it—and defend community, place, and planet. There is the home of possession, which values mostly economics. And there is the home of meaning, which cherishes place, belonging, and togetherness.

In the long run, only the home of meaning can give us strength in an era of upheaval.

✿

In the twenty-first century, both the home of possession and the home of meaning face immense threats. According to a 2021 report from the real estate company Redfin, nearly 40 percent of Utah homes are in danger of burning down, and in California, $627 billion worth of real estate is at high fire risk. Even under a relatively conservative estimate from the First Street Foundation, the United States already has $20 billion in expected real estate losses from flooding every year: in other words, the real estate industry estimates that this much damage is already happening every year right now. In thirty years, because of sea level rise that number could rise to $34 billion.

Hundreds of millions of people will face damage, part with belongings, and in many cases, lose the safety and basic shelter of a home in the next few decades. The fundamental question of how to house and shelter the people who are evicted by climate change could easily turn already gnawing crises—including both the housing and the global refugee crises—into even larger catastrophes that could swallow up the global economy and destabilize whole countries and regions.

In the face of all of these economic losses, it would be easy to overlook how the loss of social, spiritual, or cultural bonds to home and place will also affect so many of us. And you might therefore think that questions about the home of meaning would become insignificant or altogether irrelevant. You might even imagine it would be a liability to have a home of meaning in a time of climate change—it might lead someone to avoid making necessary changes to the way they live, or resist relocating when a threat arrives at their door.

Scholarship on the subject suggests the opposite is true. Glenn Albrecht, the Australian word-maker who came up with the term *solastalgia*, has noticed that anyone who takes action to restore or protect their home landscape, or anyone with a strong sense of both community and personal empowerment, tends to overcome the sadness of witnessing environmental damage at home. He calls this phenomenon *soliphilia*, from the French *solidaire* (interdependent) and the Greek *philia* (love).* "Soliphilia is manifest in the interdependent solidarity and the wholeness or unity needed between people to overcome the

---

* Chinese American geographer Yi-Fu Tuan came up with a similar concept in 1990 called *topophilia*, "the affective bond between people and place." But while Tuan's word remains in the realm of pure emotion, Albrecht's *soliphilia* tries to capture the additional sense that the love of place could motivate people to take action.

alienation and disempowerment present in contemporary political decision-making," he writes. "Soliphilia introduces the notion of political commitment to the saving of loved home environments at all scales, from the local to the global."

Of course, love of home does not lead everyone to become wiser or more civic-minded. But at least some research has bolstered Albrecht's observations. Consider one study, for instance, from the Indian state of Odisha—a place at the edge of the Bay of Bengal, troubled in recent years by extreme cyclones, heat waves, flooding, and an eroding coastline. A survey there found that families with higher "place attachment" (measured through statements like "because my forefathers were staying here, this place is very important for me") were more likely to take steps to ready themselves for a disastrous flood. In other studies, in the western United States, Canada, and Australia, people who loved the place they lived in or felt connected to their community were more likely to participate in community activities to prevent wildfires and better equipped to recover from disasters.

Hurricane (or Superstorm) Sandy, which slammed against the eastern United States in the fall of 2012, remains one of the most place-altering catastrophes this country has experienced. The storm damaged two hundred thousand homes and affected people in twenty-four states. New Jersey, New York, and Connecticut took the hardest hit. In Mid-Atlantic coastal communities, many people abandoned their flood-soaked, wind-battered homes and never returned. Those who stayed struggled for years to clean up and repair.

You might think this would also be a kind of place-detaching disaster—one that would sever the bonds between people and home—but that is not quite correct.

First, consider the people who stayed. One set of places that became important after Hurricane Sandy was a series of community gardens, many built in lower-income neighborhoods on formerly abandoned lots in New York City from the late twentieth century onward. Three researchers visited these neighborhoods to find out how people had coped with Sandy. After the floods, the gardens helped some people find their way. Gardeners looked out for each other and their neighbors, especially people who were stranded with no electricity or food. As soon as the floodwaters receded, a garden in Queens became "a place where people knew that they could go . . . go and get warm, they had a fire going and people started bringing food. And then people started seeing it as a drop-off point," one resident told the researchers. A few dozen people ate "homemade chili over an open fire two days after . . . when the National Guard [couldn't] even get through yet." At another location, in Coney Island, gardeners said one of their neighbors, an undocumented resident who was afraid to ask for any kind of official public aid, fed herself and her family almost entirely from the harvests of the garden in the season after Sandy. Even here—in the aftermath of one of the most punishing American disasters of all time—grabbing a hold of place and community helped survivors to steady themselves.

Then there are the people who left.

Take the case of Staten Island, striped with working-class neighborhoods with a generations-long sense of local history and connectedness: "There were clambakes in the summer, and the neighborhood kids played soccer together at night under the streetlights," writes Elizabeth Rush in *Rising: Dispatches from the New American Shore*. After Sandy, many residents chose to relocate. They did so with an extraordinarily organized, collective voice. "Despite their love for the place that had long defined them," Rush explains, "residents of nine local communities began begging the state government to bulldoze their homes and allow the land to return to tidal marsh." In her visits to and reflections on one of these places, Oakwood Beach, Rush describes neighbors gathering together to make the decision to leave, neighbors who loved this place and cared for one another.

In a survey led by a psychology Ph.D. student, residents of Oakwood Beach were slightly more likely to accept the buyout if they experienced a greater sense of "connection and caring" (answering yes to statements like "people in my community feel like they belong to the community"). A collective decision to relocate is often called "managed retreat," a controversial term that, to some, implies giving up.

But Rush sees strength in the way this community handled disaster—"an example for the rest of us to follow. . . . They're less victims than agents."

You can love a place and community and decide to let it go. This decision may also be an act of love.

ꙮ

When bell hooks was a girl, spending time in her grandparents' home, "no one talked about the Earth as our mother. . . . The Earth, they taught me, like all of nature, could be life giving but it could also threaten and take life, hence the need for respect for the power of one's natural habitat." For hooks, home was connected to nature but stood outside it, a place of stability: "A true home is . . . where growth is nurtured, where there is constancy."

After returning to Kentucky, bell hooks bought property in the hills and in the city. Each time she purchased real estate, she also strove to create a home of meaning. "I have bought homes to share with others, even to give away." A place of quiet reflection to "hear divine voices speak." A place of resistance and building community. A house where she could claim her right to space as a Black woman in a predominantly white neighborhood. A house with a porch looking out on the street, where she could greet others, a porch for "making contact—a place where one can be seen." All were part of hooks's effort to foster the "beloved community" that Dr. Martin Luther King, Jr., articulated, a community where all are cared for and safe.

I don't know what kind of "place attachment" will be appropriate for the rest of the twenty-first century. We may no longer be able to realize all of hooks's dreams. We live in an inconstant time, and the home of safety may be far harder to find. I don't know whether and where it will be practical to attach oneself. Some of us will inevitably have to move. We will retreat from the places we have known, in moments both managed and scattered, driven by motives that range from personal economics to self-preservation.

But in all the stories I have encountered about disaster, I am struck that so many of the leaders who emerge, the people who pull a community back up from the wreckage, are people of place. They are people who build homes of meaning and not just homes of possession. And all of them seem, consciously or unconsciously, to have a sense of the collective, the tiny unseen roots that extend from person to person to ecosystem and that allow us to draw up again from the Earth and regrow—even after disaster and dislocation.

In this moment of uprooting, certain kinds of attachments could doom us—parochialism and factionalism, racism, political divisiveness, and an elitist insistence on leaving the vulnerable to face harm alone.

But love of home is a different kind of motive and emotion altogether. This kind of attachment can enlarge our sense of self, remind us that our circle of care extends far beyond the walls we live in—to place, to beauty, to human communities, to planet, to generations past and present, to histories both triumphant and scarred. And I think that the home of meaning has a central role to play in this struggle about whether we fight climate change or succumb to catastrophe. I think we solve this crisis by remembering not just who we are, but where we belong, and what matters.

CHAPTER 9

—

LIVING WITH WATER

home waters, n. *the area of sea around one's own country.*
home wind, n. *a wind blowing toward one's home or country.*

There are at least two ways to swamp a coastal city or town.
One is a storm. Among types of storms, tropical cyclones, also known as hurricanes, are the heavyweights. They develop over ocean waters that are warmer than about 80 degrees Fahrenheit, and heat and the process of convection supply the energy to stir them. On a warming planet, these are already becoming more devilish and destructive, with even faster winds and heavier rains. Other kinds of storms are also growing more severe, like nor'easters, which can drop blizzards, heavy rain, and colossal waves along the Atlantic Coast. So are heavy inland rainstorms.

A second is via high tides. The moon follows an elliptical path around the Earth, and sometimes she tugs more forcefully on the oceans than at other times. The highest of tides are sometimes colloquially called *king tides*. Very high tides can happen, for instance, when the moon aligns with the sun—so that both are pulling on the oceans—and the moon is simultaneously at *perigee*, a word that sounds like a ballet move but refers to moments when the moon is closest to the Earth. Add some extra rain or swell to a king tide, and the water can overrun sewage treatment systems, seawalls, and other structures engineered to handle water and avert floods under average conditions.

Shallow flooding from tides and small storms is also called *nuisance flooding*, the water that makes a deep pond between the road and the storefront, the water that burbles up through the storm drains or seeps into the engines of parked cars and ruins them.

But I would argue that this is the wrong term. This flooding isn't just a bother but an omen, like the first raindrops in a deluge.

Early on a Monday in May 2019, Andrea Dutton, a renowned geologist and oceanographer, told a group of people in St. Augustine what such flooding portended. The Keeping History Above Water conference—the event coordinated by Leslee Keys of Flagler College and Marty Hylton of the University of Florida—convened on the ground floor of the Casa Monica Resort, a nineteenth-century architectural marvel formerly owned by Standard Oil magnate Henry Flagler, about a block from city hall. The conference was intended partly to draw experts to St. Augustine to help with its high-water problems.

The crowd sat around round linen-covered tables and sipped coffee. Dutton wore a scarf decorated with red and blue bands. She had the same red and blue image projected on a slide, and she unfurled the scarf for the audience, like a magic trick. *What you're looking at here is a color scale*, she said. It was a visual representation of the planet's temperatures over more than a century and a half: all the warmer-temperature red bands were at the far (recent) end of the scale. *You can wear it to your next dinner party*, she said, *so that you can go talk to your friends about this and show them, look, this is real and it's happening.* This was only partly a joke. A few slides later, she reached an image labeled *sea level commitment*, as if to tell them they had all made a pact with the sea—not such a far-fetched analogy. The image showed a trend line depicting the change in the rate of sea level rise over time. From several thousand years ago through the time of the shell-mound cultures until the modern era, the rate was relatively flat and slow. *This is the time period in which we built on our coastlines. We got comfortable there, unfortunately*, she said.

But from modernity onward the graph jutted upward rapidly, the waters surging into several possible but uncertain future states: the sea could keep rising at a faster or slower rate, depending on how much carbon humans chose to keep sending into the atmosphere.

Global mean sea level has risen about eight to nine inches since the year 1880, about a decade before Rudolf Diesel patented the engine that bears his name and that helped set us on this fossil fuel–based, high-carbon, high-water journey. Even this change has reverberated all around the coast already, boosting the erosive power of waves, which smack against the landscape. Already the sea is punishing the land, a problem especially noticeable in places like barrier islands, which protect the coasts. (The Davis Shores neighborhood of St. Augustine—where city historic preservationist Jenny Wolfe used to live—stands on such an island, Anastasia Island, and to the south, its beaches are eroding, and the county has to regularly truck in sand to keep them intact.)

In this part of the twenty-first century, the global rate of sea level rise is two to three times faster than it was when the shell mounds were built: about an inch every eight to twelve years. Moreover, because the oceans are not a giant bathtub but are shaped by currents, tectonic shifts, and other forces that alter topography (such as land subsidence, i.e., the sinking of land), in some places the rate of rise is even faster than in others. A few years previously, Andrea Dutton and some of her colleagues discovered that, between 2011 and 2015, ocean waters from Cape Hatteras National Seashore in North Carolina south to Miami, Florida, rose six times faster still than the global average, or about three inches in just four years. A few inches sounds small until it is the difference between, say, the water level below your tide gate, your doorsill, or your street surface and the level above, the tipping point between dry and wet. It sounds small until you notice, for instance, that the frequency of tidal flooding along the U.S. coastline has doubled in the past three decades.

But these troubling observations were not the most alarming part of Andrea Dutton's presentation. Geologists are the historians of the Earth, and like scholars in the fields of history and preservation, she had studied the past in order to understand the present and the future. She had estimated past sea levels by traveling around the world and examining fossils of coral (which grows, of course, at the coastal edge). She was part of a global collaboration of scientists who were using this method to determine how high the water was between ice ages—at three million years ago and again at 400,000 and 125,000 years ago, the final date occurring after humans had arrived but before we had built much of anything durable. The scientists had connected this work to models and measurements of past temperatures (such as those derived from studying fossilized sea creatures—foraminifera, which store details about ancient climates in the chemistry of their intricately patterned, microscopic shells). And the story she read in the old reefs was troubling, because it appeared that it had taken just one degree Celsius (or nearly two degrees Fahrenheit) of warming, averaged over the entire planet, for the corals to climb up twenty feet or more—onto what is currently dry land. For that to happen, the planet's vast ice sheets in Greenland and Antarctica must have shed a lot of ice.

And all this evidence from the past suggested that some shocking things lie in our future: since the beginning of the Industrial Revolution, humans have already warmed the Earth by about one degree Celsius. *We may have already warmed up the planet enough to commit ourselves to something like twenty feet of sea level rise*, she explained.

But the scientist meant to move on to other business right away. *All right, this is the depressing moment in the talk*, she said, to dispense with the shock—another magic trick—and a murmur of uneasy laughter passed through the room. *Call me Depressing Dr. Dutton, or do whatever you need to do to get through it. The good news is that this is not going to happen overnight.*

It would take more than a century, probably much more, for twenty feet of additional seawater to arrive at anyone's doorstep or floodgate. The trouble was, no one could predict the exact pace of change and whether it would arrive gradually or in bursts. The future of the sea would depend on a great number of variables, and the greatest of these would be choices made by people in the twenty-first century.

Dutton revealed another graph, with a set of rainbow-colored pathways, each representing a different level of carbon emissions and consequent rise in the oceans by 2100—the possibilities ranged from less than two feet to eight feet. Which one would we choose? How much carbon would we keep emitting? How much would we warm the Earth ultimately? How quickly would we melt the ice and augment the ocean? How much time would we buy a place like St. Augustine or Annapolis or Miami or Manhattan? How many generations would be able to hold on to these places? And when they were abandoned, would people still be able to keep their histories and their ancestors close to them, as the builders of the shell mounds had? Would they still be able to renew the world? These were not just hypothetical questions.

A day after Dutton's talk, about a dozen people from the conference, including a couple of St. Augustine city officials, staff from the National Park Service, and engineers from the U.S. Naval Academy in Annapolis, walked along the waterfront to examine the series of seawalls that had been built there over the city's history and discuss plans for new flood control measures. They strolled together to the old Spanish fort. A Naval Academy professor had agreed to lead his students in researching a series of engineering strategies—such as a breakwater, a levee, or a living shoreline made of natural material—to protect the seventeenth-century structure and the grounds around it from floods. Still, it was easy to see that any barrier would hold the tide for only so many years. *We've really got our work cut out for us*, said the city public works director.

⚛

I drove to the conference from Tampa through a heavy rainstorm that pounded my rental car and dropped sheets of water across the highway. A few hours after my arrival, I joined a trolley tour offered by Jenny Wolfe and a local archaeologist. Trolley tours were a standard local attraction; a fleet of them circled St. Augustine constantly, filling the air with a running narrative about various eras of the city's history and making stops at miscellaneous attractions like America's first wax museum, the chocolate factory, and the Spanish Military Hospital Museum, a reconstruction of an eighteenth-century hospital.

But Jenny and her colleague were offering a different kind of tour: half about history, half about flooding, it was called "Heritage at Risk." We looped through the damp, narrow streets, many of them gathering puddles.

Some of the most fragile places are cemeteries and archaeological sites.

St. Augustine has a shallow water table. In some spots, you can strike groundwater a couple of feet beneath the soil surface. Among other things, it is hard to dig a grave here without it filling with water. Jenny pointed to the Tovar House, one of a few houses in St. Augustine from the eighteenth century. Archaeologists had combed through the site, looking for clues about how the house had been used and modified over time, and had wanted to leave the dig open as an active exhibit. "But during the process, the archaeologists were dealing with rising floodwaters from the water table," she explained, so they had to close the site to protect the building's foundation. The story of this place must either be documented or lost altogether. When higher waters start to reach the building more often, it cannot easily be relocated—the old foundation is too fragile. The story of this place must be documented, so it is not lost altogether in a future flood.

St. Augustine is a treasured place. It is full of epic stories and oddities and attractions, but I didn't get a sense of it as a community or a home until the last day of the conference, when Jenny Wolfe borrowed a vehicle from the city's fleet and assembled an ad hoc tour for four historic preservationists from out of town. She planned to spotlight the city's posthurricane reconstruction and renovation, and she invited me to come along.

Jenny pulled up beside the Casa Monica in a giant white Chevrolet truck. Her hair loose, her feet in gray moccasins, she was wearing blue jeans and a royal-blue polo shirt embroidered with the city crest (a coat of arms, originally drawn up for St. Augustine in 1715). This was half-business, half-social. She knew all four women taking the tour from various professional gatherings and events, and they cracked jokes and fired questions at her. Jenny proceeded slowly down the waterfront boulevard, Avenida Menendez, named for the city's founder, Pedro Menéndez de Avilés. She stopped the truck in front of a row of houses, most from the late nineteenth and early twentieth centuries, whose first floors stood at varying heights above the street, some elevated more than others in anticipation of floods.

Here was a wood-shingle building that had been elevated an extra two feet—to about ten feet above the tide line. It was still under construction, the walls covered in black tar paper. Beneath it, the contractors had discovered coquina piers of unknown vintage. Here was a Mediterranean

Revival house with hollow clay tiles and arched doorways. The first floor had filled with water during both Irma and Matthew. The owner refused to evacuate during the second of these storms. "He has video where he's standing on his porch, and the water is just sloshing around," Jenny recalled. Rather than give up on the house, he decided to convert his first floor into a waterproofed porch and live upstairs.

Here, at the front of another house, stood a wall maintained by a private resident, a replacement of an older broken wall that was made of coquina and concrete block: the water from Hurricane Irma had retreated so quickly that the pressure sucked it toward the sea and busted it into pieces. The new wall was a couple of feet higher.

Jenny drove on, turning away from the waterfront and into Lincolnville. Oaks draped with Spanish moss leaned over the streets. Here was an empty lot where a house had been torn down, then another demolition, the site of a neighborhood community center that had been derelict and too expensive to fix after two storms and decades of deferred maintenance.

She crossed the bridge into Davis Shores, where she had lived at the time of Hurricane Matthew. She no longer called this neighborhood home. After the storm, the return to her cottage had triggered too many feelings of loss and vulnerability, and she had decided to move in with her then-boyfriend, who thereafter became her husband. Here were houses on wide, grassy lawns, then more empty lots, then a ranch elevated in defiance of the traditional low-to-the-ground style for such a house. Here were the tiny, rectangular dwellings put up after World War II. Here was an unobtrusive green city park where an archaeologist was searching for remnants of a second Spanish settlement.

She rounded a corner, and a tiny green cottage came into view, framed by a wooden fence. "This is where I used to live," she said.

"It's super cute," I exclaimed.

Jenny looked wistful for a moment. "I lost the place where I felt very content."

꙳

What do we lose when places like St. Augustine—places dense with ghosts and relics and remembrances—fall within the grasp of salt water?

Did it matter, I asked myself, to remember what had transpired here over the centuries? I didn't like these questions, but I knew others would ask them, powerful people with money, people who could make decisions about what mattered and what didn't, about what to salvage and what to abandon, and about who would keep their homes and their histories and who would lose them. "We all hated to take U.S. history, right?" Leslee Keys told me, in a tone of half-jest, half-alarm. "Well, don't worry. You won't be seeing any of that." I remembered years ago walking Boston's Freedom Trail with my brother past the house once owned by Paul Revere—who famously rode from there to Lexington in 1775 to warn American Revolutionary soldiers that the British were on the march. Some spots along that trail are already at risk of floods and will become more so as the sea rises. Places like this wouldn't disappear from textbooks, but some of them would become inaccessible, and we would lose the ability to revisit them.

History has always been a contentious project. But the stale, broken-spined history books I remember from my school classroom were not at all like the experience of encountering raw, cacophonous, unfiltered history—the struggles, the strangeness, the misdeeds and crimes, inventions and ingenuity that often speak in shockingly direct ways to the present condition. Heritage allows people to find belonging in a place, to claim it as their own and gather strength from the lessons of the past. But written history can easily gloss over complexities and rob people of their stories, alienating or marginalizing some in order to make others feel comfortable or powerful. "We have always been a pluralist nation, with a past far richer and stranger than we choose to recall," writes *New Yorker* journalist Kathryn Schulz in an article recounting the history of tamales. Social and racial reconciliation—the restoration of dignity to people who have been wronged—always requires wrestling with ghosts. But reengaging with history in this way often means returning to the landscape and to the places where past events occurred. Sometimes even the tiniest details matter when making sense of a community's origins—a broken piece of pottery, a cannonball, an etching on a wall, a corroded metal button, a bead, a bit of paint, flecks of rock or ash or pigment in old layers of earth help us locate people from the past who might otherwise have been erased, people deemed ordinary or inconsequential in their time but who later became a clue or even a momentous symbol. If we are not careful, when we lose a place like St. Augustine—especially if we do not safeguard the records and evidence of its existence—we will forfeit some of our ability to recollect, reclaim neglected stories, and correct our mistakes. To lose the past is to let go of possible futures as well.

I spent a week in St. Augustine after the conference, wandering through layers of history. The people who settled colonial Spanish Florida were culturally diverse. And there were glimpses of pluralism here, both deliberate and accidental, well before it was any kind of American ideal. On the side of a building along one colonial avenue, for instance, I found a plaque dedicated to "the memory of the 400 Greeks who arrived in St. Augustine, took on fresh supplies, then journeyed south to help settle the colony of New Smyrna, Florida. After ten difficult years, the survivors of that colony sought refuge in St. Augustine . . . the first permanent settlement of Greeks on the continent." The building is the oldest surviving Greek Orthodox house of worship in the United States.

But the most extraordinary story I encountered was already submerged and buried under a combination of water and earth, land and salt marsh. About two and a half miles from the downtown, Fort Mose Historic State Park (pronounced *Fort Mos-ay*) is understated compared to many of St. Augustine's attractions—a green L-shape on the map, not advertised by flashy billboards.

An archaeologist I met at the conference put me in touch with one of the park's chief defenders and advocates, Thomas Jackson, who agreed to meet me one afternoon at a coffee shop off one of the main highways through the area—in the strip mall zone just beyond city limits, southwest and across the river from the historic district. Thomas had wire-framed glasses and a voice like a cello, low and mellifluous with a warm drawl. He grew up in St. Augustine and remembered visiting the Castillo de San Marcos as a kid on Easter Sundays, and some of the older people would murmur, "We had a fort, too." He would hear it a handful of times in the Black community in St. Augustine, but he didn't fully understand its meaning until later.

Over time, rain and wind and decay concealed the evidence, but

Fort Mose was the first legally recognized free Black community in what is now the United States. "Freedom seekers came down from the Carolinas," Thomas explained to me as we sat at a café table, "and made their way here to Spanish Florida."

Fort Mose was a product of the bravery and perseverance of those who escaped slavery and the opportunism of the Spanish colonial government. "Black history is so intertwined with Spanish history, and the story is not told," Thomas continued, tapping the table with his hand emphatically, "especially in English-speaking society."

Spanish slavery was brutal, but, in its legal code, Spain treated slavery as an "unnatural condition" and "established mechanisms by which slaves might transform themselves from bondsmen into free vassals," writes historian Jane Landers. When Pedro Menéndez and his crew founded St. Augustine in 1565, the sailors-turned-settlers included both white Spaniards and people of African descent, some who were free and some who were enslaved. Then in 1687, when eight Black men, two women, and a child fled from their captors in St. George, Carolina, in a boat and landed in St. Augustine, the men were given paid jobs building the Castillo and working as blacksmiths and the women as domestics. When an English officer arrived to try to apprehend them, the governor of Spanish Florida refused to release them and sought input from the king. Eventually, in 1693, the king of Spain issued an official edict on such refugees from slavery, "giving liberty to all . . . the men as well as the women." Conveniently, this would also bring new laborers and soldiers to the Spanish colonies and destabilize Spain's rivals, Landers notes.

But it was a guarantee in writing only, and colonial leaders were often loath to enforce it. Fort Mose might never have existed without the unwavering determination of one West African man, who would take the Spanish name Francisco Menéndez. Menéndez escaped British slavery and fought against British colonial forces with the Indigenous Yamassee Nation but was forced back into slavery when he came to St. Augustine. He became captain of St. Augustine's Black militia while still enslaved and had to petition the governor of the colony for freedom for himself and other fugitives, which was granted in 1738.

Menéndez's efforts resulted in the establishment of Gracia Real de Santa Teresa de Mose (meaning the "royal grace of St. Teresa of Mose"), a community of initially about forty people. He became its leader.

"They could live free at Mose, as long as the able-bodied men joined the militia, and everybody in the community became Catholic," Thomas said. "Catholicism was the official religion of the Spanish crown, so that was a requirement. And the militia would help defend the city from the north." The place evolved into a multicultural society—the refugees had roots in a number of different African cultures such as the Mandinga, Mina, Kongo, and Carabalí, and some intermarried with Florida's Indigenous communities. The residents grew crops, served as blacksmiths, set up their own retail shops selling provisions, worked on construction projects in St. Augustine, and received rations and supplies from the Spanish colonial government. They built a fort, similar in shape to the Castillo de San Marcos, but made of earth and palm logs, with prickly pears and yucca (also nicknamed "Spanish bayonet") planted around the edges to deter intruders. Fort Mose was burned down by the British two years later; then a second fort was rebuilt a dozen years after that.

When the Spanish ceded Florida to the British in 1763 at the end of the Seven Years' War, the inhabitants of Mose mostly fled to Cuba and the fort was abandoned. But Fort Mose remained in the memories and stories of the community's descendants, and Black Americans' connection to this place endured. "We're almost sure that the inhabitants of Mose who left with the Spanish in 1863 and moved to Cuba—some of them moved back," Thomas told me.

He felt that the legacy of Mose resonated through the entire arc of Florida's Black history: "I think there's a direct connection between the Mose community and Catholicism, and the Lincolnville community," which was established in 1866, the year after the Civil War ended. The influence of the Catholic Church continued to thread through this storyline. In the early twentieth century, white Catholic nuns were arrested in Lincolnville for running a school for Black students. The neighborhood became a key battleground in the American civil rights movement and was the site of a series of important protests, led by Dr. Martin Luther King, Jr., that helped galvanize the passing of the Civil Rights Act of 1964.

Thomas Jackson's grandfather had come to Lincolnville in the 1920s, and Thomas attended school and church in the neighborhood as a kid. When Thomas asked why his family chose this place, his father had said, "We had been there before," though Thomas is still trying to find out what that meant.

But it would take centuries for this community to be able to reclaim both this story and the place where it happened. In the nineteenth century, a civil engineer hired by Standard Oil founder Henry Flagler obliviously dredged sand from the area that held the ruins of the forts, partially marring the site: Flagler wanted to fill a tidal creek downtown, so he could build his fancy hotels atop land that had been marsh. In the 1960s, a military historian bought the property suspected to hold Fort Mose's remains. But it wasn't systematically excavated until the mid-1980s, by a team led by archaeologist Kathleen Deagan and supported by the archival research of Jane Landers. With aerial photos, they found the imprint of the second fort. Then their excavations uncovered wooden posts, the edges of earthen walls, and a "smaller oval or circular wood and thatch structure . . . which may have been residential," the scholars wrote. Thomas Jackson watched what was being uncovered at the site. "I started getting involved with Fort Mose once I realized that there is a story that needs to be told." After the dig was complete, the state of Florida purchased the land that held the archaeological site. In 1996, Thomas and other locals set up a citizen advocacy group that prevented an adjacent parcel on higher ground from becoming a condo development. Both properties became part of Fort Mose Historic State Park, and the upland now houses an interpretive center.

Jackson volunteered to accompany me to Fort Mose. And three days after our coffee conversation, we strolled the park's boardwalk to the edge of the salt marsh. "When we first started, none of this was out here," he explained. "We had to pretty much tell the story in the thickets."

To make the experience more tangible, Thomas had learned to fire a musket. He had acquired a costume several years ago in the style of an eighteenth-century Spanish colonial militiaman, sewn by a woman in St. Augustine. He practiced the Spanish military drill regularly both at Fort Mose and with a group from the Castillo de San Marcos that performs musket-firings for the public. And over the years, he and a group of locals have held annual reenactments of a 1740 battle against the British at Mose. They have also organized regular events in which they play the parts of Black militiamen, priests, the Spanish governor, and other characters from the era. Members of the public could listen to each character tell a story about the journey from the Carolinas to Fort Mose—sometimes along the same path where Thomas and I were now standing. One of his favorite parts to play had been Mose founder Francisco Menéndez: he admired what a resourceful character Menéndez had been—warrior, sailor, speaker of multiple languages.

That day on the boardwalk, Thomas Jackson was dapper in a white polo shirt and black walking shoes, and equipped with a stylish long black umbrella for whatever the gray sky might unleash. Looking out into the river and marsh, I saw no visible traces of human history, just a lushness of cedars and oaks and palmettos, an anole lizard scurrying along one of the railings, fiddler crabs running through the mud, a heron carving through the sky, and an orchestra of birds chirruping and insects singing. Farther out into the marsh was a rookery with a breeding colony of wood storks. In the distance, the land rose up into a small, low island. A series of blue and white signs, cracked and heavily weathered with some of the letters smudged, described the vista before us. ALL THAT REMAINS OF FORT MOSE IS UNDERGROUND—ON THE ISLAND BEFORE YOU, AND IN THE SURROUNDING SALT MARSH.

Now it was simply home to the plants and animals of the marsh. JUST 250 YEARS AGO, DURING THE OCCUPATION OF FORT MOSE, THE AREA SURROUNDING THIS DOCK WAS DRY LAND USED FOR FARMS, RANCHES, AND FORTS, announced the sign. GLOBAL CLIMATE CHANGE IS ALSO HAVING AN IMPACT. WHAT WOULD HAPPEN TO OUR LOCAL COASTLINES IF THE WEST ANTARCTIC ICE SHEET MELTED, RAISING GLOBAL SEA LEVELS BY AS MUCH AS 20 FEET? The signs were about ten years old, Thomas said.

When the water rises even farther and the site becomes inaccessible even for research, the story of Fort Mose will have to live in the retelling, repeated by people like Thomas Jackson and those who come after him—people who believe that memory matters, that stories help keep us grounded and alive and give us a way to feel like we still belong in this unruly, unpredictable world.

꙳

In the fall of 2019, Hurricane Dorian whirled through the Atlantic, making landfall in the Bahamas and at Cape Hatteras, North Carolina. This hurricane dampened St. Augustine but didn't inflict the same level of damage as Irma or Matthew. According to the local newspaper, the *St. Augustine Record*, it mostly drenched the "commonly flooded areas." But a few feet of water filled parts of Davis Shores.

By now, the city government had a detailed plan for building infrastructure to manage somewhat higher waters, but it could happen only in a piecemeal manner, as all such bureaucratic efforts must. One part of the plan was to try to keep storm surges out of the city and its underground drainage networks. Twenty miles of stormwater pipes run under St. Augustine, some of them fifty years old or more. During Hurricane Matthew, the seawall at the bayfront had performed decently, but floodwater had risen up through the storm drains. So the tiniest first step was to install new valves in drains that would flow in only one direction, outward, preventing the sea from climbing up through the storm sewers during a flood and then seeping into the streets. At the south edge of the historic core of St. Augustine, the streets ended at Lake Maria Sanchez (a remnant of the creek that Henry Flagler once decided to fill in order to build the city center). The city would install a new pump station there and a bulkhead to keep stormwater from pouring in via the lake. The streets around city hall would also be getting major drainage upgrades.

Another part of St. Augustine's strategy was to set aside land that was becoming waterlogged much more often. On Anastasia Island, two pieces of flood-prone land went into retirement—one south of Davis Shores and one within that neighborhood. The former would stop an apartment complex from popping up on a site that gets partly drenched more than two dozen times per year—nuisance floods averted. The latter held a historic 1940s house, the old home of a turpentine farmer. It was one of the lowest-lying properties in that part of the neighborhood, and all around it lived residents who were sick of flooding, restless, and impatient with the city. The project became controversial. *It was quite a battle*, admitted Jessica Beach, a city stormwater engineer who became St. Augustine's chief resilience officer in early 2021. (A local news blog accused the city of using flood control as "a cover story . . . to create a park in one commissioner's neighborhood.") Still, the city government ultimately acquired the land; helped haul the farmhouse to the Florida Agricultural Museum about twenty miles south; filled and elevated the entire site, turning it into a barrier to control flooding; and added some park benches. They named it Coquina Park. It did not solve flood problems, but Beach said it fixed one of the neighborhood's weakest (or perhaps wettest) links.

Meanwhile, one of the city commissioners led a task force to change the building codes, with new guidance on proper drainage when elevating a building to move it out of the flood zone—and what not to build, including giant gravel foundation pads, which could change your neighbor's hydrology and send all the floodwaters next door.

These were all pieces of the much larger scheme to raise the city up: the streets, the seawalls, the barriers, the pump stations that were part of the wastewater treatment system. Piece by piece, project by project, until as much of the city as possible was higher than the flood level of Hurricane Matthew, which was also the estimated flood height of a more minor storm after about a foot and a half of sea level rise.

By and large, other cities were trying the same tactics. Raising, pumping, lifting, moving. St. Augustine had less money to work with than some. The Lake Maria Sanchez project alone would cost $30 million, and the wastewater treatment improvements another $14 million, high price tags for a fifteen-thousand-person city even after help from the state and the federal government. Miami Beach—with six times the population of St. Augustine, glammy and wealthy and dazzling with celebrities—had lifted three miles of its roads to the tune of $41 million and was spending $400 million on pump stations.

None of these efforts will ever be enough to save everything. The giant old hotels, the cemeteries, the archaeological remains, the delicate colonial coquina foundations—these can't be moved or raised, and only some can withstand a regular soaking.

Even the most ambitious engineering fixes may or may not protect this place through the end of the century. A previous St. Augustine mayor had traveled to the Netherlands in 2018 to speak with Dutch engineers about whether a major engineering project could protect the city—perhaps a seawall around the whole city or tide gates at the inlet. She had met with Senator Marco Rubio to ask for help funding a study with the Army Corps of Engineers to consider possibilities. But she had to resign in 2019 after facing serious health problems. The city public works department has carried the torch and continues to look into this possibility. But such federal funding is highly competitive. In the end, a big engineering project might never be feasible for a little town like St. Augustine. And Floridians may never have an appetite, even a tolerance, for the massive collective public solutions that Europe can pursue.

Meanwhile, the waters kept rising relentlessly, charging into the city again and again. On more than a dozen days every year, you needed boots to get to your car downtown.

In September 2020, a nor'easter arrived at the same time as a king tide, and nearly the whole city flooded for three days, almost as intensely as during a hurricane. Fierce winds blew water over the seawall at the waterfront. Floodwaters rose into much of the city, including the streets of Davis Shores around Coquina Park, into the garages, under the doors, up to the knees of people who lived there.

Later that fall, Jessica Beach began holding small gatherings on behalf of the city government, neighborhood by neighborhood, to talk with people about water.

Some locals were angry. From a man who had come to Davis Shores with his wife in 2019 to start a family: *If a hurricane is coming, that's fine.*

That's an inherent risk of living a hundred yards from the Intracoastal and five hundred yards from the ocean. It's just the normal stuff, you know, a little storm, a little bit of rain, the tide, the wind; you shouldn't flood out during that. The problem is water comes into places that people live. So we need to find a way to prevent people from being affected by said water. Because people choose to live here. You know? I'm never going to get rid of this home.

Some were worried about equity. From a Lincolnville resident: *A throwaway house. That's how we feel about ours. If this house just drowned and floated away, we'll be okay. But not everybody can do that.*

Some were shocked by the flooding and planned to sell and depart.

But some asked about how to raise their houses. Some asked for help. Some had been there for generations. Some didn't want to leave. Some worried about their neighbors—elderly civil rights veterans, for instance, some of whom couldn't easily afford to go elsewhere; families with roots back to the Spanish colonial era. Some developers and home buyers were moving west to higher ground, snatching up properties sometimes at the lower-income edge of the city, potentially pushing other residents out in the process.

What will be saved? Who and what will get to relocate? Who will be left to fend for themselves? Will this be the usual game where money wins and displaces the people who lack it? Or will someone set up better rules, a clearer path out of the rising water, one that can lift most people to safety?

St. Augustine has decisions to make about what to do when all this water rushes in.

They are the same decisions that every community along every coast will ultimately need to make.

🐚

In all these battles against the tide, I notice it is hard for people to talk out loud about how much we will inevitably lose, even in the rosiest scenarios.

"I can't really think about, you know, the 2100 scene—or whatever the projection is that we're going to be so severely impacted that our city may not be accessible," Jenny Wolfe admitted when I spoke to her in the spring of 2021.* "But we're still trying to think within that range of the lifetime of the mortgage—so like fifteen to thirty years—to put it into bite-size pieces."

To think beyond this span is almost profane. It is almost unbearable. Who can blame anyone for avoiding the thought?

Only a few people would ever say it bluntly out loud to me. One was Lisa Craig, the former Annapolis preservationist who had gone on to launch a consulting company for historic communities that were in danger of drowning, literally and figuratively. (Leslee Keys, who was no longer full-time at Flagler College, had been working with Lisa on this venture—a pair of realists trying to get aging cities to plan for the climate disasters advancing toward them.)

"In the big picture of it all, Annapolis is going to be underwater," she said, "as will certain portions of St. Augustine, as will certain portions of Nantucket." She could have gone on to list nearly every coastal city in the country. "There is just no way to stop and slow this unless they decide they're going to put up these massive seawalls, which nobody is entertaining."

Every little tide valve, every lifted seawall, every raised house only purchases time, not a permanent fix.

But time is necessary. It is important. If done right, with luck and thoughtfulness, buying time might shield people from the impacts just long enough. It might allow a community to help the vulnerable and the people least able to move on their own. It might let people consider their options, collect themselves and their families, gather their belongings and memories, and move up and inland. Done wrong or thoughtlessly, the process of adaptation could lose cultures, histories, people, homes. And every community will do it alternately right and wrong—some saves, some losses, some failures.

In the long run, the most powerful way to buy respite from the sea is not just with infrastructure and engineering but also by reducing carbon. If the planet runs hot, if nothing is done to slow the pumping

---

* Later that year, Jenny Wolfe would leave her city government role to take a position with an architecture firm focused on cultural preservation but would remain in St. Augustine, still devoting her energy to protecting historic sites from threats such as flooding.

of carbon into the air, to reduce the rate of ice melt, to prevent the hurricanes from raging more fiercely and more often, the water will rise faster. The engineering will fail faster. You need to tackle both. Without engineering for sea level rise, St. Augustine will have a shorter life, a dimmer story, marked by greater losses. But a worse fate will await people at the coasts everywhere if the world keeps burning fossil fuels—less than two versus eight feet of sea level rise within a lifetime. The first situation might just be manageable. The other will surely be catastrophic.

St. Augustine had never adopted a formal policy about carbon emissions. The mayor who visited the Netherlands, Nancy Shaver, had insisted on avoiding the term *climate change* altogether. "We're kind of a practical city . . . we don't really have a program around CO_2," she had told a public radio reporter. "We don't do those things." She said this without irony, like so many people who choose the "practicality" of politics over the uncompromising truth of atmospheric chemistry and physics. In early 2021, St. Augustine's state senator, Travis Hutson, introduced a series of bills making it more difficult for local governments in Florida to transition from fossil fuels to renewable energy sources. All three were signed into law in June 2021.

श्री

Sometimes the past leads us to more creative and more realistic solutions than the most fanciful, most ingenious, most technologically rich futurism. The past is what brings us home, the story of who we've been and where we came from.

But somehow even St. Augustine is failing to grasp the lessons of time—the behavior of water for millennia past, the things our ancestors knew, the evidence scientists have found recorded in the layers of rock. The most powerful lesson of history, when we encounter it raw and unfiltered, is its truthfulness—its realism about the arc of change and the things that drive it.

Written in the millions of clamshells piled and fossilized along the coast is a story about the challenge of making a home beside water.

We could read this story again; we could remember it. It could remind us of what the sea can do, and how we need to renew the world.

CHAPTER 10

A SAFE SPACE

Part of what defines *home* is a set of nested boundaries. The exterior of one's own house or shelter: walls, a door, a gate, even a meager tent flap. A neighborhood. A city limit. A state or national border. We mark the edge that differentiates the place where we belong from the space beyond.

But the process of drawing edges and borders—between private and shared, home and not-home, here and elsewhere—is often fraught. Some efforts at boundary-delineation are actually land grabs (as in the numerous instances in U.S. history in which Indigenous lands were divided into lots and given away to European American settlers for farming). Some cruelly demarcate the edges of discrimination (as in racial housing covenants). "BUILD THE WALL" was the rallying cry of the Trump years, a line separating who is part of America and who is not, a symbol of nativism and nationalism. Others are efforts to separate humans from nature, boundaries that can be all too easy to breach (as in the levees between the Mississippi and the neighborhoods of New Orleans that failed during Hurricane Katrina).

Still it is possible and necessary to set ethical limits around human space and movement—fence land for dogs or livestock to roam; demarcate the edges of parklands or schoolyards; calculate the boundaries of a floodplain and attempt to keep home building out of the path of deluges;

map the basin of a drinking-water reservoir to ensure that poisons and contaminants do not ooze into it.

In an era of climate change, we need also to recognize one ultimate limit, the outer edge of the Earth's atmosphere.* For all practical intents, the edges of this round planet are the boundaries of human life for the foreseeable future.

More than a decade ago, scientists revealed the results of a set of models that determined how to maintain safe conditions on Earth for human life. The first of these studies, published in 2008, was straightforwardly focused on carbon dioxide. The second, in 2009—with the evocative title "A Safe Operating Space for Humanity"—defined thresholds for several aspects of the global human home, including the acidity of the oceans and the rate of species extinction. What were the boundaries of a world in which humans could continue to operate—grow food, have access to adequate water, and face a level of risk (such as from disasters like flooding or hurricane or drought) that could still be manageable and survivable? What lines must not be crossed? James Hansen, then based at NASA—one of the first scientists to forthrightly warn Americans, in the 1980s, that climate change would be a major existential crisis—was a coauthor of both papers.

The original definition of *threshold* is tactile: "The piece of timber or stone which lies below the bottom of a door, and has to be crossed in entering a house." In science, a threshold describes an abrupt boundary between one condition and another. One moment you are in one state of existence, and the next, you have entered an entirely different terrain or set of circumstances. Ice melts into water. An object under strain, like an overloaded and sagging bridge, finally collapses. If you cross some thresholds, it is possible to return. A bridge can be repaired. A person with an illness may sometimes recover and return to a state of

---

* There are those, such as Elon Musk, who believe that to escape climate change, we need to set up a new human society on, say, Mars. But as Bill McKibben points out, "The single most inhospitable cubic meter of the Earth's surface—some waste of Saharan sand, some rocky Himalayan outcrop—is a thousand times more hospitable than the most appealing corner of Mars or Jupiter." Currently, there are no other planets within our reach that can sustain human life over a long period of time. In spite of fanciful, science-fictional ideas that might suggest otherwise, any future scenario that sends human refugees fleeing an uninhabitable Earth is likely to be brutal.

full health. If you cross others, the door closes behind you. Among the starkest of these crossings is the end of life—once a heart stops beating for long enough, we cannot pull someone across the threshold from death back into life—at least not with any existing medical technology.

The "safe operating space" Hansen and his colleagues calculated for atmospheric carbon dioxide—the planet where humans could reasonably continue to pursue life and civilization—was 350 parts per million. But the specific number itself was less important than another piece of information. The carbon dioxide concentration in 2009 had already reached 387 parts per million. According to Hansen and his colleagues' analyses, humans had not merely pushed the planet past the threshold. We had left home altogether and were beginning to run toward the edge of a precipice.

Some doors have already closed behind us. (The seas will likely keep rising irreversibly for centuries, though human decisions about carbon emissions still influence how rapid or catastrophic this process becomes.) Still, the scientists suggested that we might reverse course and walk back through the door to a safer Earth.

To do this, we as a species and a global society would need to collectively share the atmospheric space. We would need to come up with limits and boundaries at every level—in our communities, within our respective countries, and through international agreements—to protect our planetary home.

☙

Unfortunately, many of our society's most enduring notions about the setting of boundaries and the sharing of resources within them come from "the dismal science" (aka, economics) and social Darwinism (a distortion of evolutionary theory that proposes that human cultural groups should tussle for "survival of the fittest" in order to better the human species), a grim combination indeed. Because these ideas are so entrenched, it's hard for many of us to imagine any positive outcome from the face-off between human desires and ecological limits.

In 1833, the British economist and professor William Forster Lloyd published *Two Lectures on the Checks to Population*, his ruminations on the usefulness or alleged folly of childbearing among the poor, given cold economic calculations about the food supply. Most of his ideas weren't particularly original but echoed the influential (though also controversial) scholar Thomas Malthus—who had offered up similar analyses of population growth three decades earlier. Lloyd mixed in his own metaphors about communal grazing, in which pastoralists let their sheep congregate on shared or public pasture. His reasoning wasn't based on empirical evidence. Rather, he conjured a hypothetical case of abused pastoral land: "Why are the cattle on a common so puny and stunted?" he wrote. "Why is the common itself so bare-worn, and cropped so differently from the adjoining inclosures?" The cause of this sorry condition: human nature. According to Lloyd, anyone who kept a herd on a shared pasture would want to reap the benefit of adding just one more animal than their neighbors but would only suffer a fraction of the consequent cost of overgrazing and losing some of the land's productivity. So people would inevitably mishandle shared property—each rancher or herder putting extra cattle out to pasture until the fodder is scarce, the land ruined, and the animals half-starved. The solution, he insisted, was privatization. "The common reasons for the establishment of private property in land are deduced from the necessity of offering to individuals sufficient motives for cultivating the ground." Under private ownership, each household would manage its own pasture and bear the full cost of its degradation, and therefore have greater incentive not to tear the place up, thought Lloyd.

Lloyd's name might have slipped into obscurity had it not been dredged up, a century later, by a biologist named Garrett Hardin, who taught for thirty years at the University of California, Santa Barbara.

In 1968, writing in the journal *Science*, Hardin penned one of the most influential essays of the twenty-first century, "The Tragedy of the Commons," based partly on Lloyd's writing about pastures. Hardin postulated that selfish human nature would always befoul shared resources. And he extended Lloyd's grazing metaphor to the atmosphere and the oceans. All of these "commons" would inevitably be trashed if people were left to their own devices. "Ruin is the destination toward which all men rush," he wrote, "each pursuing his own best interest."

The only solution, in Hardin's thinking, was either privatization (capitalism) or heavy-handed, top-down, even draconian intervention (such as centralized Soviet-style socialism)—especially with regard to overpopulation.

Hardin's ideas about human scarcity and the commons were also directly linked to his own misanthropy. Social Darwinism, which also argues that some cultural groups are inherently "fitter" than others, is the basis of a number of blatantly racist ideas and ideologies—including eugenics, a racist and ableist notion that humans could improve our species if we only allowed "desirable" people to have children. In some of the less-frequently quoted passages of his 1968 paper, Hardin's words were tinged with the logic of both social Darwinism and eugenics: "It seems to me that, if there are to be differences in individual inheritance, legal possession should be perfectly correlated with biological inheritance—that those who are biologically more fit to be the custodians of property and power should legally inherit more." In other words, the "desirables" of society should inherit more of the Earth than the "undesirables." Elsewhere in his writing, he wasn't so subtle about his racist and nativist views. He railed against the United Nations' Universal Declaration of Human Rights. In later work, he compared rich nations to lifeboats with finite resources and capacity, arguing that letting those from poor nations aboard would swamp the boat. He argued in favor of forced sterilization. He insisted that rich nations should not deliver food aid to poor nations, because it would encourage them to "breed." He went on to serve on the board of a white nationalist publishing outlet, where he contributed commentaries sometimes steeped in Islamophobia and fears that Muslims would "outbreed" whites and cause the "genocide" of Northern-European-based cultures simply by outnumbering them. "A multiethnic society is a disaster," he opined.

But a large number of scholars tried to peel apart Hardin's academic ideas and separate them from the ugliness of his racial and cultural views. Politely stripped and whitewashed, the "tragedy of the commons" quickly became conventional wisdom. A 1985 essay in *American Zoologist* on environmental education suggested Hardin's essay should be "required reading for all students . . . and, if I had my way, for all human beings." The essay has been cited more than forty-five thousand times in academic literature alone since it was first published and still appears today—without irony or critique—in TED talks, contemporary business and financial writing, environmental analyses, and scientific discussions about evolutionary biology.

We carry in our minds this misanthropic story that humans can never responsibly share a home or a pasture on this blue watery planet—that selfishly trashing our surroundings and fighting over what remains is our birthright.

As a result, many of us imagine we can save ourselves only by building high walls around our own particular plot of ground and preparing to fight anyone who might trespass.

Too often, we cling to this grim and questionable logic. But it is impossible to permanently wall off the sky or the rain or the swelling ocean. We are inextricably bound to one another by the common Earth we all live on.

ß

Around the same time that Garrett Hardin was contemplating tragedy, a political scientist named Elinor Scott (better known later on as Elinor Ostrom) undertook a quiet, scholarly pilgrimage through the Los Angeles area to learn how people there used groundwater, an important water source for dozens of cities in the region.

Born in Los Angeles and raised there through the Great Depression mostly by her mother, a musician (after her father, a set designer, left the family when she was young), Elinor learned to make do with limited means. But her mother clamored to get her admitted to an elite high school, where she was surrounded by wealthier students. She became the first in her family to get a college degree—even though her mother "saw no reason whatsoever for me to do that." At the time, even fewer people could imagine the usefulness of a woman pursuing an advanced degree. But Elinor was passionate about learning and enrolled in UCLA's graduate program in political science in 1957—and divorced her first husband in large part because he objected. There she studied with a professor named Vincent Ostrom, whom she married a few years later. Vincent was an expert in Western resources and had studied strategies for sharing various kinds of commons, such as water, while avoiding "patterns of mutual destruction," as he explained in a documentary later in his life.

Groundwater was another kind of commons. From the 1920s onward, people living above the Central and West Coast Groundwater Basins of Los Angeles—underground reservoirs stretching across 420 square miles, from beneath Long Beach (near the Port of Los Angeles) up to Hollywood—had been using groundwater a bit too voraciously, faster than the rainfall could replenish it. Wells were deepened and powerful pumps installed to slurp from the sinking water table. Seawater trickled in—the salt penetrating freshwater supplies and creeping inland, threatening the entire basin. In the 1940s, school and park lawns started to die when they were watered from salty parts of the aquifer. But by the 1960s, the crisis had mostly been averted, the aquifer saved from destruction. (Four decades later, the water levels in the western aquifer had actually risen about thirty feet.) Elinor wanted to understand how. She was searching for answers to the same question as Hardin: How do people manage a common resource? However, she was more interested in gathering evidence than in holding forth. She sifted through memos and correspondence, sat in on meetings, and interviewed water users throughout the area.

In this case a large part of the problem turned out to be, in fact, privatization—an ineffective and convoluted system of laws that created incentives for people to overdraft and waste water in order to claim private rights to use it. But after the U.S. Geological Survey and the Los Angeles Flood Control District reported that salt could soon wreck the whole basin, various water companies, industrial users, and landowners started an association to agree to constrain their water use. Voters set up what's called a "watermaster," the Water Replenishment District of Southern California, to watch over all these enterprises.

One could easily take a dim view of the murky bureaucracy (described in somewhat painful detail in Elinor Ostrom's 1990 book, *Governing the Commons*) required to manage the region's water supplies. The whole process was legally costly, although, as Ostrom points out, an order of magnitude less expensive than losing all of that groundwater.

The result was no bright, green, perfect utopia. (Essayist and former city official D. J. Waldie writes that questionable practices have leached into the LA regional water boards in more recent years, and local water politics still need reform.) But it has been no tragedy either. People saw the possibility for ruin on the horizon and stopped. Today, LA's approach to groundwater management may also help the region cope with climate change. LA's water supply relies heavily on the Colorado River Aqueduct—a massive engineering project that draws water across the desert to the city. But that water is fed by a now-dwindling mountain snowpack. The Metropolitan Water District of Southern California is now considering erecting a wastewater recycling plant that would help replenish nearby groundwater basins. Those basins, still around because of the groundwater that was saved and protected decades ago, give parts of LA an alternative source of water as the Colorado becomes less reliable. LA's response to its past problems has made it far easier for the region to manage present and future strains on its water supply.

And notably, the resource remains shared, not divided (as Hardin might have insisted was necessary). "No one 'owns' the basins themselves," writes Ostrom. She calls the management of LA's aquifers "polycentric."

Polycentric, not iron-fisted or heavy-handed but grassroots, democratic, collaborative, cooperative, and messy. Complex like nature itself, many parts involved. Like the ripples in a lake surface after rain, waves emanating from many droplets. A cacophonous flock of birds seen from a great distance, suddenly a self-organizing pattern of wingbeats and trills. A murmuration. Circles of responsibility and care overlapping one another. People elbowing each other noisily to reach a system, a working solution that will have to change and change again.

In 1973, five years after the publication of "The Tragedy of the Commons," Elinor and Vincent and colleagues set up a research center at Indiana University in Bloomington, where they had both taken posts in the political science department, to study what actually happens when people manage common property, sometimes for many generations, even centuries. They collected examples from around the world—communally managed forests in Japan and Mexico and Nepal, shared cattle pastures in the Swiss Alps, cooperative irrigation societies in the Philippines and Spain.*

In 2009—the moment when climate scientists announced that humans were no longer living in a "safe operating space"—Elinor Ostrom became the first and, at the time of this writing, still the only woman to win the Nobel Prize in Economic Sciences. Her win was a shock to those in the field of economics. Business and finance journalist John Cassidy, commenting in the *New Yorker*, called her "positively obscure," and murmurs of disapproval buzzed among the practitioners of the dismal science.

But if you read her work closely—in between all of the descriptions of labyrinthine systems for handling watersheds or fishing or community gardens—you can see something revolutionary. Elinor and Vincent Ostrom and their colleagues have amassed a vast body of evidence that people can in fact drink from the same well, literally and figuratively, or fish from the same lake, or take up space under the same sky without detriment. But doing so always requires some complex housekeeping—to ensure that everything is transparent, that anyone who takes more than their share faces consequences, that conflicts can be resolved. Communal resources also have "clearly defined boundaries," writes Elinor; everyone has to know the limits, the walls, the edges, the rules of their shared space and their particular rights and roles within it. And these systems are tailored and adapted to the particular conditions, cultures, and needs of a specific place—fundamentally grassroots, as diverse as the regions they operate in. Each of these systems springs from a sense of shared belonging—a sense of home.

In the years since, Elinor Ostrom has been popularized as the woman who "disproved" the tragedy of the commons. But Elinor herself took issue with this characterization. Hardin "was addressing a problem of

---

* Such irrigation systems exist in a number of places that were influenced by Spanish culture, including the acequias of the American Southwest. These sophisticated, centuries-old networks of irrigation channels are some of the most interesting examples of technology shared across human cultures, developed in the Middle East thousands of years ago, borrowed by the Spanish, brought to North America by Spanish Catholic missionaries. Indigenous communities such as the Pueblo had independently developed similar water systems and integrated the methods of acequia maintenance into their own land management practices. Today acequias are still communally managed in northern New Mexico, are considered a vital part of Indigenous culture in the region, and are recognized as governmental units under New Mexico law.

considerable significance that we need to take seriously," she said in an interview with *YES!* magazine. "It's just that he went too far. He said people could never manage the commons well." Tragedies can still result when people choose to compete rather than communicate and cooperate, to fight over a resource instead of resolving to conserve it together.

But if Ostrom didn't rule out all of Hardin's ideas, at least she showed us that we have a choice. We don't have to write that tragedy. We don't have to divide the Earth into billions of little allotments or fiefdoms and put up walls to keep our neighbors out of our personal space. Nor do we need to build a dystopia. No iron-fisted ruler, no draconian policies or denial of human rights will save the planet we live on. In fact, either of those strategies could almost ensure ruination.*

After careful observation, Ostrom suggested that what we need to do is much harder, much more complicated, much more diffuse and diverse, and much more democratic.

꙰

The atmosphere of the Earth is also a commons, not unlike a pasture or a groundwater basin. But it is both invisible and global, vaster than our imaginations and therefore bewildering to try to manage.

The climate crisis is not really a tragedy of the commons, however. We do not each have an equal share of the sky (as Doria Robinson has pointed out) nor have we had collective say over what happens to it. Climate change is another kind of calamity—largely caused, not solved, by the actions of certain powerful private enterprises, and by inequality, a lack of transparency, and outright denial and deception. According to one analysis, since the mid-nineteenth century, just ninety companies have borne the responsibility for two-thirds of greenhouse-gas emissions

---

* Naomi Klein writes in *On Fire: The (Burning) Case for a Green New Deal*, "The reality is that Soviet-era state socialism was a disaster for the climate. It devoured resources with as much enthusiasm as capitalism, and spewed waste just as recklessly. . . . And while some point to the dizzying expansion of China's renewable energy programs to argue that only centrally controlled regimes can get the green job done, China's command-and-control economy continues to be harnessed to wage an all-out war with nature, through massively disruptive mega-dams, superhighways, and extraction-based energy projects, particularly coal."

(and the company that has done the most, that has taken up the largest share of the atmosphere, is none other than Chevron, the company operating the Richmond oil refinery). Every year, some of these same companies spend hundreds of millions of dollars lobbying against laws that would set boundaries and limits on the emissions that cause climate change. Since at least the late 1970s—before the public understood what climate change was—fossil fuel companies and energy moguls have known that their emissions could drive the planet toward dangerous levels of warming. But instead of acting responsibly, they underwrote disinformation campaigns designed to confuse the public about the scientific findings. This colossal theft from the global public, this extraordinary act of deception, has been documented by reams of evidence from the world's leading journalists and academics.* In social science, the technical term for a freeloader or a mooch who robs the public is a *free rider*. The fossil fuel industry seems to want a free and endless ride around the planet, the unfettered opportunity to use this atmosphere to turn as much profit as possible and leave everyone else to wrangle with the long-term costs.

It is up to the rest of us to hold the boundaries, build the metaphorical fences, and rein in the problem. In the same month that she won the Nobel, Elinor Ostrom also wrote up some of her own ideas for the World Bank about how this kind of regulatory process might work for carbon emissions. Her discussion of the subject was not so different from the story she'd told about any other shared resource. In essence, there would need to be polycentric solutions—at every level, in cities, towns, regions, states, and nations, tailored to the particular places where we all belong. And at every level, a climate-master of sorts (like a

---

* In 2015, the *Los Angeles Times* and Inside Climate News published a detailed series of investigations revealing that Exxon and other major oil companies conducted their own sophisticated scientific research on climate change in the 1970s and 1980s, then deliberately buried their findings. "Exxon documents show that top corporate managers were aware of their scientists' early conclusions about carbon dioxide's impact on the climate," three reporters for Inside Climate News wrote. "They reveal that scientists warned management that policy changes to address climate change might affect profitability. After a decade of frank internal discussions on global warming and conducting unbiased studies on it, Exxon changed direction in 1989 and spent more than 20 years discrediting the research its own scientists had once confirmed." These investigations have led to a series of ongoing lawsuits accusing Exxon of misleading the public and shareholders.

watermaster) would have to stop the free riders. In February 2012, four months before her death from pancreatic cancer, she published another set of thoughts in an academic paper calling for solutions at many scales to address climate change. "A global strategy is frequently posited as the only strategy needed," she wrote. But "if not backed up by a variety of efforts at national, regional, and local levels," global solutions "are not guaranteed to work effectively." The work of safeguarding a livable planet would require a proliferation of such efforts, and people might have to push hard from the grass roots before the world would follow.

To date, governments around the world, under heavy political pressure from these same polluting corporations, have not taken much action against fossil fuel companies nor implemented serious-enough rules on carbon dioxide. Even the Paris Climate Agreement, heralded as one of the world's most promising political accomplishments on climate change, doesn't establish real means of halting a free rider or of setting well-enforced atmospheric boundaries for the countries involved.

The burden of trying to hold these decision-makers to account has been taken up primarily by a growing number of grassroots activists around the world. In 2008, a group of young activists at Middlebury College in Vermont, along with climate journalist and author Bill McKibben, took James Hansen's threshold for keeping the planet safe as the name of their organization, 350.org, now one of the largest and most recognized climate justice networks in the world. Their initial stated goal was to "make sure everyone knows the target so that our political leaders feel real pressure to act."

Since then, the diversity, creativity, and ferocity of the climate movement has been awe-inspiring. It has also drawn strength and power from the ethic of home expressed by communities like the Standing Rock Sioux Tribe, whose residents camped through the frigid North Dakota winter of 2016 to 2017 to try to prevent another oil conduit—the Dakota Access Pipeline—from crossing their ancestral lands. (The protest wasn't successful at halting the pipeline project during the Trump administration, but it spurred national outrage. In 2020, legal actions brought by tribes forced the Army Corps of Engineers to complete a more extensive environmental review.) The movement has also been buoyed by diffuse localized efforts—from solar on rooftops to community gardens to green building initiatives—and has begun to find solidarity with larger political protest movements, including Black Lives Matter. In the summer of 2020, the Sunrise Movement, a climate justice network led by Gen Z, made that alliance abundantly clear in a plea on its website: "We must see the fights for racial justice and climate action as two fronts of the same fight. If our society valued Black, Brown, and Indigenous lives, we wouldn't be in the mess we're in with climate change in the first place."

Simultaneously, the climate movement has flowered on a global stage. In September 2019, millions of people joined a strike comprising thousands of events around the world—inspired by the Swedish teen activist Greta Thunberg—to demand that governments take ambitious steps to combat climate change.

But even a movement this powerful has not yet been enough. More than a decade after 350 became "the most important number on the planet," in the words of Bill McKibben, the measurement of atmospheric carbon at the Mauna Loa Observatory on the Big Island of Hawai'i reached nearly 420 parts per million. By the time you read this, it may be higher still.

The free riders are still driving us inexorably beyond the limits of what is safe, and I wonder if too many of us are waiting for some iron fist of global government to save us. I wonder what it would mean if we all fully grasped that this is a fight for our homes and our safety. What if more of us stepped forward to defend the space above us, the ground beneath us? What if we took charge everywhere, to create new rules, to draw boundaries around the places we care about and insist together that they cannot be crossed? We hold this one planetary home in common, and it is both unethical and disastrous to continue to allow a few bad actors to decide what happens under our shared sky.

CHAPTER 11

—

TO MOVE HOME

out of house and home (idiom). *evicted; no longer having a place to live.*

Where the Alaskan village of Newtok meets the Ningliq River, the erosion line looks less like a riverbank and more like a cliff edge, like something sliced apart in an earthquake. On top of the cliff is tundra, knitted together with grass. Scramble down about eight or ten feet to the bottom: here you reach the deconstructed muck, where water and land have merged into a viscous material with a texture something like quicksand. You can walk on this surface if you don't linger long enough for your boots to sink. In some places, a trapped lens of liquid makes the muck undulate underfoot like a waterbed. From this vantage point, you can see the dangling roots of grass above you and the exposed subsurfaces of the soil, including a cross section of permafrost, the former icy bedrock of this land now thawing and dripping and decaying. Bits of trash and lost belongings fall into the muck here—broken boat hulls, wiring, bedsprings, forgotten furniture. Look hard and you can also see the detritus of millennia past, such as bones from a mammoth—small finds like vertebrae or teeth, and sometimes whole tusks worth substantial cash if you can get ahold of a fossil hunter or manage to haul the bones to Anchorage. But the muck is no longer land in any real sense—it is the erasure of a place.

On two days in late September 2018, Andrew John, nephew of the late tribal administrator, Tom John, walked to the edge of the erosion line in Newtok. He unspooled a yellow and white measuring tape and recorded the distance between it and the nearest houses in anticipation of a brewing storm. Andrew John was preparing a report for Patrick LeMay, an Anchorage engineer who had been helping the village negotiate with the Federal Emergency Management Agency, which will sometimes offer financial assistance, in the form of a buyout, for homeowners in the path of flooding.

The next day, Andrew walked again through the yellow-green grass and the ever-sodden ground and staked wooden posts every ten feet from the corner of the house nearest to the erosion. The village needed to know: How much time did these families have left?

He started from the house where Lisa Charles had spent her childhood summers and her teen and young adult years, her late grandmother's place—cranberry red with white trim and mounted on wooden pilings, the facade faded and mottled where the rain and wind had stripped the paint from the graying wood. By then, Nathan Tom, her uncle, lived there with other relatives, while Lisa, her husband, Jeff, and their kids were still in the blue-gray house farther from the river. On September 26, just fifty-eight feet of land stood between that house and the cliff edge.

As the storm passed through the community, Andrew John collected detailed notes: *The storm began late in the evening of October 3rd. . . . The winds were from the southwest 25 to 30 mph and gusting as high as 45 mph . . . we lost 20 feet of shoreline from Nathan Tom's home.* Even after the storm had departed, high winds kept pummeling the landscape, *blowing 20–30 mph and gusting anywhere between 40–50 mph depending on the current location of the storms passing out in the Bering Sea.*

On October 12, he brought his findings to the residents of the old red house and asked them to *start thinking about packing important documents, family heirlooms, photos, and things that cannot be replaced.* The next day, the twenty-foot marker beside the house fell into the river. The erosion line was just a few paces away, and another storm of similar ferocity could easily devour the remaining earth and topple the house. The day after, he helped the family move into a place beside the school normally reserved for teachers but presently vacant. Four days later, on October 16, the village cut off the electricity to the old home.

Andrew John is the double cousin of Lisa Charles—they are the grandchildren of two siblings from one family who married two siblings from another family. He had grown up in Newtok, graduated high school in 1999, and eventually signed up for the Marine Corps. This experience may be how he picked up a slight drawl and a formal manner, responding *Yes, ma'am* or *sir* when asked a question. When I eventually met him, I noticed he also had something of a military aesthetic, often wearing army green and camouflage rain boots, though this was also a common rural Alaskan habit. He told me he had done tours in Afghanistan and one in Iraq, had studied for a time at the University of Alaska, in both Fairbanks and Anchorage, first focusing on mechanical engineering and eventually pursuing a degree in process technology, a course of study that can lead to oil industry jobs.

Shortly after his uncle's death in 2017, Andrew had accepted Tom John's former role as tribal administrator. The job brought him full circle, back to his home village.

His new post was at a desk in a rectangular metal-roofed building on a high steel frame, known to everyone as "the brown building." His office was on one side of a large, dingy conference room.

In a small office on the other side sat a colleague of his late uncle named Romy Cadiente. A small-framed man with a black walrus mustache and well-defined furrows in his forehead and laugh lines at the corners of his eyes, Romy worked dogged hours under the fluorescent lights of this building, on long conference calls and in meetings, trying to get help for the village through any available means. On the wall near his desk, he had hung a piece of paper printed in large font with an inspirational quote from Tom John: *As you go through your day, be sure to guard against negative thinking, whether about your life or about other people.* The rest of the quote evoked Tom John's devout Catholicism,

Instead fill your mind with good, noble, and holy thoughts, and his love of the land, *contemplate the beauty of the natural world and let your heart be lifted up as you do*. In this spirit, Romy was a purveyor of optimism, as if it were not merely an attitude but a thing you could produce in prodigious quantities, enough especially to offer extra to anyone who might need inspiration to imagine that the village had a rosy future waiting on the opposite side of the Ningliq River. On phone calls and in emails, his speech was peppered with superlatives, *awesome, neat, lucky, grateful, blessed.*

Romy was one of the few outsiders to take up permanent residence in Newtok. A Native Hawaiian, he had accompanied a friend on a fishing trip to this region, the Yukon-Kuskokwim Delta, two decades previously, when he met and fell for a woman named Charlene Carl, a former high school classmate of Lisa Charles and the daughter of George Carl, one of the founders of the newer Newtok Village Council. After the pair corresponded for a couple of years, Charlene eventually went to Oklahoma to live with him. They moved back to Newtok in 2010 to help with the relocation. In the beginning, Romy was out of his element here. He joked that the first time he saw a moose, *I'm like, that's a weird-looking horse*, and the first time he came to Newtok, *I was like, I don't see any roads*. The winter weather was also a shock. *I have never seen snow going horizontal. Snowdrifts would cover the homes. You walk outside; I swear your body just goes numb even with all of these fifteen million layers.* But certain things about the community resonated with his Hawaiian Catholic upbringing. *We both believe in the same God. We live from the sea. Our values are consistent, taking care of the ocean that gives us life.* The rhythmic patterns of the Yup'ik language sounded like a song to him. He eventually took the role of relocation coordinator and was often the face of Newtok in discussions with government agencies and funders.

In 2018, the community was finally able to string together several million dollars in grant money. They hired the Alaska Native Tribal Health Consortium, a nonprofit organization with experience running development projects and health clinics all over rural Alaska, to oversee the first stage of the move. Four more houses went up. The Mertarvik Evacuation

Center (known as *the MEC* and pronounced *meck*) was finally finished and could now function not just as an emergency shelter during a severe storm but also as a temporary launchpad for the move. They built a construction camp with twenty-four beds, a mess hall, a shower, and a laundry facility. Still, there were no phone lines, no power, no reliable heat, no sanitation. Mertarvik still wasn't ready to serve a fully functioning community.

But after the storm arrived in early October that year, it was clear that the community would need to leave as soon as possible. The sky and the river had dictated the terms.

⚓

Newtok remains one of the most famous climate-migrant communities in North America. But by the time the river reached the village, the community was hardly alone, and over the years, the global ranks of climate migrants have continued to swell. Just in Alaska, dozens of other remote rural communities face major risks from flooding, erosion, or permafrost collapse—problems all made worse by climate change.

All over the United States, people have been quietly packing up and moving away from coastlines and shorelines. Over the last three decades, in more than a thousand counties in forty-nine states, homeowners have accepted buyouts from FEMA so they can remove themselves from the paths of recurring floods. Increasingly, these kinds of relocations are driven by climate change—as the sea level rises, the land erodes and, in some regions, far more rain pours onto river floodplains. FEMA's program is sometimes controversial: no one wants to feel like the federal government is pressuring them to leave a beloved home. But according to an analysis published by the National Institute of Building Sciences, the American public would save money if FEMA issued far more buyouts, instead of allowing people to rebuild under the National Flood Insurance Program—especially in areas that suffer repeated and increasingly severe flooding.

Where a relocation has threatened the integrity of a culture or a way of life, some communities have tried to stick together—as Newtok has. In Louisiana, an Indigenous community on a rapidly shrinking island named Isle de Jean Charles (in the region that formed the basis for the fictional movie *Beasts of the Southern Wild*) has worked for years to organize a relocation to higher land. On the Olympic Peninsula in Washington State, the Quileute Tribe secured a land exchange from Congress in 2012 so it can flee risks from both tsunamis and sea level rise.

Elsewhere in the world, the Republic of Maldives—a nation situated on a group of islands in the Indian Ocean—has tried to prepare for the moment when rising seas inevitably swamp most of its low-lying land base. Between 2008 and 2012, then president of the Maldives Mohamed Nasheed sought an organized way for his fellow citizens to relocate collectively to another country such as Sri Lanka, India, or Australia, so that they would not, in his words, be "living in tents for decades." But Nasheed was ousted in a military coup in 2012, and tragically, the question of what will happen to his compatriots remains unresolved.

And then there are the hidden ranks of the displaced within and beyond American borders. According to some estimates, the diaspora that permanently departed New Orleans, after Hurricane Katrina struck in 2005, numbers in the tens of thousands. In 2020, the Internal Displacement Monitoring Centre, a research organization established by the Norwegian Refugee Council, estimates that 30.7 million people worldwide, including 1.7 million Americans, had to flee their homes because of weather-related disasters. Not all of these are related to climate change, but the organization predicts that the number of people around the world forced out of their homes by flooding will rise by 50 percent with each extra degree Celsius the planet warms.

Climate migration is especially challenging to track: often more than one factor drives someone from their home, such as when a drought or a flood worsens an existing economic crisis. "There are no reliable estimates of climate-change induced migration," writes the International Organization for Migration. But some experts have tried to project what may come: globally, in a worst-case scenario, the population of climate-change refugees could rise from twenty million or so to as high as one billion people. Already, a portion of the vast population of migrants that has been detained at the U.S.-Mexico border in recent years is fleeing the disasters of climate change—such as the increasingly severe droughts, hurricanes, and floods that have plagued countries like Guatemala, Honduras, and El Salvador. Globally, there are millions more stories of people who have tried to find a new home after the old one vanished or became untenable. Most of these stories go unheard.

In some ways, Newtok has to stand for all of them.

The stakes are high. The knowledge that resides in Newtok—the insight and skill its people have collected over generations—could also erode away if village residents become scattered in urban areas such as Anchorage or Fairbanks. Newtok's success or failure could determine the willingness of sympathetic parties, like funders and government agencies, to lend a hand to any other community in a similar plight. And to allow Alaskan communities to disappear would mean the partial erasure of cultures that have, for centuries, beat back the forces that tried to extinguish or assimilate their lifeways. "What we were really scared about was dislocation, because these people were here for generations," Romy Cadiente said. "This Mother Nature is their medicine, is their guidance. And if you lose any of that stuff, you essentially kill these people that have been here for thousands of years."

The people of Newtok needed to find their way home—intact and together.

❧

The spring and summer seasons of 2019 on the Y-K Delta were disorienting for a variety of reasons. For one, the weather was profoundly askew. High temperatures in February and March broke records, and two men died on the Kuskokwim River after falling through thin ice. In midsummer, a mass of warm air arrived and squatted over Alaska, carrying with it an ominous heat wave and generating some of the warmest temperatures on record in the state. On the average summer day in the Y-K Delta, temperatures reach the mid-50s. But that July, about one hundred miles from Newtok, Bethel logged 91 degrees (although equipment problems at the weather station made this an unconfirmed record). In Chevak, about fifty miles northwest, the thermometer reached the low 80s.

Such unusual warmth boded poorly for the many Alaskans who had built their entire lives on hunting, fishing, and living from the land—and for the other humans sharing the planet with them. It meant that the Far North was continuing to heat up far too quickly, and this development would further distort weather patterns at other latitudes. (The warming of the Arctic and the resulting wrinkles in the jet stream would be partly to blame for the freak 2020 cold-weather disaster in Texas, for instance.)

Despite the unsettling weather, Newtok pushed forward with its plans. First, the community had to begin disassembling the old houses—which was simultaneously a FEMA requirement and a necessary safety precaution (a collapsing house is dangerous).

In the spring of 2019, the house that had belonged to Lisa's grandmother came down. The village had no heavy equipment—which couldn't feasibly be dragged across the already unstable ground—so a demolition crew, nicknamed the demo team, disassembled it by hand. Board by board, shingle by shingle, they pulled the house apart and stacked the pieces on the tundra. Though she had prepared for this moment for years, finally seeing the deconstruction of her childhood home was a shock for Lisa. *I almost started crying. I was like, oh, my God, they're doing this.*

She was not the only person to mourn the loss. A couple months later, Romy's partner, Charlene, began packing up her Uncle Pete's place, a seafoam-blue house that had belonged previously to her own grandmother. *All those precious teacups and glass teakettles that belonged to my late grandma, Elsie. I packed them in newspaper, all that time remembering Grandma, how she used to sit on the couch and just listen to everyone, her drinking tea with her beautiful teacups and telling us stories about when she was little with her older sister. I tried very hard to hear her voice, how she used to sound.* That house came down in May.

Then three more in the summer. Katie Ayuluk—a twenty-five-year-old who had been raising two small daughters in a house near Pete's—watched her family place come down. That house had been so close to the edge that the family's steam bath, an unpainted square hut, had been teetering on collapse, and they had hauled it back from the erosion edge. *When the storm hit, our house would start shaking, depending on which way the wind was blowing. And every time the waves hit the land, the water would splash our window.* She had been tormented by recurring nightmares in which *our land fell in water, and the only place we could go to was our boats.* Katie knew it was necessary to dismantle the house, but she was still grief-stricken. *I walked out of that house crying. I get more anxiety now.* She disliked her new, temporary accommodations in a building behind the school.

By August, it was Charlene and Romy's turn—a crowded red three-bedroom house that they shared with fourteen other people, including her parents, siblings, and nieces and nephews. Charlene's mother and father had to leave the village for a medical appointment, and Romy was busy organizing the relocation. So Charlene and her sister did most of the packing. While she collected shoes, a small television, some electrical wiring, the yarn with which she crocheted, and various papers and books, Charlene queued up pop songs on her iPod and turned up the volume— Kelly Clarkson's "Stronger" and Alessia Cara's "Scars to Your Beautiful." Intermittently, she cried.

She wanted to be happy about the idea of moving but instead felt ambivalent. Later that fall, when the birds began to leave, she felt sad. *The birds make it not so lonesome,* she said. But she had loved it here, even in the harshest moments. *In winter, it's so quiet you can feel the presence of God.*

☙

The question that haunted the residents of Newtok was, would the new village be ready in time? And the answer had to be yes; any other scenario was nearly unthinkable. *You just try, try, try, and you hope and pray,* said Romy.

But there were still numerous complications to reckon with. For one thing, any arctic or subarctic construction process must surmount extra design obstacles that don't exist at warmer latitudes. Many basic architecture and infrastructure designs, from a house blueprint to the setup of a septic system, are based on a set of assumptions about the conditions of the world—including predictable weather patterns, temperature ranges, cycles of freeze and thaw, and stability of land—that have never really applied to Alaska and may no longer be realistic in any locale in the stormier twenty-first century. And there are simple practical issues that must be handled differently in the Far North. For instance, sewage systems are exorbitant to build in a place as remote as Newtok, and even after the investment is made, conditions make them prone to cracking and leaking. In the absence of other solutions, many Alaskan villages, Newtok included, have had to use "honey buckets," as in hauling a bucket of human waste out of the house and dumping it in the only place available, the river. In a storm, the waste sometimes washed back into the village. It was a public health disaster in the making, but there had been no other real options. "There have been two choices for Alaska," said Aaron Cooke, the architect with the Fairbanks-based Cold Climate Housing Research Center who had camped out and led the crew that built Mertarvik's prototype house in 2016. "You get a multimillion-dollar [water and sewer] system that no one, once it's built, can afford to maintain. Or you get a bucket to poop in. What is the holy grail? Everybody's looking for something in between."

By the summer of 2019, Newtok and its various advisors, after years of research and trial and, especially, error had finally come up with the means to overcome both sanitation troubles and housing design issues. Eleven houses stood at Mertarvik (though the first three with their mold troubles were unlivable). When the 2016 prototype proved to be sturdy and mold-free, the Newtok Village Council contracted the Cold Climate Housing Research Center to design thirteen more hyperefficient houses and serve as housing gurus on technical matters. Cold Climate and the Alaska Native Tribal Health Consortium together came up with a simple waterless toilet and a gravity-fed water-treatment tank that could go inside each house.

With the blueprints on hand, the construction of Mertarvik finally shifted into high speed that summer—like a time-lapse film of the settlement of a frontier town, shrunk from centuries or decades into months. First, the military returned in much larger numbers than previous seasons. Nearly two hundred men and women from the Air Force Reserve, the Army National Guard, the Air National Guard, and the Marine Forces Reserve came through between June and late August. They arrived on charter planes and Black Hawk helicopters and shuttled people and supplies back and forth. Lumber, steel, glass, and other heavy materials came by barge. The people of Newtok welcomed them by hosting multiple receptions with traditional foods like dried fish and *akutaq*—a dish made of wild berries and sometimes nicknamed "Eskimo* ice cream"—and attempting to teach the military crews Yup'ik dances. This produced a strange tableau in the cavernous main hall of the MEC, with men and women in combat fatigues awkwardly wiggling their arms beside Yup'ik people who tried to illustrate how to gracefully wave dance fans. But all involved were moved by the exchange. At one of the dances, Romy took a video from the indoor balcony and watched them in awe and genuine gratitude. *I was thinking to myself, these people came from all over Alaska, just to help this little tiny community. It was really overwhelming. And then the sun was going down. I get goose bumps every time I see that video, because, you know, in spite of everything that's going on in America, all divided, here all these people came together just for one reason.*

The military crews agreed to build four houses. On top of this, Newtok hired the Ukpeaġvik Iñupiat Corporation, an Alaska Native company based in the North Slope, to put up nine more (which would raise the total number of habitable houses to twenty-one by the end of the season). Everyone had to work with precision, since materials were not easily replaceable; the nearest lumberyard was hundreds of miles away. "We can't have very many mistakes," Senior Master Sergeant Denver Long said in one of a series of promotional videos the military produced about the work there.

Mertarvik turned into an active construction site, with crews running bulldozers, loaders, and excavators, hauling gravel and lumber, hammering roofs together, mounting doors and windows. Many New-

---

* There is some debate about both the origins of the word *Eskimo* and whether it is an acceptable term for Alaska Natives and other arctic and subarctic peoples. Linguists at the Alaska Native Language Center at the University of Alaska Fairbanks believe the word comes from the Montagnais (Innu) word *ayaškimew* and means "netter of snowshoes." Archaeologists still use *Paleo-Eskimo* as a technical term for some of the earliest human inhabitants of the Arctic. The people of Newtok commonly refer to "Eskimo food" and "Eskimo dancing," and those whom I asked said they consider the word unproblematic. But other sources suggest *Eskimo* is derogatory or at very least a colonizer's word. I use it here only in reference to common phrases spoken by Newtok residents.

tok residents worked on the new site. Lisa Charles's oldest daughter, Ashley, got a job in housekeeping at the construction camp.

A cousin of Jeff's worked on the road crew. And at first, it was just a paycheck for hard work such as loading rock and gravel into trucks. But eventually, *I started to realize I'm making a village for my people*, he thought. *I can tell my future grandchildren that I was there to build the place that they're living in.*

The plan was for a third of Newtok (about 140 people) to move over to Mertarvik in the fall of 2019, where there would also be a school and health clinic, a new diesel power plant and electrical distribution system, and a simple village water treatment system. For safety reasons, no one could visit the construction site without permission. On a clear day, you could often see Mertarvik from Newtok. Romy would sometimes watch the construction work through binoculars, so he could know what was happening without always making the journey. But it could also sometimes seem like a mirage, a dream far away, one that might or might not be realized.

☙

In late September 2019, toward the end of the construction season, the Newtok Village Council allowed me to return and stay in the school, as I had a few years before. I planned to spend nearly two weeks there this time, and I readied myself for another unpredictable journey in storm season.

I left for the airport before sunrise, and on my stop in Anchorage, Romy Cadiente called to warn me about inclement weather over the Y-K Delta. But when I tried to reschedule my flight, the airline agents seemed to know nothing about the conditions over the remote tundra region—as if the place existed in some alternate world—and I had to give up and take my chances. While it was raining and gray when I arrived in Bethel, the little propeller plane still took off with me and about a half-dozen other passengers. From the window, I gazed down at the matrix of ponds and river channels and the deep green and brown land suspended between them. The landscape looked more delicate than I had remembered. *It's like the land is incidental or merely the boundary of river*, I wrote in my notebook.

On arrival, the agent who worked for the regional airline hauled me and my bags on the back of an ATV to the school. Romy and Charlene met me at the entrance and helped me carry my belongings into a storage room full of cabinets and stacks of chairs, where I would sleep on the floor again.

"Notice anything different?" Romy said, moments after setting down my bags. And I realized I had been too travel-weary to make sense of what I was seeing. "Look out the front door," he instructed.

And there it was, the river less than the length of a soccer field away, and in front of us bare, muddy ground. "The water is closer," I said. "Yeah, where are all the houses?" said Charlene, and for a moment I laughed awkwardly and stared at the scene in front of me as if it were a hallucination. Then I realized they had lost their own house, and I was not the only person living in the school.

They walked me down the hall and into the classroom where they had taken up residence, which was also where Charlene taught grades three to six by day. The walls were papered with phrases in both Yup'ik and English. The shelves were full of science and geography textbooks. On a table by the window, the couple had stashed a pink electric teakettle and a microwave. They had hid the rest of their stuff. "I don't want anyone to know I'm living here," Charlene told me, especially her students. She offered to show me where to find the athletic mats in the school gymnasium, which we could pull out and sleep on every night. "I cannot stand sleeping on this hard floor. Golly! Kill your back," exclaimed Romy.

They made me cups of tea. Romy seemed to alternate between enthusiasm and exhaustion. He leaned over a school desk in an olive T-shirt, his hair slightly mussed, a silver cross dangling around his neck. "You actually came at a pivotal time because we're going to start moving people over to Mertarvik."

It was a Saturday night, but he told me he might take a boat to the new village site in the morning, depending on the tides and winds. "Wanna go?" he asked.

But Romy did not take me to Mertarvik the next day. He had too much to do to supervise a journalist at an active construction site and departed quickly in the middle of the day on his own.

In his absence, I picked wild cranberries, collecting a bag of the jewel-like tart red fruits under the afternoon sun with Charlene and her sister. Charlene pointed out the pond from which Newtok pumped its water. Until recently, the community had used another pond just south of this one. But frequent floods were dirtying the water supply, so they had moved the pump.

The next day, Romy left in the morning without me again. Meanwhile, Andrew John brought Newtok's Catholic priest—a tall, angular, garrulous man from Poland—across the water to say a prayer inside each of the houses and sprinkle them with holy water. I saw Andrew in the brown building on their return. He gazed at me blankly. "Who are you with again?" he said. There had been parades of reporters through the village all summer and a documentary crew on location for weeks, and the community was tired of outsiders.

I began a waiting game, much like the people who lived here (only the stakes for me were far lower). The following day, the weather turned, and the wide Ningliq River was too dangerous to cross in high winds.

I stopped by the house of Bernice John, Tom John's widow, at the center of the village, about a quarter mile from the erosion line and near a pond edged with grass and willows. She had taken in a pair of puppies—fuzzy, droop eared, and only a foot long from nose to tail— who climbed onto my boots when I approached the door. I spotted a musk ox skull in her backyard, though it wasn't clear where it had come from. Inside, the house was much the same as it had been four years previously. A profusion of leafy houseplants and clothes hung from the ceiling. A jumble of storage tubs was stacked against the wall. Beds with foam mattresses ran from one end of the one-room dwelling to the other. Nine people still lived here. But now there was a picture of her late husband, Tom John, hanging above her bed beside a small cross.

Bernice no longer held a post on the village council and was now spending much of her time taking care of a grandson, an infant with large, dark eyes who was named after Tom John. Bernice, the baby, and her adult daughter (the baby's aunt) all sprawled across the same bed, and I asked if Bernice would miss her house, the one her husband had built two and a half decades ago.

"No, it's too small and moldy," she replied matter-of-factly. "Plus my foundation's getting bad." The permafrost decay was causing houses to sink and shift, even those that stood some distance from the erosion edge. In a recent storm, an entire house had slid off its foundations, though its occupants had still not abandoned it—the village was too crowded. "When it's windy, my house is like a loose tooth," Bernice quipped. But she was optimistic about Mertarvik. She hoped the first generation to grow up there, little Tom John's generation, would be smart and technologically savvy and well educated—and also still practice the old traditions of living on this land.

On other days, I visited Jeff and Lisa Charles's house. Once for a dinner of traditional food—both dried and boiled pieces of salmon, dried herring, frozen whitefish eaten uncooked like sashimi, seal oil, boiled cabbage and turnips, fatty strips of beluga whale skin, *akutaq* for dessert, some wild-cranberry pancakes, and an infusion of Labrador tea, which grows wild on the Y-K Delta. I sat next to her uncle George, the one who had moved out of her grandmother's house, and across from two of their friends with a small infant. The house was unchanged, but Lisa had adopted three silly, fluffy, toy-size dogs, one who had just delivered a litter of puppies, and had a fourth pooch who lived outside. One of her daughters had also tamed a seagull, whom she fed by hand outside the window. The house brimmed with activity, and it was hard for me to imagine that they would leave it so soon. But Lisa and Jeff were scheduled to move to the other side of the water in less than a month.

The next night I came by Jeff and Lisa's again for tea and a long conversation about the history of the relocation, about Newtok, about fishing and other miscellany.

The television was tuned to the news on both visits, and in New York City, the United Nations was holding a series of meetings on climate change. The Anchorage news station was running both national and Alaskan stories about climate change.

That night the news turned our conversation to global environmental issues. That summer, dead animals had washed up on Alaskan shores and riverbanks—whales, birds, seals, and salmon. Scientists called it a "multi-species mortality event" (almost like a term used in military combat to drain the emotion from tragedy). They couldn't attribute the exact cause of death but pointed to overheated ocean waters as a likely culprit. "It makes me worry," Lisa said slowly. "Like the elders used to say, famine will come again," repeating the story I had heard from Tom John's mother on my previous visit. "The elders used to say it will get really warm here. They're like, one day there will be no more winters. So you could say it's started to come true. The elders used to say the weather follows the people."

A few days before, Swedish teen activist Greta Thunberg had marched with tens of thousands of protesters in Manhattan as part of the worldwide September climate strikes. "This is an emergency. Our house is on fire," Thunberg had said. "And it's not just the young people's house. We all live here. It affects all of us."

Lisa said she liked what Thunberg had to say. "She's so young, and she's so smart," she reflected. I would hear the same from others. The kids who were marching in the streets made Newtok somehow less remote, part of something larger.

I stayed at the Charleses' house until well after dark that night. There was a storm brewing. But I hadn't realized how quickly the weather had turned until I began walking back to the school. The wind was so strong that I felt as if I had plunged into a fierce current of water and struggled just to move forward. A few times, the wind snatched my breath away, and I gulped and gasped, trying to gather more air into my lungs. The wind bellowed and sang.

When I finally returned to the school and settled into my sleeping bag, I could feel the entire edifice shudder and tremble, bullied by the wind. Later that night, the building began to rock slightly back and forth. All night I heard the sound of metal clanging against metal, as if someone were trying to break in.

The next day (four days after my arrival in Newtok), another teacher who had been living temporarily in a trailer near the erosion line moved her family into the classroom next door. Every night through the rest of my stay, she sang or played music or videos for her grandchildren, and sometimes I heard Yup'ik songs, sometimes old country melodies.

"The erosion used to be four steps from my door," she told another teacher, illustrating the distance by walking across the floor. "Now it's two steps."

⯘

I continued waiting for a chance to visit Mertarvik, but the conditions weren't right. I asked Romy every day, until the question started to seem petulant. But the tide had to be high enough for anyone to get out of Newtok easily—otherwise the boats could get stuck in the eroded silt. The wind had to be calm. And most importantly, Romy had to be ready. No one could distract from the process of preparing the new village for its first inhabitants.

Meanwhile, visitors came in and out of the room I stayed in—a U.S. census worker, a group of people from the Alaska Native Tribal Health Consortium who were training locals about how to maintain the water-less toilets in the new houses, a pair of communications workers who would shore up Newtok's satellite dishes and try to set up a connection in Mertarvik.

On Friday, nearly a full week after my arrival, some Newtok residents, including a couple of teachers, held a farewell dance at the school for the first group of people who would soon move to the new village site. A small band played guitar, bass, and drums and sang decades-old country songs, while an audience gathered on the bleachers and little girls spun around the dance floor, giggling.

On Saturday, I visited Charlene's parents, Lucy and George Carl, who were living in teacher housing. George, a self-trained carpenter, had built many of the houses in the village and helped found the Newtok Village Council. He had grown up in a sod house, then a log cabin, then eventually his uncle had enough money to buy lumber for a stick-frame house, after getting a job in the fishing industry in Bristol Bay. He said he had never been too excited about Mertarvik, because it had no easy access to small, quiet streams and backwaters, only the wide, untamable Ningliq, which took longer to freeze in the winter. "But I don't mind getting away from this old village now because it's not too safe anymore," he reflected.

On Sunday morning, I watched a family return from a hunting trip and unload armloads of moose meat, already cut and skinned. They looked celebratory, grinning at the world at large. But the rest of the village was so quiet it seemed like it was holding its breath. It felt like we were all waiting.

It was Monday morning, the last day of September, when Romy finally announced he could bring a boatful of visitors to Mertarvik—a group of filmmakers, me, and an Alaskan photographer who had arrived over the weekend to shoot some images with me.

In the afternoon, we followed him out the school door and along the muddy boardwalks, past another red house that was being demolished, to a metal outboard motorboat the size of a small truck that had parked at the water's edge. Romy walked quickly and with energy, in blue jeans and a brown zipper vest, a bright orange construction vest on top of that, a hat stitched with an image of the Hawaiian islands, bits of grass clinging to his neoprene boots, his shoulders hunched slightly forward like someone carrying a heavy load. A rare, vividly blue sky hung over us, and we stood while the demo team loaded heavy stacks of wood panels onto the boat, some bare, some painted, the bones of Newtok's dismantled houses. The wood could be repurposed for storage sheds, fish-drying racks, and huts for steam baths. The community was sending everything that could be salvaged across the water, and no boat trip could be wasted on just a few journalists. We sat on top of the panels, Romy in the middle, a driver at the back with a group of Newtok construction workers. The boat engine roared to life, and brown water sloshed in our wake. We passed an area of tidal mudflats and followed the shoreline. Gulls sat atop the choppy water, so fat they looked like sailboats. An arctic fox perched on the side of a muddy bluff, watching us.

Then Mertarvik came into focus—blue and hazy, then sunlit and solid. Romy summoned me to the middle of the boat, leaned in, and explained what I was seeing in the distance. Twenty-one houses lined the bluff, in bright shades of garnet, sky blue, sage green, and tawny brown. Beside them, a teal building, the Mertarvik Evacuation Center.

Above this were a landfill and a silver-gray rock quarry; next to those, a wide gravel runway. In the sunshine, Romy's eyes became crescent-shaped, a half-squint, half-smile. "There's a lighting system over there where the planes can land at night also," he shouted over the engine. Above it all was the unbroken tundra, the expanses where musk oxen gathered in the winter, the fields of wild berries. "That's our little village," he said, a broad grin spreading across his face.

He told me he had ambitious dreams of what it could become—full-size planes landing here, a hotel, bigger barges—a transportation hub like Bethel. But for now this would be enough. "All of the transformers are already up; power lines are up. Look at that; they already have the porches and stuff on some of the houses. Beautiful, eh?"

The boat landed. I could tell even from the first footsteps how much more solid this place felt. A place coming together instead of falling apart. The roads were dry and free of mud, even though it had rained the previous day.

We walked the length of the first street of houses, each perched on a gravel pad with wide wooden porches facing the water and Newtok in the far distance. We entered a brown house at the end—George Carl's house, the new home of Charlene's family. Inside, the house was simple. The water tank perched demurely near the front door. The walls were about fourteen inches thick, full of high-value insulation, the windows triple-paned to keep out the cold and the storms. It was empty but clean and warm, full of light and fresh air. (The house looked *nunaniq*, a Newtok resident told me later that evening, when I showed her pictures of the place on my smartphone; she said the word meant a combination of "happy" and "wow" in Yup'ik.)

After leaving the house, we crossed a meadow full of Labrador tea, smelling a little like citrus and cedar, and toured the high-ceilinged, airy halls of the evacuation center. The main hall, where they had taught the military construction crew how to dance, was so bright and open that it looked like a concert venue. This would also be the temporary school, Romy said, where Charlene would teach after they moved here.

From there, we walked up the hill. Romy pointed out the utility poles, straight and sturdy and recently strung with electrical wires. He paused in the road and turned his face toward the vista of houses and water.

"Stop," he said suddenly. "Isn't it pretty?"

❧

I wanted to believe the story that Newtok would have a blessed future. I wanted to believe the community could chart a path out of the disasters of the twenty-first century. I wanted to believe they could find answers that would help the rest of us, and we could all hang on to the histories and identities that made us who we are, even as the planet grew stormier and more precarious.

I had begun to think that the optimism Romy produced wasn't just for others but for himself as well, something he needed to keep going. Late that night after the Mertarvik tour, I sat with the photographer in a classroom, and Romy came in to talk with us. He asked what we had thought. In previous conversations, I had tried to congratulate him on the years of work he had put into this effort. The success, the fact that Mertarvik was a real place, belonged partly to him. And he had demurred. "I'm just the paper boy," he had said, and scoffed.

But this evening, something shifted in his eyes. His face flushed with emotion, and his voice became hoarse. "I just hope I'm doing the right thing," he said quietly. "I care about these people so much. I see so much." Then a tear rolled down his cheek, and he turned away. "I'm sorry, I'm sorry," he whispered.

I left two days later on another partly stormy afternoon.

Over the next month, George and Lucy Carl, Charlene Carl, Romy Cadiente, Bernice John, Lisa and Jeff Charles, and eighteen other households arrived in Mertarvik not as visitors but, for the first time, as residents.

When I called George Carl in November, he spoke of Mertarvik as if it were a set of brand-new shoes—well-made but not yet broken in. Cell phones worked only sporadically, and the store hadn't opened yet. George complained that there was little to do; the place was too quiet.

Lisa and Jeff and their kids had brought most of their stuff over with their own boat. Four freezers had also been hauled across the river for them—for storing food the family gathered, fished, and hunted. Lisa said her kids missed their cousins, and one of her daughters kept lobbying to visit Newtok so they wouldn't forget her.

Romy and Charlene remained itinerant, taking a room temporarily in the MEC, then moving a few months later to a trailer on-site. But all of it was better than living on a classroom floor.

Cold weather set in, and the snow fell. Storms gathered day after day, but the houses at Mertarvik stayed warm and the land steady. In November, the Y-K Delta entered the season of freeze-up, when boats can no longer travel the waterways but the ice isn't yet thick enough for snow machines.

The two villages sat apart from each other, old and new, temporarily separated, waiting for the ice to bridge them and for the year to end and another to begin, with its cycles of cold and thaw.

               ᴓ

A good story—a love story between people and the land they live on—should have ended there, with the heroes sailing across the water to their newer, happier lives. But the twenty-first century offers messier plotlines.

When the pandemic began in early 2020, I was haunted by something a group of nurses had told me in Newtok. The village held screenings for tuberculosis while I was there—a disease rare in the Lower 48 but common in rural Alaska. The crowded, poorly ventilated, multigenerational houses in Newtok had created conditions for respiratory bacteria and viruses to spread more easily. I imagined how the coronavirus could quickly invade and raise havoc in a tiny, close-knit place like this.

However, that is not what happened. Communities in rural Alaska have not forgotten how outsiders brought devastating illnesses before—including the deadly influenza pandemic (the Spanish flu) between 1918 and 1920. Many parts of rural Alaska enacted some of the strictest pandemic lockdowns and travel restrictions in the nation. Over the next year, Newtok had only one case of Covid, a teacher who quarantined and didn't spread the disease to anyone else.*

Because of these precautions, to some degree life in Newtok and Mertarvik proceeded as normal. Even so, a year of isolation is stressful, especially in a community already facing crisis. Many of the outside helpers of Newtok—advisors, workers, and experts from various government agencies and nonprofits—simply couldn't make the long journey. And Newtok slipped out of the national headlines while other troubles occupied America's attention.

In the summer of 2020, the Newtok Village Council chose to apply federal Covid recovery funds—money that could otherwise have gone to individual households—to construction costs at Mertarvik. But not everyone was pleased with this decision. Rumors and complaints about the council and the relocation process circulated on Newtok's Facebook forums.

Then, in the still-eroding village site on the other side of the Ningliq River, the old diesel power plant died in late August and couldn't be repaired for about a month. Some people cranked up their personal generators, but burning fuel day after day was costly. It was hard especially to keep powering the freezers—and eventually much of the fish the Newtok families had caught and frozen, the moose meat they had bagged, the berries they'd handpicked from the tundra and set aside for the winter, turned to rot. This was the food they had put by to make it through the long winter, a nutritional and cultural necessity for survival, especially the months when ice and rain make movement across the region difficult and grocery shipments unreliable. When other Y-K Delta communities heard about their plight, some sent donations of meat by air or barge.

But even after the power plant was repaired, the damage to the social fabric of the community was not mended. A breakdown, disagreements over money, matters of real estate, festering tensions between families, long histories, old wounds—these are the ingredients of a fight.

I never received any satisfying answers about what happened next—

---

* Sadly, Newtok was not able to keep the pandemic out forever, and, in August 2021, the community reported thirty-two cases of Covid.

simply that mounting stress, bickering, and bitterness seemed to squall through the community and uproot some long-standing relationships. In October, I heard from Charlene that Romy had reached his limit and resigned, and the two of them abruptly left Mertarvik. The following spring, I learned that Andrew John was no longer employed as tribal administrator, and community leaders had told him not to speak with media about the village's relocation.

But when I called Bernice John, who had taken a post as an "elder advisor" to the relocation, I heard little beyond down-to-earth cheerfulness. "With this pandemic, it's sort of slowing everything down," she said in her usual singsong voice. "But then it wouldn't stop the relocation. It's a little bumpy road but then we've passed through these bumpy roads before, you know," she insisted.

A day later, I spoke with Patrick LeMay, the Anchorage engineer who had been helping with the FEMA buyout process a couple years before. In 2020, he had moved to Mertarvik to work with Romy, spending six months apart from his family so he could assist with the next round of construction ("It was a longer deployment than probably my shortest Marine Corps deployment," he told me). He quarantined for two weeks on arrival in May. The village hired a crew of about fifteen people to put up houses and four more to build roads, and hammered away until November. LeMay had taken the principles of Mertarvik's previous housing designs and slimmed the costs by swapping out Cold Climate's custom technologies with items that could be purchased from familiar building suppliers—like an off-the-shelf composting toilet and a less expensive air exchanger to ventilate the thickly insulated houses. He estimated that he'd shrunk the price tag of each house to less than half of the previous generation but that they had about the same top-notch efficiency rating. To help make the village more energy efficient, he installed heat exchangers, which would transfer any "recovered heat" emitted by Mertarvik's diesel power plant to the water pipes, to keep them from freezing in the winter. (He told me the community's eventual goal would be to free the new village from diesel power as much as possible—with some combination of solar and battery storage.) The only significant problems he encountered were the disruptions that rippled out through economic supply chains—lumber stores that shut down because of the pandemic, a water tank unavailable for weeks. Nine nearly finished new houses stood in Mertarvik, twenty-nine gravel pads ready for more, and another half mile of one-lane road (mostly designed for ATVs to reach the front doors of their owners).

He sent me photographs taken by drone—of a road carved like a stem into the russet-colored tundra vegetation, little squares of houses leafing out from it, each covered in pale-green polymer moisture barrier before the siding was put up. Romy left in October, and Patrick stayed on until November, still under contract with the village. "We had an absolutely amazing year," he said gleefully. "We got a lot of work done." (This felt partly genuine and partly an effort to persuade me. Then again, I was fully aware of how essential it was to produce optimism.)

Later in the year, he reported back that the crews had finished a duplex at Mertarvik, and work had begun on a new airstrip.

Assuming what had been built proved to be durable, you could call this effort significant progress. Mertarvik was "a messy-assed success," Aaron Cooke, the Cold Climate architect, told me. "The reason I feel optimistic is they've got a foothold. Mertarvik is a real place. It's not a plan. It's a place."

I also hoped Mertarvik would outlast these crises and skirmishes. I remembered a sign that I had seen posted on the wall of the brown building. CHIEF IS A TITLE NOT GIVEN TO ELDERS IN OUR CULTURE. . . . CHIEF IS A KASS'AQ [white person's or outsider's] TERM FOR LEAD ELDER. ALL WERE EQUAL. In the long memory of Newtok, somewhere there were stories and instructions about how to act together rather than become divided, how to resolve differences, and how to stake out a home and face down a disaster. If the people of Newtok made it over to Mertarvik intact, it would be because they loved their land and their community too much to allow any one setback to overwhelm them.

This is the combination of love and perseverance we will all need to cultivate in this tumultuous moment on Earth.

CHAPTER 12

—

TO CLEAN HOUSE

bring something home to (idiom): *1. to impress upon or make clear to. 2. to fasten the blame (for something) on (someone).*

The bronchi, the branched air passages of the human lungs, are shaped somewhat like vegetables—like upside-down stems of celery or broccoli, rooted at the throat, flowering into hundreds of millions of alveoli, the sacs that exchange oxygen and carbon dioxide between the air and the bloodstream. Like the stems of plants, these structures serve as transport systems, carrying material from the world outside into the inner world of our bodies.

In the five hours that the Richmond refinery fumed and smoldered on August 6, 2012, the black smoke that spread across the San Francisco Bay Area potentially reached into hundreds of thousands of throats and lungs.

The plume was visible for miles.

Ash fell on all the fields and the soil and the food plants that Urban Tilth had carefully planted in the ground. At times it seemed to Adam Boisvert like a snowfall, ash flaking from the sky onto the cars and the sidewalks, the hanging laundry, the playground equipment. There was also a dark, filmy, gooey material on many surfaces the next day, Doria

Robinson noticed. The Urban Tilthers all wondered what might enter the stems and the xylem, the circulatory systems of the vegetables they had grown all summer. What might seep into the ground in the next rain? What might pulse into the waterways and be digested by the oysters of the San Francisco Bay? A ripple of toxins? A poison inhaled into the urban ecosystem and carried outward into the world at large?

Late that night, Chevron announced it would host a town hall–style meeting for Richmond residents the following evening. (On Twitter, someone launched a parody account called @Ch3vronPR and quipped, *In PR school, we had nightmares about refineries blowing up in our cities so we built them all in yours. #ChevronFire #ThoughtsDuringSchool.*)

Adam and one of his Tilther housemates biked to the Richmond school garden to survey the state of the beds. The day was mild, in the 60s, and the smoke had dispersed. But the signs of the fire were obvious to Adam. *Everything had ash all over it.* The beds at the school had collected the fallout. Ash on the plants, ash on the soil. The pair pulled tomatoes and squash out by the roots and collected them.

Tania headed to the Greenway Garden and Berryland, where about a half-dozen teen gardeners waited for her, for one of their last days of the summer apprenticeship program. The teens had helped tend food here all summer long, and the gardens were full and ripe, with raspberries and strawberries, the undulating leaves of purple tree collards, beans, basil, cilantro, zucchini, corn, and fat squash. This was supposed to be their second-to-last day of planting, and they were scheduled to have a gardener graduation later that week, an annual rite of passage for each class of apprentices with a party that included a meal of their own hand-grown vegetables. Instead, Tania had to send them away empty-handed. She warned them not to touch the beds. *We don't know if the soil is toxic.*

One girl was especially incensed. *Why do we even grow food?* she shouted. Arms crossed. *Why did we do all of this?* Tania looked at this young, angry face, not so different from the person she had been as a teen, and felt that the girl was just used to fighting, fighting for her life. *She had been here before on this road of hopelessness,* Tania thought.

The Tilthers called an emergency meeting to consider what they would do. In that moment, everything was up for discussion, including whether the organization should stop its work altogether. *There was a big question around, like, is it safe to grow food in Richmond?* Tania remembered. *Like, are we feeding people toxic food because there's all this stuff that's falling from the sky?*

She was also grief-stricken. *It's one of our little pockets of joy, like a little oasis that we create in the community—that's how I feel every time I've helped start a garden. Just to think that we would not be able to do that anymore was another level of heartbreak.*

The Tilthers decided there had to be a reckoning. Since its founding, Urban Tilth hadn't been an overtly political group. Local leaders, including the mayor, had at times invited Doria to run for political office. But she preferred to have her hands in the literal dirt and in the tangible work of growing the community. Doria was forthright: *We grow food. We invite people to change the way that they live their lives.* But this approach to transforming the city would never work, she realized, if a large corporation could poison all of their efforts without consequence.

The staff of Urban Tilth decided they would stage both a press conference and a protest. *We were really mad, as a group,* Tania said.

When the evening arrived, the Urban Tilthers gathered with around five hundred other residents and local advocacy and public service groups in the plaza outside the Richmond Civic Center carrying flags and signs with slogans like PEOPLE'S HEALTH NOT CORPORATE WEALTH. The Tilthers wore face masks and bandannas over their noses and mouths, a symbol of unbreathable air, and carried wheelbarrows and trash cans and plastic tubs full of the vegetables that they and the youth had grown. Urban Tilth also held a press conference there. Tania and Adam each gave interviews to journalists. *What are you growing right now?* a radio reporter asked Tania. As she tried to form the words, she cast a glance at Adam, standing a few feet away as he spoke on camera to a pair of documentarians. When their eyes met, they both began to cry.

Doria also spoke to the media, and she had a firm message for the oil company. *We're not asking, we're telling Chevron, they have to be accountable,* she said. *They have to pay. Our entire program is in jeopardy.*

At 6:00 P.M. on August 7, 2012, about twenty-four hours after the fire first erupted, doors opened to a three-thousand-seat auditorium where a group of city, county, and regional agency officials, along with Chevron's spokespeople, sat in a row in front of a mustard-yellow curtain. The room filled quickly, and reporters lurked at the margins, taking video footage and photographs. The head of a Richmond-based charitable foundation* ran the meeting: she tried to defuse the heat from the room before the questions began. *Those feeling angry, please stand up,*† she instructed the room. Nearly the entire crowd leaped to its feet. They booed the general manager of the refinery as he tried to apologize to the city, and they grew restless as the speakers told the crowd they were still gathering information, still considering how to update local warning systems, still figuring out what had gone wrong.

At the end, the audience was invited to stand at a mic at the front of the room and ask questions. A long line of questioners formed that snaked around the auditorium. One man ran toward the stage carrying a CHEVRON OUT OF RICHMOND sign and shouting the same words until the moderator persuaded him to settle into a seat.

The Tilthers waited, and when their turn came, they carried the by-now-wilted vegetables to the stage, along with a sign that read NOTHING THAT IS POISONED CAN GROW.

Is this contaminated? the group shouted at the panel. And then they began to chant, *Tell us now! Tell us now!* until other voices joined the chorus and nearly the whole audience was chanting and stomping their collective feet. But the panel never responded and moved to other topics. Some of the Tilther youth headed to the microphone and vented their anger, cursing at the panel. Doria would hear criticism later about their strident tone, but *we did not control them.* She felt it was important for the youth to have space for their feelings. *That blowup was due to a*

---

* This foundation—then the Richmond Community Foundation and now named RCF Connects—has in at least one case accepted funding from Chevron. In 2014, Chevron chose RCF Connects to lead an economic development initiative.

† Since 2009, students at the UC Berkeley Graduate School of Journalism have run a news lab called *Richmond Confidential* to report on events in the city. Over the years, *RC* reporters have done deep investigations into Chevron and Richmond politics, and the most detailed account of the August 7 meeting comes from *RC.* Some details that appear here are drawn from *RC*'s reporting and from KQED, regional public radio.

lot of years, maybe even generations, of taking it, years of putting up with poverty, discrimination, and pollution.

Meanwhile, the regional air quality district announced that levels of "potentially toxic pollutants" measured at the scene of the fire were "well under their reference exposure levels" and "not a significant health concern."

Three days later, that agency backpedaled and announced it had found elevated levels of acrolein, a smelly chemical also present in second-hand smoke and car exhaust. Much later, an independent academic analysis concluded that no one had actually collected reliable information about pollution during the fire. The closest air monitors took no samples until after the fire was over. A year after the accident, Chevron would install a series of fence line air monitors (a bargain it had struck with Richmond in 2010 when the city threatened to raise its taxes). But no one who lived through the 2012 fire would ever know exactly what they had inhaled. During and just after the accident, fifteen thousand people from surrounding communities turned up at hospitals and doctors' offices, complaining of chest pain, shortness of breath, sore throats, and headaches.

In the days after the protest, Doria received hate mail from those who didn't agree with her stance. *This woman said that we were liars and we weren't from Richmond. Somebody even wrote twice—they were just so mad.* She was criticized by some of her friends and supporters. *People were, like, how unruly! But the young folks were heated. They had worked so hard.*

At that moment, there was no way we could stay quiet.

ॐ

From 2012 onward, I would make periodic visits to Richmond. I would drive through the flatlands, photographing the patterns that striped through the city's grid—abandoned lots beside tiny residential bungalows on sparsely treed streets; unauthorized art galleries of graffiti along walkways, alleys, and industrial corridors, some involving elaborate murals, some scrawled haphazardly; the monumental infrastructure of the port, shipyards, and railyards. I would visit the Rosie the Riveter WWII Home Front National Historical Park and eat lunch on the patio at an adjacent swanky restaurant, both housed in the old Ford assembly plant. On the other side of the harbor channel, I toured a vintage naval ship docked at the edge of Shipyard No. 3. Here an archipelago of interpretive signs spanned the vast expanse of concrete, paying tribute to the women and men who had helped build wartime vessels—and their stories of triumph over discrimination in a time of prosperity. "At first the shipyards and other war industries attempted to operate only with white men," writes historian Richard Rothstein, "but as the war dragged on, unable to find a sufficient number to meet their military orders, they were forced to hire white women, then Black men, and eventually Black women as well." Richmond often bills itself as the home of Rosie—the bicep-flexing, blue-shirted white female worker depicted in U.S. government posters to recruit women for jobs in the defense industries in the 1940s—the symbol of mighty women taking traditional male roles in the workforce. But the histories inscribed here on the shipyard signs said nothing about the soul-crushing economic collapse that came in the decades after the war, nothing about the incredible labor of growing a twenty-first-century city out of old dirt and cracked pavement.

Since my first visit here, something about this city had taken up residence in my thoughts, and I wanted to understand it. Richmond represented a twentieth-century industrial moment of optimism that has since been shattered. And it was not so different from any other such place in America—the hollowed-out manufacturing centers of the Rust Belt; the hurricane-battered oil and gas communities near the Gulf of Mexico. Left-behind places, haunted by toxicity and decaying infrastructure, by a globalized economy that had eviscerated America's ability to make most things for itself. You could imagine both optimistic and dystopian ends to their stories—one in which such communities lumbered awkwardly but hopefully toward a greener economy, one where wealth might be diffuse and local and spread across many households, like solar panels on rooftops. You could imagine another where such places dried up or sank into polluted backwaters.

The oil industry has underwritten so much of Richmond's past and present. According to various estimates, between one-tenth and one-third of the city's annual revenue comes from Chevron.* The company is a major philanthropic force, propping up many community organizations. Its supporters knew what kind of bargain they were making. Its opponents especially knew what they were risking. To stand against the refinery required tenacity. But for people like Doria, any benefit Chevron had ever offered to the community was far outweighed by the costs it exacted. Profit at the expense of people's health and future had no lasting value.

In the fall of 2012, I visited Doria at her yellow house. There were bicycles stacked by the door, along with a pile of old vinyl records, and sets of bags and coats I had to step over to enter her living room. "So many backpacks and bikes around here," she said offhandedly. It seemed like a theme: she wore a yellow T-shirt emblazoned with the words THE ORIGINAL RIDE OR DIE CHICK. Doria had always loved bikes, she told me. Urban Tilth's first garden was, after all, on a bicycle path.

By this time, she had begun dating a bicycle enthusiast, entrepreneur, and artist named Najari Smith (who would later become a life partner). Najari had moved to the Bay Area and eventually made his way to Richmond after years of financial struggles in New York City with jobs that never paid him enough to afford rent (forcing him first to live with piles of roommates and, at one point, leaving him homeless, sleeping on trains before reporting to the office). When he and Doria met, he was working as a graphic designer and serving as a volunteer on a city bicycle and pedestrian advisory committee. In the summer of 2012, they had launched a cyclist advocacy group called Rich City Rides. Like Urban Tilth, the goal wasn't just to fix bicycles (or grow vegetables) but to use these activities to foster something a little more subversive—to make people healthy and energized enough that they might also demand better living conditions in their city. Doria was also running yoga classes by donation downtown, focused on stress reduction. "Part of my personal mission is trying to find other like minds who are just isolated somewhere in the city, who just need to know that

---

* In 2020, according to the Othering and Belonging Institute, Chevron taxes made up nearly 24 percent of the city's general fund revenue. However, the company has disputed city efforts to raise its taxes and also pays below-market property taxes, a real estate tax break enshrined in a 1970s-era California law.

they're not alone and that we also want change and we're tired of being bullied," she mused. Bullied by Chevron, she meant.

She made me a cup of tea, and we sat in her living room at a wooden table by the window. Tobias, the cat, squatted beside her.

Just weeks after the fire, she was strikingly optimistic. "It's a really exciting time to be in Richmond," she said. That year, the city government had also adopted a new plan that focused all of its decisions on health—its policies now had to take the wellness of city residents in mind. "We have to think through what type of developments we want, what type of industry we want. We want to actively invite the kinds of things that will improve quality of life," she said. Implicitly this could mean that food gardens would be a high priority and industries that might pollute them would eventually become unsuitable. Doria had an answer to the question that had loomed over her and her staff during the fire. The gardens and farms would stay.

But she also wasn't comfortable continuing to plant without knowing the soil was safe from contamination. Urban Tilth didn't offer food to anyone for nine months after the 2012 fire. The organization couldn't seek any compensation from Chevron to cover any cleanup costs, because the Tilthers had no legal proof of any damage. No one had gathered baseline soil data at the farm and garden sites, and it was therefore nearly impossible to demonstrate harm. Over the next couple of years, Urban Tilth tore out and rebuilt the fruit and vegetable beds on the Richmond Greenway. They had to fundraise to cover all of the replacements. They eventually imported fresh soil and installed sturdier beds—made of pine lumber instead of old driftwood and reclaimed wood—and planted them again.

🌱

There is nothing good that can be said about one of the largest industrial disasters in California history, about the injuries sustained by six oil workers who were on-site during the 2012 Richmond refinery fire, about the money the city had to pour into emergency response, about the terror felt by locals who watched the flames tower into the sky that night, about the company pleading guilty to six counts of criminal neglect. There is no silver lining, no path to grace, no revealing plot point in some grand happy parable.

But such a moment sometimes offers up rueful clarity.

In the past, refinery incidents and accidents had been an endemic affliction for Richmond—a problem that the rest of the world mostly chose to ignore.

But by the 2010s the climate justice movement had grown larger, stronger, and more confrontational. After many years that had yielded little progress on climate change, national environmental organizations set their sights on fossil fuel companies. They knew by then that the oil and gas industry had tried to get a free ride, a pass on its responsibilities to the public—by using its lobbying might to disrupt political efforts to regulate carbon emissions from fossil fuels, which are the largest contributor to climate change. They knew that fossil fuel money had also funded organizations like the Heartland Institute, devoted to disinformation, infamous for its efforts to discredit scientists and sow doubt about research and findings on climate change.* They also understood that if the industry continued to pursue more sources of oil, coal, and gas—such as those extracted from fracked gas and from tar sands—the climate crisis would become even more dire.

In response, even long-established, staid environmental organizations got riled up. In February 2013, for instance, the Sierra Club broke its 120-year prohibition against civil disobedience to join protests against the Keystone XL pipeline; four years later, the organization's board changed its stance on this type of activism altogether. Climate activists became more willing to engage in a sort of direct combat—including civil disobedience to confront the industry on its own turf. (Analysts even warned that activists had become effective enough to get

---

* Harvard University historians Geoffrey Supran and Naomi Oreskes have also documented oil companies' strategies for misleading the public. Oreskes described her findings in a congressional committee hearing in 2019: "Rather than accept the science and alter its business model accordingly, [the industry] made the fateful decision to fight the facts. For more than thirty years, the fossil fuel industry and its allies have denied the truth about anthropogenic global warming. They have systematically misled the American people, and contributed to delay in acting on the issue, by discounted [sic] and disparaging climate science, mispresenting scientific findings, and attempting to discredit climate scientists," through misleading advertisements, by funding third-party groups that produce propaganda, by supporting trade associations that run climate-denialist media campaigns, by attacking scientists directly, and by lobbying against regulations.

in the way of planned fuel-extraction projects.) The 2012 fire brought Richmond into the spotlight and drew the attention of national networks of activists.

In August 2013, three days before the one-year anniversary of the fire, hundreds of out-of-towners arrived in Richmond and joined locals for a rally near the BART station. It had all been organized by a coalition that included Communities for a Better Environment (the environmental justice organization that had been fighting Chevron in the courts and had an office in downtown Richmond) and the San Francisco chapter of Idle No More, a network of Indigenous activists. Bill McKibben, normally based in Vermont, attended in person, along with then Richmond mayor Gayle McLaughlin. A motley coalition of supporters assembled with them—including nurses and local labor unions. Some marchers carried sunflowers. Some toted brightly painted signs with slogans like ¡SÍ A LA ENERGÍA LIMPIA! ¡NO MÁS TÓXICOS! (YES TO CLEAN ENERGY/NO MORE POISONS) and WE HAVE A RIGHT TO GROW HEALTHY. Yellow flags with the words STOP CLIMATE CHAOS shook in the breeze. Organizers set up a PA system at the front of the rally, and Doria Robinson gave a speech. She held a bright blue microphone to her mouth and addressed the crowd, her voice full-throated with emotion as she told the story of the previous summer. *It made us actually open our eyes even further—I mean, we knew, but we didn't know, right?—to the need that we have to stand up as Richmond residents on the front line to Chevron*, she shouted.

McLaughlin, now in the middle of her second term, had announced the day before that the city was suing Chevron to force the company to take extra safety measures. A round-faced woman with a treble voice that seemed to still carry a little of the tone acquired from schoolteacher jobs of her past, she told the crowd, *The community doesn't deserve to be traumatized.*

And when McKibben took the mic, he expanded the geography of this story. *Chevron is a really bad actor. OK? . . . Ask the people in Canada fighting their fracking. Ask the people in Ecuador who have had to live with their waste. When they get it here to refine it, they're a bad actor. . . . And they are bad, bad actors on this planet. They have nine billion barrels of oil in their reserves. OK? If they burn most of those, then we cannot deal with climate change.*

Afterward, at least 2,500 ralliers marched to the refinery. They painted a giant yellow sunflower on the street in front of the gates. Idle No More activists led them in a round dance, a lively circular performance of unity. Then, according to activists' accounts, the protesters approached the cops, who rounded up those who trespassed, until there were no more plastic handcuffs and officers had to send for more. In the end, the police arrested more than two hundred people.

Doria did not participate in the march. She had mixed feelings about the out-of-towners. *The city becomes an arena of battle*, she reflected later, and there was a disconnect between the activists who march *through the neighborhoods where the people who are actually hurting from all this stuff live* and the residents who *don't even know what the hell is going on.* But after the march, Urban Tilth held its own event, a fossil-fuel-free party on the Greenway where the out-of-towners and locals mixed and members of the community had a chance to talk about their experiences with the fire. This was the more significant part of the day for Doria, a moment when the loneliness of Richmond seemed to ebb away and people could feel their connections with one another. Her sense of mission had shifted over the past year: she felt she was no longer just growing food but fighting against a corporation for her own and her neighbors' survival. And she realized this was no small battle: it was global. But it was critical, she thought, to empower the voices of people who had felt the hardest blows from fossil fuels and had the most at stake.

Shortly after the protest, Doria received an international invitation. Richmond's oil-industry opposition had come to the attention of officials in Ecuador, where a group of Amazonian communities had sued over contamination left behind by the oil-drilling ventures of Texaco, a company acquired by Chevron in 2001. The incident was sometimes nicknamed Amazon Chernobyl. Texaco had spilled about sixteen billion gallons of toxic wastewater in the rain forest, according to the environmental group Amazon Watch. The company had filled hundreds of pits with a stew of petroleum wastes including polycyclic aromatic hydrocarbons, or PAHs, carcinogens and by-products of combustion later also found in trace amounts in Richmond's soils. According to one report commissioned by a human rights group in 1993—one year after Texaco ceased production in this region of Ecuador—residents near these polluted sites had a high risk of cancer, and people exposed directly to oil contamination experienced "elevated rates of fungal infection, dermatitis, headache, and nausea." Communities in the region had been trying to get compensation from Texaco and then Chevron for two decades. In 2007, then president Rafael Correa had ridden into office with the support of Ecuador's Indigenous communities. During his administration, Ecuador changed its constitution to recognize the rights of nature. Correa also entreated international governments to pay billions of dollars to protect the rain forest if the country would, in turn, agree to stop drilling for oil there. In 2011, an Ecuadoran court awarded $18 billion to the Amazonian plaintiffs, though the amount was eventually dropped to $9.5 billion. But the company refused to pay up and moved its assets out of the country. Employing a legal team that literally numbered in the hundreds, the company also sued the plaintiffs' New York–based lawyer, Steven Donziger, for fraud and racketeering, and a Manhattan court scheduled hearings in that case in the fall of 2013.* After the Richmond protest made international news, Correa's administration reached out to Mayor McLaughlin, offering to fund a visit in which she and other Richmonders could meet the people who lived beside the now-abandoned Amazonian oil-waste pits.

Correa's relationship with progressives in both the United States and Latin America had already started to sour as his administration opened other Indigenous lands, including parts of a national park, to oil extraction. He would become an even more controversial figure for, among other things, using the legal system to harass and prosecute both Indigenous and environmental activists, and after he left office in 2017, he was found guilty on corruption charges. Arguably the trip may have been a publicity stunt for the Ecuadoran president. But McLaughlin's

---

* In 2014, a U.S. judge declared the Ecuadoran court's decision against Chevron invalid. Donziger was found guilty and sentenced to house arrest, a ruling that was contested by the Office of the United Nations High Commissioner for Human Rights, which called for an investigation. In 2020, twenty-nine Nobel laureates wrote an open letter accusing Chevron of harassing and defaming the lawyer. In 2021, Donziger was also found guilty of criminal contempt of court.

intent was clearer. *My main purpose was to see it for myself, come back, educate the community, build solidarity*, McLaughlin said, years later, and she asked Doria Robinson to accompany her, along with a journalist from a regional newspaper. Over a week, they went to Quito to meet with Correa then took a flight to Lago Agrio to see the oil pits. Nearly everywhere McLaughlin was followed by television cameras and radio broadcasters. In the Amazon, the group had police escorts. The Ecuadoran officials staged a press conference at one of the now-defunct oil wells. Doria kept a blog, where she described the scene.

Many of the contaminated pits were covered with a thin layer of dirt by Chevron-Texaco before they left Ecuador as a remediation measure. This soil acts like a gelatin-like cover. When you step on it it feels like walking on a water-bed with small holes where water seeps through. . . .

Reporters milled around exploring the pit for themselves, stepping on its strange surface, commenting on the rank smell in the air, taking photographs and film footage.

In the late morning, Correa himself made a dramatic entrance—wearing blue jeans and yellow rubber boots, with the theme from *Star Wars* blaring behind him—and walked into the oil pit to announce a campaign named the Dirty Hand of Chevron, calling for a worldwide boycott against the company. After this spectacle, Mayor McLaughlin and two mayors of Amazonian towns walked together to the oil pit, dipped gloved fingers in, and held up their oil-coated hands for a somber photograph.

In the days that followed, the Richmond contingent heard stories from families in the region who had suffered from cancers or birth defects that they blamed on oil pollution, and McLaughlin vowed to build relationships with communities here and in other countries that had complaints against Chevron. For Doria, Ecuador felt like *a House of Mirrors for visiting Richmond residents*, she wrote in a blog entry. *The resemblance to Richmond California's economic and political landscape is uncanny.* She pondered climate change often during the trip. Sometimes the overpowering scale of the problem would hit her—and the sheer size and economic might of the corporations that dominated the world's oil, gas, and coal reserves. *It's difficult to preserve space for hope*, she reflected.

Chevron later funded a billboard advertisement criticizing McLaughlin for this trip and others like it (such as to a sister city in Cuba). IF YOU SEE GAYLE MCLAUGHLIN, TELL HER TO CALL RICHMOND, one ad read in large lettering, and GAYLE MCLAUGHLIN: TRAVELING THE WORLD, IGNORING RICHMOND, read another, along with an unflattering image of her with a puckered face and a strangely disproportionate hand raised to her chin.

※

In 2014, I traveled to Richmond twice, in both spring and autumn. It was an election year for city officials, and the community was wrestling with questions about who and what it might become, its split personalities facing off. There was the old oil company town and the newer green face of the Left Coast. There was also an underlying socioeconomic clash—those who had rooted here for generations versus newcomers, some of whom represented new wealth and its influence on property values, the tendency of the rich to drive up the market and price out the poor. Various sides would claim they were more authentically Richmond, each insisting that it had the community's best interests at heart. (Doria observed that there was also an older philosophical strain, one she could trace through her family and the civil rights movement—about lifting people up and mending community collectively—but this was rarely well represented in politics at any level.)

Politics, identity, and economy intersect in complex ways. Arguably, all identities have politics, and allegiance to a community or heritage can be radically empowering when the intent is to lift up people who have been marginalized. The phrase *identity politics* originated in the 1970s with a group of Black feminists who sought an approach to fighting oppression that was rooted in the needs of their community: "We believe that the most profound and potentially the most radical politics come directly out of our own identity," they wrote. In popular usage, the phrase sometimes morphs into a weapon, a pejorative wielded mostly against the Left. But Black Lives Matter cofounder Alicia Garza argues that the idea remains relevant to present-day racial justice movements: "Identity politics says that no longer should we be expected to fight against someone else's oppression without fighting against our own."

Identity is also about home, where you imagine you come from, where you choose to plant yourself, and what you will build in the space around you. But not all identities are a path to empowerment. And there are places where industries are themselves an identity. Sometimes the story looms as large as or larger than the economics. Coal country is more than a place; it is also a story about heritage. And it is, of course, normal to have pride in the things that you or your parents or grandparents have built in a place. But in a time when fossil fuel pollution puts everyone's future at risk, it's a trap to cling too fiercely to the industries of the past.

In the fossil fuel politics of Richmond, some residents felt Chevron was always the "other," always a towering externality bearing down monstrously on their lives. But some, it seems, felt that the city and the refinery were inseparable siblings, or that the oil company was a justifiable, if messy, means to an end. Even after the 2012 disaster, a number of community leaders believed that the company's money might be the only way to ameliorate the poverty found too commonly here and to pay for public needs—and simultaneously to fund salaries and campaigns. Scores of Richmond community service organizations have long been supported by Chevron philanthropy.

Chevron, meanwhile, had a specific agenda in Richmond that election season. For the past several years, the company had been pushing for approval of a plan to *modernize* the Richmond refinery—or to *expand* it, depending on which verb seemed more apt to one's view of the situation. In 2008, a previous city council, more sympathetic to the company's aims, had approved one version of Chevron's modernization plan. (The councilors who voted in favor earned the epithet the Chevron Five for their consistent support of the company.) Then Communities for a Better Environment, the West County Toxics Coalition, and the Asian Pacific Environmental Network led a lawsuit against the city council over the plan, insisting that its environmental impacts hadn't been rigorously studied. A judge halted the project in 2009. In 2014, the company was trying again with a scaled-back version. But one of the goals of both plans seemed obvious to those who analyzed it: to process high-sulfur crude, otherwise known as sour or dirty crude, a category that included fracked oil. Kamala Harris, then serving as the attorney general of California, wrote a ten-page letter criticizing the company's plan. Dirty crude was more corrosive and more likely to create repeats of the 2012 fire and to increase the refinery's carbon emissions. "Given that the residents of Richmond are already facing some of the highest pollution burdens in California," Harris wrote, the city's environmental impact review needed to "analyze whether additional pollution will contribute significantly to the community's existing public health problems."

In response, Chevron aggressively pleaded its case to Richmond's citizens and tried to sway city politics. The company started its own news site called the Richmond Standard in January that year, and its consultants helped launch another "citizen journalist" venture, called Radio Free Richmond, a reference to the cold war–era radio programs broadcast by the United States to Soviet-occupied countries, perhaps playful or perhaps suggesting that Richmond needed liberating from the Left. Chevron also bought up most of the billboard space in the center of the city and ran a combination of feel-good public relations messages and ads for political candidates who supported its interests. During the year, the company also spent about $3 million on political campaigning aimed at trying to influence local elections in Richmond that November. Gayle McLaughlin had reached the end of her term limit, and Chevron hoped no one with similar anticorporate political positions would succeed her. The company backed a candidate named Nat Bates, who was serving on the city council. Peppered across town were billboards featuring Bates's face, with the slogan BUILDING THE RICHMOND WE NEED, HONORING THE RICHMOND WE LOVE, and beneath, in smaller print, was a list of ad sponsors, including "Major Funding by Chevron."

One morning, I drove to the Richmond Civic Center to talk with Bates at his office. The Civic Center is Richmond's city hall and also a community gathering space, an arts center, a plaza, a modernist collection of brick buildings designed in the 1940s to project the image of an industrial city on the rise. I parked nearby and walked past a senior center. On one curved wall of this building was a florid, intricate mural depicting various chapters of Richmond's history—including one of the original Rosie the Riveter workers, the oldest national park ranger in the country, and the Japanese gardeners who had run a series of commercial rose nurseries in Richmond in the early twentieth century. The last panel featured Doria Robinson, larger-than-life yet unpretentious, even when essentialized as art—one eyebrow slightly arched, wearing a wry smile. Behind her, a crew of Tilthers, and in the background, crop rows of flowers and greenery; at the far horizon, the refinery; and to the west, an assembly of activists with protest signs bearing messages like CLEAN UP CHEVRON.

"We will show Standard Oil's refineries transform into solar plants and green industry," the artist wrote in a statement describing the mural. The city council had funded the artwork, but Nat Bates had cast an opposing vote. This was not his vision of Richmond.

He had also recently fulminated against the bicycle infrastructure Doria and her partner, Najari, had helped push the city to install. Bates insisted that cyclists were almost nonexistent in Richmond and that bike lanes were "popping up all over" and "creating all kind of havoc." (Najari argued the opposite, that many people who couldn't afford cars needed to ride safely, and later that day, when I was walking downtown, two young men traveled past me on bikes, one on the sidewalk trying awkwardly to balance on a too-small frame, the other in the street. The second saw my camera and stopped to ask if I was a reporter and if something was going on in the city.)

But Nat Bates and Doria Robinson had grown from the same roots. He belonged to her grandparents' generation. From a southern migrant family, Bates was raised by a single mom who brought him to Richmond from Texas and who got a job with the Santa Fe Railroad in the 1940s. In the 1950s, he'd had a brief career as a professional baseball player in Canada for two seasons. Then, back in Richmond, he was a probation officer for thirty-five years. In 1967, he became one of the first Black leaders elected to Richmond's city council. He had already served two terms as mayor in the 1970s. Recently, he had been invited to the White House, and some of his campaign ads featured a picture of him standing beside President Barack Obama.

Then in his eighties, he was known as a local gadfly, especially to his opponents on the city council. He was sometimes caricatured, accused of being "in Chevron's pocket," and in turn, he was often provocative toward his opponents, accusing them of holding a radical agenda.

I also recognized his innate charisma and charm. He greeted me in the hallway with a handshake and led me to his office, where I sat across from him at his desk. He had large weathered hands and a small well-groomed mustache and wore a shiny silver and gold watch and chunky rectangular glasses. A prescription bottle protruded from his pocket. He spoke in a soft, warm drawl and was unabashed about his support of Chevron. In his view, it was simply part of an economic reality.

Did he have any concerns about Richmond being so reliant on the revenue of one company, I asked? "Well, in life, you play with the cards they dealt you, right?" he said, as if I had asked a philosophical question instead of an economic one. "I guarantee you there are a lot of cities that would love, just love to have Chevron USA in their city," he said. He called the fire "troublesome," but insisted that the city should do whatever it could to move the modernization plan forward and get the refinery running at full capacity again, "get all those jobs back online." But he favored any flavor of economic development: he envisioned Richmond becoming a hub for importing electric cars and said he was happy to see a solar panel company at the waterfront.

When I asked about climate change, Bates demurred: he repeated the arguments that have been fabricated by climate disinformation groups. "When you have two or three or four different scientists come forward with different opinions about the necessity, and the effects of fossil fuels and climate control and a host of other scientific issues, obviously, we have to try to . . . arrive at a decision that, hopefully is in the best interests of the community," he said. I couldn't tell whether he was cagey or truly misinformed—practically speaking, there is a wildly strong consensus among scientists. A moment later he walked back his position and acknowledged that the Obama administration had pledged to tackle climate change, and he wanted to be supportive.

Critics of Bates have raised questions about the legality of his connections to Chevron and the political action committees that have supported his campaigns. Perhaps Bates's enthusiasm for Chevron was purely self-interest. But in the moment, I assumed that he was sincere and that he could not imagine a desirable reality in which the refinery was not central to the city's functioning and identity. In a stump speech, Bates later insisted that anyone who didn't accept Chevron's support was "a damn fool."

And Bates wasn't entirely wrong: the city couldn't yet afford to lose Chevron's business.

A few days later, I stopped by a job-training program called RichmondBUILD. Both Gayle McLaughlin and Nat Bates had touted RichmondBUILD as a city accomplishment—McLaughlin because it trained for green jobs, including solar installation and the gritty work of cleaning up old industrial sites; Bates because any kind of job creation was a net positive in his estimate.

The program was held inside a stucco yellow building about a block from the railroad tracks. Inside, on a concrete floor, the students were completing a series of physical agility tests—crawling into a confined space, scaling up tall ladders, hauling weighty buckets—to prove they could handle such labor. The program manager, Fred Lucero, was a pragmatist. Green jobs, he told me, were fine; they made for a good story. But anyone who came from intergenerational poverty in Richmond just needed to work. "I emphasize going to work," Lucero told me. "We serve a population that is tough, that's been through a lot."

I pulled three students aside and asked about their aspirations—one wanted to be a carpenter, one a park ranger or any other sort of green job. And the third said she hoped to work at the refinery. Her name was Breonna, and she had a tattoo on one arm shaped like an infinity symbol, with the words LOVE THE LIFE YOU LIVE. LIVE THE LIFE YOU LOVE. She hadn't liked her previous job at Walmart. "That's not for me," she said. I asked if she worried that a fire or accident might strike the plant again. "Personally, I think fear is a choice in life," she told me. "I just stay positive."

There were plenty of others in this town who had made a similar choice, people who needed to play the cards they were dealt.

But in the long term, it was hard to see how this community, or any community, would survive if we all continued to tether our future, our choices, our energy, our economy, our homes, to such an industry. "This image that Nat Bates keeps about Richmond," Doria told me later, "it's not a healthy one. This corporation throws money from time to time, and everybody has to run and catch it." But she also respected Bates's lifetime contributions to the city. "There's this huge danger in throwing away the people who fought hard for justice, who came before us, because the justice they were fighting for doesn't look like the justice we're fighting for today." Doria thought Bates had failed to understand the root causes of the community's suffering. (Nat Bates and Doria Robinson would never see eye to eye politically, but a few years later, he would drop by one of Urban Tilth's farms to observe their work and offer to make some calls to find a person who could loan them some heavy machinery.)

In July 2014, the Richmond city council voted to approve Chevron's modernization plan. No one opposed but the mayor and vice-mayor abstained, protesting that they had not discussed the details thoroughly. Doria's mother attended and gave an interview to a small newspaper run by a Richmond teacher and his students. *I've been in Richmond almost sixty years*, she told the reporter. *This company does not care about this community.*

Chevron continued its public relations blitz, called Richmond Proud, through Election Day. The campaign won awards from the PR industry. But it also backfired, provoking the ire and ridicule of national media. David Horsey, a cartoonist and editorial writer for the *Los Angeles Times*, wrote that the company deserved "this year's top prize for brazen conduct."

Senator Bernie Sanders stopped in Richmond to a give a speech. *You're seeing right here, in this small city, unlimited sums of money from one of the largest corporations in America, who says, "How dare you ordinary people—working class people, people of color, young people—how dare you think you have the right to run your city government?"* he homilized to an audience of several hundred, who rewarded him with a standing ovation.

In November, every candidate Chevron had thrown its weight behind lost their bid. Nat Bates's opponent won by a landslide. Bates would eventually return to a seat on the city council. But it was clear that the oil industry couldn't reliably buy an election here—not even with an outpouring of money and advertising—and a large number of Richmond voters no longer wanted their home to be a company town. They had made a different choice.

ℬ

No matter what happened in the political landscape, for Urban Tilth it was always necessary to reckon with the dirt.

Dirt is an ecosystem. Peer through a microscope, and soil can be a forest of half-composed stems and leaves with filigreed veins, shiny mites and little bugs, worms and their glistening trails, netted root hairs, strands and filaments of fungi, and various microbes and molds, a furious mix of rot and rebirth. Soil is also a kind of historical record—it hangs on to waste and files it into layers of accumulated dust and particles. Soil recalls the things that happened on its surface. In some kinds of soil layers you can find the char of millennia-old fires or volcanic events. Soil can also be forgiving: in experiments around the world, fungi and microbes have sometimes been able to munch through toxic substances and break them down, even clear away some residues of oil spills.

In the past, Urban Tilth had relied on new soil with no memory, manufactured by nurseries, donated and hauled to its gardens. But before the 2012 accident, at the middle school up on the hill, the young farming apprentices had planted directly into the earth. And in the long run, it would be hard to grow a lot of food in Richmond if the soil always needed replacing and importing. The question remained: Could the soil endemic to Richmond ever be safe enough?

There's never been a lot of study on what happens to plants and dirt just after industrial accidents. But there were reasons to worry about the soils in Richmond. More than a decade earlier, for instance, after an industrial fire at a warehouse, some British scientists had detected alarming levels of PAHs, the same class of chemicals found in the Ecuadoran oil pits. The scientists determined that the level of

PAHs was 70 times higher than normal in grass shoots and 370 times higher in soil.

Joshua (Josh) Arnold, a soil science student at the University of California, Berkeley, had just moved to Richmond at the time of the refinery fire. *I remember just being absolutely astounded that there was a refinery there! I thought I was moving to the Bay Area that's super environmentally friendly.* Two years later, Josh, who had just graduated, and two others, an organic gardening teacher and a Ph.D. student in soil science, offered to help other Richmonders test the soil. They recruited people from all over the city to volunteer their yards and dug samples from nineteen spots in Richmond. They sent bags of dirt away to labs, searching for trace amounts of carcinogens like xylene, benzene, and toluene, found in crude oil—and PAHs. Although these are common contaminants, it is surprisingly difficult to find any standardized safety rules specifically for residential soils. There's no straightforward consensus about what concentration of these chemicals the average row of backyard tomatoes or roses can hold or what might still be safe if, say, a child eats some dirt or a dog tracks garden mud through the living room. So the students cobbled together a list of health guidelines from a motley range of sources. When the results came back, they found few things of great concern—traces of hydrocarbons and xylene. The students recommended earthworms and compost as an antidote to any lingering contamination. And at the middle school farm, the Tilthers aggressively layered on compost and cover crops and followed the guidelines offered by the soil science students until Doria felt safe to let teens plant vegetables there again.

This act of soil renewal prepared Urban Tilth for a far more ambitious project to reclaim abandoned land at the edge of the city.

Richmond sits at the northwest tip of Contra Costa County, which curves east along the Suisun Bay and south, behind Oakland and Berkeley—a mishmash of agriculture and industry. To the north lies a community outside city limits called North Richmond, a former agricultural outpost for Italian, Mexican, and Asian immigrants that, in the 1940s, evolved into a blue-collar community for Black industrial workers. It is somewhere between urban and rural, like a place that has fallen off the map. Muddy with a tendency to flood, crowded, shunted into the far, neglected corner of Richmond's urban consciousness, in the 2010s the community of about four thousand remained "underserved"—with high crime rates and poverty and not a lot of help from government entities—and consequently its residents were sometimes full of distrust. Garbage dumps and recycling centers, storage lots, homeless encampments, and run-down convenience stores were splotched through the area, along with an elementary school and residential houses that were too often aging and cramped.

The area was also riddled with vacant lots, and the supervisor of Contra Costa County had recently become an enthusiast of urban farming. He asked his staff to sniff out some property that might still be suitable for growing food, then chose Urban Tilth to develop a three-acre plot of land between two creeks. The organization would rent the land for the nominal fee of $500 per year. The property, three miles north of the refinery, was tiny compared to an industrial farm, but it would be large enough to demonstrate what could be possible and to allow people's imaginations to unfurl.

The chosen land parcel spread northeast from an intersection of Fred Jackson Way (named after a local civil rights activist). It lay in a historic Japanese flower-growing district; some descendants of those growers still owned the property next door and rented greenhouses to an orchid cultivator and a wholesale seedling nursery. Numerous Indigenous Huchiun Ohlone villages had stood near and around the land long before North Richmond was settled by migrants from elsewhere. A local association of Black cowboys had run horses there. But industry had never touched it, and the first soil tests came back clean.

It took two years for the Tilthers to negotiate with the county—to rezone the land, get permits, and draw up a lease.

When Doria first laid eyes on the place, it was a jungle of weeds higher than her head, in some spots twice as high, with a few trees scattered within them, and some weeping willows slumped by the street like long-haired old men.

Urban Tilth mowed the weeds down. But by the time the lease was approved, they had grown back, especially the blackberries.

Much of the coastal West suffers from an infestation of an especially aggressive variety of blackberry from Eurasia. (It's rumored that the famed botanist Luther Burbank first planted them on American soil, and from there, the thorny fruits went rogue.) While the berries are tasty, the canes sometimes grow into a colossal tangle—a Medusa with barbed tentacles sharp as knives, capable of grabbing and shredding clothing and skin.

The thicket of thorns at the farm site was roughly the size of eight basketball courts lined end to end, surrounded by other dense weeds, including enormous, fragrant, stubborn-rooted fennel. The land hadn't been fenced for four decades, and as the Tilthers explored further, they discovered it had also been an informal dump and was piled with hidden trash.

Yet Doria's optimism was unscathed. She could see it already, a vision in the weeds—the rows of vegetables, an educational center. *My eyes are permanently rose covered. I'm like, oh, man, this is the best farm ever.* By the time they began clearing the land, she and Tania Pulido had already assembled a group of local residents to help them in this seemingly quixotic process—and their dream for the site grew more elaborate—a farm stand, an amphitheater, perhaps even a café. *We just had no idea what the heck we were jumping into,* Tania said afterward.

The land itself was a cantankerous thing. First, the Tilthers rented a team of goats to chomp down the weeds. But the goats *looked at that blackberry patch, and they're like, oh, no. You will have to deal with that,* Doria said. Then they brought in large mechanized equipment—toothy mowers and mulchers, grinders and bulldozers. Underneath the weeds, they found orphaned couches, chimneys, chunks of broken asphalt, used motor oil, several discarded wallets including credit cards, rusty barbed wire, old tires. Urban Tilth hired workers from several local organizations, including a program that worked with women who were recently released from prison and eager to gain new skills. With hands and machines, these crews dragged dozens of semitrailer truckloads from the site.

But the land was like a geologic formation of waste, strata full of trash layered like fossils in rock. *We went through a process of realizing that everything we were standing on was also dumped material*, Doria remembered. With the machines, they scraped the ground surface, then walked behind and scooped up more debris by hand. *We would literally just clear a few feet at a time, trying to get all the stuff out.*

The Berkeley soil science students came to the farm to run a more detailed search for contamination. They drew a grid over the entire lot; then Urban Tilth's summer apprentices retrieved soil samples from each square of the grid. The soil was so hard that the kids had to use pickaxes in some places to extract the earth. The results came back from two different labs: there were dregs and residues of asphalt and one spot of lead, all in small amounts. The ground was also nearly devoid of life or nutrients—it was mostly crushed rock and clay. But the situation could be handled as before—with compost and earthworms and bugs.

We need to put back a ton of organic materials, Doria thought. *We just need all kinds of rot to be happening throughout the site.* Truckloads of manure arrived on the land, forty cubic yards of compost every week, donated by the same company that was contracted to handle city organic waste, along with straw—spread and raked and spread again. Over and over for months, dumping manure, turning manure, planting cover crops of buckwheat and daikon radish. Turning the soil again. Rot and renewal.

In the fall of 2016, while this process was still underway, Doria and the Tilthers decided to plant what they could. At first, they had to bring in soil and more compost to create a series of mounded rows, arranged in a circle, where they grew mustard greens, collard greens, garlic, Swiss chard, cabbage, broccoli, and cilantro.

The next year they plotted an orchard on the northwest edge and sectioned it off, ten-foot-by-ten-foot squares, each a home for a tree. About 350 volunteers showed up and laid down more layers of manure, straw, and compost. And in each square, they placed a bare-root whip, a little stick that would become an apricot, peach, persimmon, pear, or apple tree. Over the next year, the vast majority of the trees survived. Within a couple years, they were bearing fruit that was sent to members of Urban Tilth's farm share program.

About a year later, the Tilthers could finally begin planting in crop rows, in soil they were building there on the farm.

The first greenhouses went up.

It was the un-wasting of land, the reimagining of place.

It was also an act of rebellion. No one would tell the Tilthers what was and was not possible on an old, broken bit of ground.

❧

The next couple of years brought a series of troubling events. In 2017, officially, the seven-year California drought ended, which should have been a good sign, but stretches of dangerously arid, scorching conditions recurred—and scientists predicted such droughts would become ever worse, ever more intense, ever longer in duration as the state headed further into the twenty-first century. Around Labor Day that year, the San Francisco Bay Area clocked its highest-ever temperature record, 106 degrees, as a heat wave rolled into the region. The following year, heat waves seared Southern California. The heat was most dangerous, always, in places with too much concrete, too few trees, too many diesel fumes, and too much poverty, as in the Iron Triangle and in parts of Los Angeles.

In May 2018, the city of Richmond reached a settlement with Chevron—the company would pay $5 million in damages for the 2012 fire. Many locals said it was not enough. By comparison, Chevron's CEO, Michael Wirth, earned a $21 million salary that year. Meanwhile, the Richmond refinery continued to send up flares—on New Year's Eve 2017, in February 2018, in April after a pair of steam boilers malfunctioned, in June as a result of an "upset" in a process unit, said a Chevron spokesperson. The company insisted the flares were routine. Local activists said they were pollution—"episodic exposures . . . that may cause lung disease, cancer and other health problems," read a report by a scientist from Communities for a Better Environment.

In October 2018, the Intergovernmental Panel on Climate Change warned that the world's carbon emissions would need to start declining sharply within the next dozen years or sooner to avoid crossing troubling thresholds. The crisis was already here, the scientists acknowledged, and some of its consequences were unavoidable. The global average temperature would keep warming for at least twenty to thirty years, and the seas were now inevitably going to rise for centuries to come (as Andrea Dutton, with her colorful scarf, had told people in St. Augustine). But every impact could become far worse if these emissions weren't curbed quickly.

A pair of earth scientists announced that parts of the San Francisco Bay were sinking as the drought led people to pump out more groundwater, the land dropping down and the water rising up. That summer, the Richmond city council voted to declare "a Climate Emergency that threatens our city, region, state, nation, civilization, humanity, and the natural world."

In North Richmond, a team of architects studied flood control. An old stormwater pump station was failing, and the designers, the county, and a local watershed organization wanted to build a levee that would also function as a marsh, a type of strategy called green engineering, to control flooding now and in the future. But how well such projects would work is always a function of how much the rest of the world might do to reduce carbon emissions.

And 2018 was the first moment that I heard Doria Robinson's optimism waver.

"I'm not gonna kill myself tomorrow, you know. I have two kids in the world," she blurted in a phone call that summer.

She said she worried for her kids. Her daughter had a running cynical joke: *When I'm your age, I'm going to be living in a refugee camp.*

"I know we can't prevent the damage that we've already set up," she said more slowly, reaching for each word. "But I feel committed to just doing everything in my power to set up as many things as possible to make the impact less painful."

There were larger forces at work in the world that Doria had no immediate control over.

California has some of the strongest climate-change regulations in the nation. But from the beginning, they contained loopholes for the oil industry, including Chevron. The people who had the most power to deal with the problem of climate change—people with money, people with political clout—rarely spoke to those struggling in a place like Richmond, even though a community like this would always bear the brunt of failures.

Economists are often most interested in a set of high-level mechanisms—trying to push carbon emissions downward through taxes or fees or tradable permits, creating a market in which it costs money to pollute, a disincentive. The European Union has a carbon market. The eleven U.S. states that belong to the Regional Greenhouse Gas Initiative on the East Coast use market-based regulations to force emissions cuts from power plants. Such policies can produce results, but some experts say they could also fail in the long term if they don't target the oil and gas industry directly. They won't work if they leave places like Richmond stranded or allow fossil fuel companies to delay the inevitable—a transition to other forms of energy that won't damage the atmosphere.

Moreover, a refinery or a coal plant exists in a place. And as such a facility reduces its carbon output, it also generally gasps out fewer toxic chemicals, which spares the lungs and throats of people living nearby. Many environmental justice activists feel that if the world is going to clean up fossil fuels and usher in a shiny green economy, the promise of this should come to the people who have suffered all of the years of fires and asthma and fear. This should not be a world where only the rich can breathe.

One of California's cornerstone climate laws, launched in 2013, established a cap-and-trade program that relies on tradable permits to emit carbon. The permits create a market—you can buy the right to pollute. Year after year, the number of permits decreases, the cap tightens, and carbon emissions are therefore supposed to decline. The state combines this program with other policies that promote renewable energy, clean cars, and green buildings, and by some measures, these efforts have helped. The state's emissions have declined, and California beat its 2020 goals for cutting carbon. But many experts say the Golden State will have more trouble trimming its emissions in the next decade: the state needs to invest more in clean cars, say some, and it needs to get tougher on refineries, say others.

According to an investigation by ProPublica, oil industry emissions actually increased in California after the cap-and-trade program began, and companies had even banked some of their permits so that they could pollute more in later years when the restrictions became tougher. In 2017, when the law was reapproved, the state gave away millions of free carbon permits to the oil industry, more than the companies needed. Environmental justice groups like Communities for a Better Environment tried to pressure the state to place a cap on refinery emissions at existing pollution levels, so that at least the problem couldn't worsen. But the state refused. It appeared that the free riders were continuing to mooch off the public.

In the fall of 2018, the governor of California held a conference in San Francisco with representatives from all over the world and from the United Nations, a West Coast climate-change meeting at a time when Washington, DC, was denying not only the urgency but the existence of the crisis. Michael Bloomberg, the billionaire erstwhile mayor of New York City, was interrupted during his keynote address by a group of a protesters shouting, *Mother Earth is not for sale!* He scoffed at them, *Only in America could you have environmentalists protesting an environmental conference.* But the activists didn't trust billionaires like Bloomberg and were afraid of more of the same, more free rides for industry, more suffering for communities that were already bearing the highest burdens.

Outside on the streets, it wasn't a joke. That Saturday, thirty thousand people marched through San Francisco—banging drums, playing instruments, cheering, chanting, carrying signs and banners. They had come from everywhere, including people from other communities at the edges of oil and gas extraction or production, traveling long distances—such as from the Ecuadoran Amazon, Canada, and Nigeria. They also held a nearly weeklong series of events, including a summit where they traded ideas and shared expertise, played music, and made art. Doria and some of her staff at Urban Tilth and their friends and allies had spent months organizing their own response. Two hundred people visited the North Richmond Farm.

Afterward, the Tilthers felt somehow larger and more powerful than they had before. *It definitely made a huge impact on everyone,* Doria said. *We needed to make it clear that you can be unafraid; you can have a voice.*

෨

I had made plans to visit the Bay Area in April 2020. But in January and February the world went askew, and by March, cases of Covid were exploding across the country. I abruptly canceled my trip and stayed in my house in Seattle. *We are scrambling to reorganize so we can keep everyone employed through this crisis*, Doria wrote me in mid-March. She had sent some staff home on paid leave, even though it strained the organization's budget, and let a few stay on the farm, as long as they kept away from each other. On March 17, the Bay Area began a shelter-in-place order. But food enterprises, including Urban Tilth, were allowed to keep operating.

As with most other catastrophes—from a refinery explosion to a worldwide climate crisis—there is nothing good that can be said about the worst global public health disaster in a century. There is no platitude you can offer to turn the senselessness of the millions of lives lost into a lesson in grace. But when tragedy reshapes the world, the upheaval can sometimes throw certain details into relief. Like turning up soil with a shovel, you can see things that were buried before.

At the same time the pandemic began, a staff member named Rudy—hired just out of high school five years previously—passed away suddenly, not of Covid but from other causes.

His friends and colleagues at the farms and gardens reeled with grief and were overwhelmed, but they put themselves into the work of growing. In this particular crisis and city, the demand for food—especially fresh, affordable food—soared. To Doria it felt like a scramble for survival. *If we hadn't pushed so hard, we'd have just been out, like Covid would've just knocked us out. Because we insisted that we were going to make an impact, that's why we had enough infrastructure*, Doria said. She worked almost constantly. Adam Boisvert, who had joined Urban Tilth full-time the year before, started working at the North Richmond Farm almost daily—in between teaching an urban agriculture class at Richmond High. The Tilthers mounded up and planted twenty-five new crop rows in two months—a project that was originally supposed to take them two more years. (At the ends of some of the rows, they placed empty bottles of kombucha. Rudy used to drink kombucha prodigiously and absentmindedly leave the empties in the field, so the bottles here served as a memorial.)

Big green crates of food from Urban Tilth's community-supported agriculture program circulated through Richmond—citrus, greens, broccoli, tomatoes, whatever could be plucked from the vine or the ground at that moment in the season. Some of the produce came from the Tilth farms and some from a group of other farms in the region that had formed partnerships with the organization. At the beginning of April, about 80 families were ordering a weekly box of fruits and vegetables from Urban Tilth on a sliding scale from $10 up (with a larger amount subsidizing boxes for those in need); by the end of the month, it was about 250 families. They gave away 200 more to people who couldn't afford to pay. By the end of 2020, Urban Tilth had distributed more than sixty thousand pounds of produce.

At the same time, the human world quieted. Fewer cars traversed the labyrinth of highways that crisscrossed California. The air above the San Francisco Bay became measurably clearer. And in late April the price of crude oil plummeted until it was less than worthless, valued at negative $37 a barrel.

Some things rise in an emergency, and others fall.

By the summer, investigative journalist Antonia Juhasz was projecting the end of oil. "Demand has cratered, prices have collapsed, and profits are shrinking. The oil majors (giant global corporations including BP, Chevron, and Shell) are taking billions of dollars in losses while cutting tens of thousands of jobs," she wrote in a sweeping analysis in *Sierra* magazine. "It is clear that the oil industry will not recover from Covid-19 and return to its former self. What form it ultimately takes, or whether it will even survive, is now very much an open question."*

Years ago, Richmond's environmental justice activists used to say they merely wanted Chevron to clean up. But in the summer of 2020, the pandemic gave them permission to express radical desires.

---

* The price of oil, of course, rebounded in 2021 and soared in 2022. "But that doesn't mean that everything went back to where it was before," Juhasz told me in early 2022. "There is a very urgent push in a growing number of communities to break free of fossil fuels. You have to put in place the policies that allow you to shift." Moreover, the fossil fuel industry remains inherently volatile, and the movement to divest continues to grow.

Shut The Muthas Down! Andres Soto, longtime organizer with Communities for a Better Environment and radio host in Richmond, wrote on Facebook in July. His organization had released a report, authored by a longtime California oil industry researcher, called *Decommissioning California Refineries*, a 124-page treatise including diagrams and illustrations and discussions of how a refinery operates. "Machines that burn oil are going away," the report read. Refineries emitted more carbon than any other single sector in the entire state. There was simply no way for California to do anything useful about climate change if it didn't begin to shut down some of its oil refineries. "Early action to decommission refining capacity is a critical component of the least-impact, most socially just, most *feasible* paths to climate stabilization in California. . . . When should the decommissioning start? Right away."

In all the years that the world had been talking about climate change, few people had dared to ask: What would it in fact mean to entirely shut down fossil fuels, not in the abstract, but here, now, at home?

In late July, the oil and gas company Marathon abruptly announced it was closing an oil refinery in Martinez, a city just thirteen miles northeast of Richmond, and a second in New Mexico, and laying off most of the workers at both sites. "We will indefinitely idle these facilities with no plans to restart normal operations," the company explained in an official statement. In August, Phillips 66 said its refinery in Rodeo, eight miles north of Richmond, would no longer process oil but biofuels.

The news shuddered through the region. Machines that burned and refined oil were going away. But the transition process would be full of upheaval and uncertainty.

꙳

The summer of 2020 was a season of exhilaration and terror and outrage.

The pandemic hit Richmond hard. Covid was worse in communities that breathed too much air pollution, where people worked frontline jobs, where multigenerational families lived together, and in communities of color. In Richmond, all of these factors merged. That summer,

Doria lost two aunts and an uncle to Covid. Others in Urban Tilth mourned family members. *There has been so much dying. It's hurting my soul*, Doria wrote on Facebook.

In June, Richmond organized its own protests to mark the death of George Floyd in Minneapolis—including a guerrilla-art mass demonstration that created a Black Lives Matter mural, painted in yellow on the street in front of the Civic Center. A week later, Najari, Doria, and a group of artists co-organized a second mural-painting on Macdonald Avenue, a main thoroughfare chosen partly for its symbolism: it had been the site of a Ku Klux Klan march in the 1920s. The event was also a call for police reform: Doria drafted a series of demands to the city, including a "change to a culture that values Black Lives" and a reallocation of funding so that policing would focus on stopping and solving violent crimes. She also handled food. Activists held banners on either end of a two-block stretch to shut down traffic. There were several dozen painters and about two hundred people in attendance overall. Gayle McLaughlin showed up to help, stooping over the pavement with her bicycle helmet on, a mask on her face, and a paint roller in one hand. At the end, the yellow paint on the street read: REPARATIONS NOW.

Late in the season, Doria and I felt we could still meet safely if we kept our distance and spoke outdoors. I made a plan to drive down the coast and visit the Urban Tilth farms in early September, but when the date arrived, everything had gone up in flames—wildfire on the interstate, wildfire on the Pacific Coast Highway, and a cloud of toxic particulates enshrouding the West. The smoke colored the sky above the San Francisco Bay burnt orange for an entire day—the hue of molten glass, smoldering embers, rusted iron. The smoke blotted out the sun and turned everything around the region as cold as January, woolsock weather.

Everything felt deadly, as if we'd been caught between walls made of smoke.

When I called Doria again, all we could do was laugh, the visceral laughter that comes when something is so intolerable that it must also be absurd. We said that we could try again the following month, barring a "zombie outbreak."

It was almost Halloween before I could make my way there.

I showed up on a Saturday in late October. It was a California autumn day that began with fog and birdsong and one of the first rainfalls of the season. A collection of just over a dozen people had come to help relocate a greenhouse. About half of them worked for Urban Tilth. I recognized Adam instantly, even behind his face mask, and he gave me an air fist bump, our knuckles not quite making contact. Of the volunteers, a pair of women lived nearby, one had driven all the way from San Francisco, and a special ed teacher had come from Oakland. A woman named Tania (no relation to Tania Pulido), a farm manager-in-training, balanced a cheerful baby wearing a fuzzy hoodie with gray bear-ears sewn on. She told me her father had helped drive one of the machines used to clear the site a few years ago.

The farm itself was unrecognizable as the place I had seen a few years previously. A double-wide construction trailer at the front served as an office space. Beside it was a colorful sign that explained Urban Tilth's plans for the place. A VISION FOR A HEALTHY NORTH RICHMOND! it read in cheerful green lettering. Behind it a giant rainwater cistern, a green lawn with a few picnic tables, and then a wide flat rectangular area where the rows of crops had been multiplying all year long. In the rows were sunflowers. Trellises heavy with tall but mostly spent heirloom tomatoes with deep purply red skin. Kale, chard, marigolds, mustards, everything lavishly green from the California weather. Roosters strutted in a coop along the eastern edge, singing cacophonously. (All of these birds were castoffs, strays that people had dumped on the property. Retired fighting cocks, maybe, or just nuisance birds. But Urban Tilth had taken them in.)

At the front of the farm, the Tilthers had put in a wash station—a shelter with pallets and water hoses where they could clean the dirt from the crops. They were now packing more than four hundred boxes of produce and giving away fruits and vegetables to more than a hundred other families at free farm stands every week.

A mix of pop, hip-hop, R&B, and some Janet Jackson thumped on a large boom box that looked like a power tool—bright yellow with rubberized bumpers at the edges—as half the crew set to unscrewing the supports that held up the greenhouse—a tentlike, half-cylindrical structure, with a metal spine and plastic covering—and the other half took a set of heavy-duty rakes to the new location to level the ground. As they worked with their hands, they discussed topics like the virtues of watermelon radishes and how to teach kids about science and photography. It was a barn raising of sorts.

Doria showed up later in the morning and rushed me over to the "chicken mansion"—a henhouse at the back, recently built, made of reclaimed redwood with a wide porch so that it looked more like an artist's cottage but occupied by actual hens. She slid the door open and greeted the birds—"Hi, ladies"—who clucked at her agreeably.

Behind the henhouse stood a semicircular bench made of cob—an earthen mix of clay and straw, with a large brown glass jug of kombucha set on the top of it like an ornament. Beside it, a few pots of flowering currant. They were putting in a small medicinal garden here, Doria said, a symbol of healing in a time of loss.

🐦

I spent a week taking in Richmond.

It felt like a different place than the one I had first visited a decade previously. Less emptiness, more occupation, more people at home in the open space of the city.

I borrowed a bicycle from Najari's shop, Rich City Rides, and pedaled down the Greenway path through downtown. I passed a long series of murals, many vivid with scenes of nature—a pair of Black women birdwatchers, a tree with the word HONOR painted above it, waves of water and sunflowers and the phrase WE CAN STOP CLIMATE CHAOS. Between all of these, unofficial street art and murals. Some were tags, others elaborately painted and expressive. One with a woman's face with tears on her cheeks. One read, ENJOY MY RICHMOND.

The Richmond Greenway—which had been largely bare on my first visit—was now a series of parks, one after another. A central plaza with a pavilion and picnic tables, a playground, a basketball court, some plantings of native vegetation, a few sculptures, an orchard where any passerby could pick. I noticed a few new apple blossoms breaking open.

Beyond this, a local group had built a dirt-bike park with berms mounded into tracks and jumps. I passed parents watching their giggling children spin bikes through the paths. I passed dog walkers and joggers. Rusted corrugated warehouses with metal grates on the windows. Weeds blooming along the railroad tracks. The smell of flowers. The smell of urine. The smell of diesel. I rode to the end of the Greenway trail and the roaring parkway, not far from the refinery gates, then beyond. Across the railroad tracks from Chevron was a mural with the word UNITY in a geometric design of purple, orange, and green.

A couple of miles beyond that, a solar company had opened a sixty-acre farm covered in photovoltaic panels on cleaned-up industrial land leased from the oil giant. The panels turned their glinting faces to the sky—they were almost blinding.

I rode on until I reached a weedy meadow full of tents and trailers occupied by the houseless. Then a street full of semitrailer trucks. I rode until the pavement was broken, and it was clear I had reached a place no longer intended for humans and bicycles.

On my last day, I had a long conversation with Doria on the back deck of her house. She had been working sixteen-hour days and seemed overwhelmed, as if she were trying singlehandedly to hold this place back from the edge of another disaster.

But that morning we carved out a space in her backyard, apart from the commotion. We sat at a table beside pots of basil and rosemary. Her two Australian shepherd dogs, one dark brown and named Happy, one sandy-red and named Mocha, tussled enthusiastically on the lawn. Behind us stood a trapeze that her daughter had used for aerial acrobatics and the weathered wooden frame of what used to be the twins' playhouse. Crows rasped behind us. A train whistle wailed.

She sat with her hands in her pockets, wearing a knit hat the color of buttercups, and told me of her dreams for her hometown. She and Najari and some other local activists had started another initiative in Richmond to support new worker-owned cooperatives. This effort included a small-business loan fund—financed partly by a national network that underwrites community-based business and partly with money from various other foundations. The loan fund was helping a series of small Richmond-based entrepreneurial efforts get off the ground, including a mobile food truck and catering business, a company devoted to DJing and music, a landscaping business, and a laundromat. And possibly a new grocery store in North Richmond—to be built by an affordable housing developer where Tania Pulido now worked. Eventually, the activists' goal was also to focus on businesses that would more directly help with the transition away from oil—such as community solar and electric vehicles.

In the backdrop of this effort, the potential for calamity still loomed. The North Richmond Farm now had its own air monitors, so that Urban Tilth could detect a spike in contaminants. Doria knew there would be another fire. There would be another setback. "That's the hazard of living here. We may have to cease operations, like we did before, for a long period of time, but this time, we're going to have data," she said. "We can hold them accountable."

When the sun broke through the fog, she rolled out a turquoise umbrella that covered a wide wooden table. The morning waned.

I asked Doria if she ever felt disheartened.

"I just know it's going to get worse before it gets better, and I'm trying to figure out how to survive." There was a voice of urgency in her mind, driving her ceaselessly on. "Like you better try to look further into the future and put things in place. The storm's gonna hit, girl. You better get your shit together," she said. "I just keep trying to figure out how to do it more sanely, you know, so I don't end up dying. Like, I don't want to get cancer or diabetes or all these other things like stress and dysfunction.

"We try, we try, we try. And there's definitely things that are better than not having anything, but it just feels like, I don't know, spitting on a wildfire.

"And it's not true, because there's a lot of impact. It's just the scale that we need is just so much bigger."

It was always too much. It was never enough.

Like always, the city stood at the edge of promise and peril.

That same week, an eight-thousand-gallon tanker truck caught fire on the interstate in Richmond, the highway closed, and part of the city was told to shelter in place.

On November 2, 2020, the day before the election, the refinery flared because of a "power disruption" to part of the facility. Chevron issued a "level one" warning—claiming there would be no health impacts.

I wondered again, could you really dig into a place like Richmond? Could you really make a home on this tainted patch of earth and asphalt, in this moment of possibility and impossibility?

This is not separate from the question of surviving climate change.

ॐ

It is easy to feel hopeless in the face of climate change, in the shadow of corporations with incomprehensibly large amounts of money and influence, in a moment of dizzying political instability and multiple overlapping crises.

But a time of unruliness is also a moment of rearrangement—there can be sudden shifts in who holds sway and what people value.

During 2021, community activists in Richmond and some members of the city council (which gained a progressive, anti-Chevron majority in the 2020 election) continued to discuss and study what it might mean to close down the century-old oil refinery, which has dominated this place from the very beginning. Communities for a Better Environment organized a crowded virtual public meeting to discuss what a *healthy breakup* between Chevron and Richmond might look like. The online chat window filled with comments from frustrated residents decrying all the ways that Chevron was *making us feel like our illnesses are our fault; corrupt; toxic, a dance with death; paternalistic.* The participants also feared what would happen when the company departed: *when they're done with Richmond, they leave us with a toxic waste dump,* wrote one person.

But people in Richmond were also starting to realize that they, too, have influence. For one thing, the city has some authority over land and over what can be permitted within its boundaries. "The powerful have never been able to take land use planning away from local people," says Greg Karras, author of the 2020 refinery decommissioning report. That means Richmond holds some authority over Chevron and the land that the refinery sits on. It means the people have a choice about what their future looks like.

It will still take time and imagination to figure out what this city can become.

I have taken one main lesson from observing the struggles of Richmond and other communities like it: we think of power as belonging only to a select few, those who rule the world and those who own most of the wealth. But there is a kind of power that grows from the ground around you. Power can come from community. Power can come from home—from knowing that we belong to a place and a planet, and it is our collective job to grow something useful here and to create space for the generations that come after us.

Enormous, radical solutions are necessary to remedy the climate crisis: big policies, big ideas, big economic shifts. So far, the rulers and the billionaires haven't been leading us toward anything that will realistically keep us from catastrophe. So far, many of the most ambitious strides have come from the grass roots, from the marchers and protesters, the small-town mayors, the artists and teachers and musicians, the firefighters and farmers and scientists, the people who have their hands in the dirt, the people who are able to reimagine how to live. Here, in the smallest places, there are big transformations that could reverberate and ripple around the world, especially if we decided as a society to nourish such efforts, to lend our own energy to them, to dig in.

In September 2021, Urban Tilth bought the North Richmond Farm from the county. Doria had committed herself to this place long ago, but the purchase made it feel even deeper and more permanent. "There's something about saying, 'We're not going to just move,'" she told me over the phone. "We're not going to be able to screw this up, just move somewhere else and then leave our trash behind us." But the land purchase was about more than just taking responsibility. Again, she and the Tilthers were sowing seeds, literally and metaphorically. "I am unsure of the exact destination of where this farm is going to take us. But I have faith that we're seeding the kind of power our community needs to take control of our lives and to have a say in what happens in the world around us, in the environment, in the air and in the water, in our schools, in our government. I feel like these are the fertile conditions we need to grow something for ourselves."

She and her fellow farmers would keep replenishing the soil, keep putting down deep roots, keep holding tightly to home.

EPILOGUE

I wrote most of these words at home, settled at my kitchen table or on my porch or at my desk in the back room. I wrote them through a pandemic that confined me, like many people, to a smaller-scale life. For months, I barely left my neighborhood, tracing the same sidewalks, the same footpaths, day after day.

But paradoxically, as my life felt more constrained, the boundary between the global and the personal seemed even more porous than before. The world was inexorably connected by a virus. The world was quarrelsome and factional and unstable but sometimes shaken by shared angers or anxieties. The fiery racial justice protests that broke out in Minneapolis also reverberated into my city, my neighborhood. My neighbors marched down my street with their small children and chanted "Black Lives Matter" in front of a cluster of aging bungalows, at the edge of an industrial river, in view of a shipping port that sends cargo between my city and other continents. The emergencies that appeared on national news were personal. Friends fled from hurricanes and wildfires, driving hours along highways that became corridors of flame or spending entire days stranded in a line of cars leading away from the coast. A tropical storm poured sheets of water onto the houses of my relatives.

All of these disruptions blurred in my mind, especially at night. I dreamed about escaping floods. I dreamed about running from fires. I dreamed that I was walking through a dry valley while big-eyed animals stared at me. I dreamed about a fleet of fuming diesel trucks charging toward me through a narrow tunnel.

On one cluster of days in late June 2021, my city and my region of mountains and water groaned with extraordinary heat. In many places, the heat surpassed anything written down in more than a century of weather records: 108 degrees Fahrenheit in Seattle, 116 in Portland, 118 near Forks, sandwiched between the coast and the rain forest on the mossy Olympic Peninsula. The days were thirty to forty degrees above normal, as if someone had invented an entirely new season for the region. News reports described the temperatures with words like *shattered* and *broken*.

What I sensed, though, was not a shattering of reality but its suspension. Like a science fiction plot in which you are transported into an alternate world—one that looks like your own but something is off-kilter or amiss. Our house had no air-conditioning apart from a few ragged-looking ornamental plum trees that kept some of the sunlight off the roof. During the peak heat of the hottest day, my body felt leaden. I was unable to form full thoughts. My cat sprawled across the floor with an elongated spine and cast pitiful looks in my direction. I occupied myself with the constant readjustment of the three electric fans we owned, trying to channel air through the house efficiently enough to keep myself from withering. Finally, that afternoon, my husband and I drove to a reliably cold urban lake and jumped in. This wasn't merely recreation but a survival strategy. The human body can withstand some amount of scorching heat if it has time to cool, especially at night. But in this moment, water offered more relief than nighttime. At dusk, the sky hazed over with a blistering mustard-colored sunset, and the air remained in the 90s Fahrenheit until well after dark.

Just north of Seattle, transportation crews shut down multiple lanes of the interstate after they warped and buckled in the heat. At the shores of the estuary that curves between Seattle and Canada, mussels, oysters, clams, and snails clung to the rocks and died in the sun. Kelp and surf-grass bleached to white. In the days thereafter, the beaches smelled like rot. The estimated human death toll was more than 1,200 in Washington, Oregon, and British Columbia combined. We were all at risk, all inescapably bound to one another, mussel to rock to beach, human to city to water, ocean to atmosphere.

☙

By now, I hope you are reading this book from a place of safety, but I know that may not be the case. The world feels perilous. There is uncertainty in the air, at our feet, in the water, in the wind pressing against the windows.

I have told you that we will need to learn how to live in an era of storms and disruption.

I have tried to offer you stories that are like fables. Platitudes are cheap, but fables offer lessons drawn from experience and metaphor. What I found in these stories were lessons about home.

Home is powerful because it is both an intimate place and also a place of connection. To have safe homes in the twenty-first century, we cannot keep acting as if we are isolated individuals. We are not just consumers. We are not just a collection of independent bodies with separate carbon footprints. There are no sharp boundaries between our lives and the lives of others. There are no clear borders between the safety of our households and our bodies and the health of the land and the ecosystems around us.

We are building and rebuilding the world every day, even when we are not acting deliberately. We are all reworking the next acts of these stories right now, collectively.

And while many of the problems we face are global, some of the most imaginative, powerful, passionate solutions come from home. Home is a place we can act. Home is a place we can take care of. Home is a feeling that can inspire us. Home is a way for us to rethink and reimagine and remake our lives. Home asks us to adjust ourselves, to rewrite ourselves, to reconsider who we are, again and again, each time we occupy a new space or refashion an old one. We are all building these walls and roofs and lives together, on this one messy and unruly blue planet.

NOTES

The four narratives that form the backbone of this book—Okanogan County, Washington; St. Augustine, Florida, and Annapolis, Maryland; Newtok, Alaska; and Richmond, California—are based on a combination of field reporting and extensive interviews in these communities. The four essayistic chapters also rely heavily on in-person and phone interviews. Quotes, biographical information, and details of events that are not annotated in this section are drawn from personal communications. When quotes appear in reconstructed events, they are italicized—to signal that they are drawn from the past and from the memories of the people who were involved, which, like all memories, are sometimes inexact.

Front Epigraphs

vii **I want you to act as you would:** Greta Thunberg, "'Our House Is on Fire': Greta Thunberg, 16, Urges Leaders to Act on Climate," *Guardian*, January 25, 2019, https://www.theguardian .com/environment/2019/jan/25/our-house-is-on-fire-greta-thunberg16-urges-leaders-to-act -on-climate.

vii **Because we have not made our lives to fit:** Wendell Berry, ["Because we have not made our lives to fit"] from *This Day: Collected and New Sabbath Poems 1979–2012.* Copyright © 1999 by Wendell Berry. Reprinted with the permission of The Permissions Company, LLC on behalf of Counterpoint Press, counterpointpress.com.

vii **But the ethereal and timeless power:** Barry Lopez, *Arctic Dreams* (New York: Bantam Books, 1986), 411.

Prologue

2 **Home is "not a house for sale or a site for 'development'":** Wendell Berry, *The Art of Loading Brush: New Agrarian Writings* (Berkeley, CA: Counterpoint, 2017), 6.

2 **Not long ago, a group of high school science students collected tree moss in the valley:** Bellamy Pailthorp, "In South Seattle, Teens Collect Moss to Help ID Air Quality," KNKX Public Radio, June 29, 2020, https://www.knkx.org/news/2020-06-29/in-south-seattle-teens-collect -moss-to-help-id-air-quality; "Duwamish Valley Clean Air Program Moss Study Community Fact Sheet," Duwamish River Cleanup Coalition, accessed October 18, 2021, https://www .duwamishcleanup.org/moss-study.

4 **In the summer of 2021, one in three Americans:** Sarah Kaplan and Andrew Ba Tran, "Nearly

in 3 Americans Experienced a Weather Disaster This Summer," *Washington Post*, September 4, 2021, https://www.washingtonpost.com/climate-environment/2021/09/04/climate-disaster-hurricane-ida/.

4 **Elsewhere, the Italian island of Sicily:** Gaia Pianigiani, "Sicily Registers Record-High Temperature as Heat Wave Sweeps Italian Island," *New York Times*, August 12, 2021, https://www.nytimes.com/2021/08/12/world/europe/sicily-record-high-temperature-119-degrees.html.

4 **Villagers on the Greek island of Evia:** Lefteris Papadimas, "Greek Villagers Try to Save Homes as Fire Crews Brace for Winds Whipping Flames," Reuters, August 10, 2021, https://www.reuters.com/world/europe/greek-villagers-try-save-homes-fire-crews-brace-winds-whipping-flames-2021-08-10/.

4 **At the same time, a vicious drought hung over Angola:** Kaula Nhongo, "Namibia Under Pressure from Angolan Migrants Fleeing Drought," Bloomberg, October 7, 2021, https://www.bloomberg.com/news/articles/2021-10-07/namibia-under-pressure-from-angolan-migrants-fleeing-drought.

4 **The warming of the planet:** William V. Sweet, Radley Horton, Robert E. Kopp, Allegra N. LeGrande, and Anastasia Romanou, "Sea Level Rise," in *Climate Science Special Report: Fourth National Climate Assessment*, vol. 1, ed. Donald J. Wuebbles, David W. Fahey, Kathy A. Hibbard, David J. Dokken, Brooke C. Stewart, and Thomas K. Maycock, U.S. Global Change Research Program, accessed October 19, 2021, https://science2017.globalchange.gov/chapter/12/.

4 **Because of the emissions human societies have already sent:** "Summary for Policymakers," in *Climate Change 2021: The Physical Science Basis; Working Group I Contribution to the Sixth Assessment Report of the Intergovernmental Panel on Climate Change*, ed. Valérie Masson-Delmotte, Panmao Zhai, Anna Pirani, Sarah L. Connors, Clotilde Péan, Yang Chen, Leah Goldfarb, et al.

5 **The word *ecology* originates from:** ecology, n., *Oxford English Dictionary Online*, Oxford University Press, September 2021.

5 **The word *economy* grows from the same root:** economy, n., *OED Online*.

5 **But an economy should be a means:** "The Meaning of Home," Movement Generation Justice and Ecology Project, accessed October 19, 2021, https://movementgeneration.org/eco-means-home/.

6 ***Unruly* has several meanings:** unruly, n.4., *OED Online*.

7 **Many cultures recount stories in which heroes:** Christopher Booker, *The Seven Basic Plots: Why We Tell Stories* (New York: Bloomsbury, 2005).

1: The Fire

Some of the reporting in chapters 1 and 7 also appeared in "For Forest Blazes Grown Wilder, an Alternative: The 'Good Fire,'" my October 2021 article for *Undark* magazine.

11 **home, n.:** home, n.A.I. and home, n.A.I.1.a, *OED Online*.

11 **home truth, n.:** home truth, n., *Merriam-Webster*, accessed September 6, 2021, https://www.merriam-webster.com/dictionary/home%20truth.

12 **the American public had first become aware of climate change:** Philip Shabecoff, "Global Warming Has Begun, Expert Tells Senate," *New York Times*, June 24, 1988, https://www.nytimes.com/1988/06/24/us/global-warming-has-begun-expert-tells-senate.html; Andrew C. Revkin, "Years Later, Climatologist Renews His Call for Action," *New York Times*, June 23, 2008, https://www.nytimes.com/2008/06/23/science/earth/23climate.html.

12 **Scientists had predicted a possible crisis:** Benjamin Franta, "On Its 100th Birthday in 1959, Edward Teller Warned the Oil Industry About Global Warming," *Guardian*, January 1, 2018, https://www.theguardian.com/environment/climate-consensus-97-per-cent/2018/jan/01/on-its-hundredth-birthday-in-1959-edward-teller-warned-the-oil-industry-about-global-warming; Roger Revelle and Hans E. Suess, "Carbon Dioxide Exchange Between Atmosphere and Ocean and the Question of an Increase of Atmospheric CO_2 During the Past Decades," *Tellus* 9, no. 1 (1957): 18–27, https://doi.org/10.3402/tellusa.v9i1.9075.

12 **By 1992, the concentration of carbon dioxide:** "Trends in Atmospheric Concentrations

of CO_2 (Ppm), CH_4 (Ppb) and N_2O (Ppb), Between 1800 and 2017," European Environment Agency, accessed October 19, 2021, https://www.eea.europa.eu/data-and-maps/daviz/atmospheric-concentration-of-carbon-dioxide-5.

12 **Swedish scientist Svante Arrhenius is often credited:** Ian Sample, "The Father of Climate Change," *Guardian*, June 30, 2005, https://www.theguardian.com/environment/2005/jun/30/climatechange.climatechangeenvironment2; Ayana Elizabeth Johnson and Katharine K. Wilkinson, eds., *All We Can Save: Truth, Courage, and Solutions for the Climate Crisis* (New York: Random House, 2021), xvii.

12 **The planet had already warmed:** "Global Temperature," NASA, accessed October 19, 2021, https://climate.nasa.gov/vital-signs/global-temperature.

12 **though some places, like the Arctic:** J. Overpeck, K. Hughen, D. Hardy, R. Bradley, R. Case, M. Douglas, B. Finney, et al., "Arctic Environmental Change of the Last Four Centuries," *Science* 278, no. 5341 (November 14, 1997): 1251–56, https://doi.org/10.1126/science.278.5341.1251.

13 **In 1990 and 1992, the first reports:** Intergovernmental Panel on Climate Change, *Climate Change: The IPCC 1990 and 1992 Assessments*, June 1992, https://www.ipcc.ch/report/climate-change-the-ipcc-1990-and-1992-assessments/.

13 **Through her lake-sediment detective work, Susan's doctoral dissertation:** Susan J. Prichard, "Spatial and Temporal Dynamics of Fire and Vegetation Change in Thunder Creek Watershed, North Cascades National Park, Washington" (Ph.D. diss., University of Washington, 2003), https://digital.lib.washington.edu:443/researchworks/handle/1773/5601.

13 **But as climate change grew:** Scientific studies and white papers written in the 1990s predicted that climate change would increase both the frequency of fires and the area burned. Here are a few: Margaret S. Torn and Jeremy S. Fried, "Predicting the Impacts of Global Warming on Wildland Fire," *Climatic Change* 21, no. 3 (July 1, 1992): 257–74; M. D. Flannigan and C. E. Van Wagner, "Climate Change and Wildfire in Canada," *Canadian Journal of Forest Research* 21, no. 1 (January 1, 1991): 66–72; Margaret S. Torn, Evan Mills, and Jeremy S. Fried, "Will Climate Change Spark More Wildfire Damages?" LBNL Report No. 42592, Lawrence Berkeley National Laboratory, 1998.

14 **Jerry Williams, the former national director:** Michael Kodas, *Megafire: The Race to Extinguish a Deadly Epidemic of Flame* (Boston: Houghton Mifflin Harcourt, 2017); "The Mega-Fire Phenomenon: Toward a More Effective Management Model," Brookings Institution, September 15, 2005.

15 **But some had been deadly:** U.S. Forest Service, *Thirtymile Fire Investigation: Accident Investigation Factual Report and Management Evaluation Report*, October 16, 2001, https://www.fs.fed.us/t-d/lessons/documents/Thirtymile_Reports/Thirtymile-Final-Report-2.pdf; David Bowermaster, Maureen O'Hagan, and Warren Cornwall, "Thirty Mile Crew Boss Charged in 4 Fire Deaths," *Seattle Times*, December 21, 2006, https://www.seattletimes.com/seattle-news/thirty-mile-crew-boss-charged-in-4-fire-deaths/.

15 **In 2006, a heat wave radiated:** Alexander Gershunov, Daniel R. Cayan, and Sam F. Iacobellis, "The Great 2006 Heat Wave over California and Nevada: Signal of an Increasing Trend," *Journal of Climate* 22, no. 23 (December 1, 2009): 6181–203, https://doi.org/10.1175/2009JCLI2465.1.

15 **In late July in the Methow:** Susan J. Prichard and Maureen C. Kennedy, "Fuel Treatments and Landform Modify Landscape Patterns of Burn Severity in an Extreme Fire Event," *Ecological Applications* 24, no. 3 (April 1, 2014): 571–90, https://esajournals.onlinelibrary.wiley.com/doi/10.1890/13-0343.1.

15 **The smoke plume:** Susan J. Prichard, "Learning to Live with Wildfires: How Communities Can Become 'Fire-Adapted,'" *The Conversation*, July 6, 2016, http://theconversation.com/learning-to-live-with-wildfires-how-communities-can-become-fire-adapted-59508.

15 **already weakened by infestations of beetles:** Prichard and Kennedy, "Fuel Treatments and Landform."

16 **eventually grew to more than 175,000 acres:** Hal Bernton, "Forest Was Easy Prey for Raging

Tripod Fire," *Seattle Times*, September 24, 2006, https://www.seattletimes.com/seattle-news /forest-was-easy-prey-for-raging-tripod-fire/.

16 **and bigger in land area than the 150,000-acre Camp Fire:** California Department of Forestry and Fire Protection, "CAL FIRE Investigators Determine Cause of the Camp Fire," May 15, 2019, https://www.fire.ca.gov/media/5121/campfire_cause.pdf.

16 **In October, snowfall finally quenched:** Susan J. Prichard and David L. Peterson, *Landscape Analysis of Fuel Treatment Longevity and Effectiveness in the 2006 Tripod Complex Fires: Final Report to the Joint Fire Science Program*, 2013, https://www.firescience.gov/projects/09-1-01-19 /project/09-1-01-19_final_report.pdf.

17 **This region would be a refuge:** Cliff Mass, "Will the Pacific Northwest Be a Climate Refuge Under Global Warming?" *Cliff Mass Weather Blog*, July 28, 2014, https://cliffmass.blogspot.com /2014/07/will-pacific-northwest-be-climate.html; Stacy Vynne, Steve Adams, Roger Hamilton, and Bob Doppelt, *Building Climate Resiliency in the Lower Willamette Region of Western Oregon: A Report on Stakeholder Findings and Recommendations*, The Resource Innovation Group's Climate Leadership Initiative, January 2011; Hannah Hickey, "UW Study: Will Puget Sound's Population Spike Under Climate Change?" UW News, October 20, 2015, https://www.washington .edu/news/2015/10/20/uw-study-will-puget-sounds-population-spike-under-climate-change/.

17 **the city's famously snarky alternative newspaper, named it:** Bethany Jean Clement, "HOT-POCALYPSE 2014: Seattle Heatwave Survival Guide," *Stranger*, July 11, 2014, http://slog .thestranger.com/slog/archives/2014/07/11/hotpocalypse-2014-seattle-heatwave-survival-guide.

18 **That year, the *Seattle Times* posted a series of images:** Benjamin Woodard, "Check Out These Spectacular Photos of Seattle's Smoky Sunsets," July 9, 2015, https://www.seattletimes .com/seattle-news/check-out-these-spectacular-photos-of-seattles-smoky-sunsets/.

18 **"The fires' impact—the claustrophobia":** Lindy West, "We're Choking on Smoke in Seattle," *New York Times*, August 9, 2017, https://www.nytimes.com/2017/08/09/opinion/smoke-heat -seattle-climate.html.

18 **Rivers of smoke gushed:** Asia Fields and Michelle Baruchman, "Seattle Pollution Levels Surge, as Smoky Air Returns Through at Least Wednesday," *Seattle Times*, August 19, 2018, https:// www.seattletimes.com/seattle-news/environment/smoky-air-returns-to-seattle-through-at-least -wednesday/.

18 **Seattle air quality was suddenly:** Kipp Robertson, "Seattle's Air Quality Worse Than India, China," KING5, August 15, 2018, https://www.king5.com/article/weather/weather-blog /seattles-air-quality-worse-than-india-china/281-584411100.

18 **"birthplace of smokejumping":** "History," North Cascades Smokejumper Base, accessed October 19, 2021, https://www.northcascadessmokejumperbase.com/history/.

19 **A fire called the Williams Flats was burning on the Colville Reservation:** "Williams Flats Fire on Colville Reservation Still Burning, But Fully Contained," KREM, August 25, 2019, https://www.krem.com/article/news/local/wildfire/lightning-may-have-sparked-fire-burning -on-colville-indian-reservation/293-39d02b18-2e01-4666-a5c4-4767f68e93c3.

19 **there had been seven thousand lightning strikes in Washington and Oregon:** Dee Camp, "Thunderstorms Touch Off Several Fires," *Omak-Okanogan County Chronicle*, https://www .omakchronicle.com/free/thunderstorms-touch-off-several-fires/article_f388cbf2-bdf3-11e9-974c -9bc5f6cccf37.html; Bill Gabbert, "More Than 7,000 Lightning Strikes in Washington and Oregon Saturday," Wildfire Today, August 11, 2019, https://wildfiretoday.com/2019/08/11/more-than -7000-lightning-strikes-in-washington-and-oregon-saturday/.

20 **The history of smokejumping was closely entwined:** Jordan Fisher Smith, *Engineering Eden: The True Story of a Violent Death, a Trial, and the Fight over Controlling Nature* (New York: Crown, 2016).

20 **the only all-Black airborne unit:** Linton Weeks, "How Black Smokejumpers Helped Save the American West," *NPR History Dept.* (blog), National Public Radio, January 22, 2015, https:// www.npr.org/sections/npr-history-dept/2015/01/22/376973981/how-black-smokejumpers -helped-save-the-american-west.

20 **By 1935, the Forest Service's official policy:** Diane M. Smith, "Sustainability and Wildland Fire: The Origins of Forest Service Wildland Fire Research," U.S. Forest Service, FS-1085, May 2017; Susan J. Husari and Kevin S. McKelvey, "Fire-Management Policies and Programs," in *Sierra Nevada Ecosystem Project: Final Report to Congress, vol. II, Assessments and Scientific Basis for Management Options* (Davis: University of California, Centers for Water and Wildland Resources), 1996.

20 **But in the last decade, the Forest Service has invested:** "Suppression Costs," National Interagency Fire Center, accessed February 14, 2022, https://www.nifc.gov/fire-information/statistics /suppression-costs.

20 **Engineers eventually developed:** "Planes," U.S. Forest Service, December 12, 2016, http:// www.fs.usda.gov/managing-land/fire/planes.

21 **But some others had previously been clear-cut:** Prichard and Peterson, "Landscape Analysis."

21 **When early anthropologists and historians tried to estimate the population:** Alexander Koch, Chris Brierley, Mark M. Maslin, and Simon L. Lewis, "Earth System Impacts of the European Arrival and Great Dying in the Americas After 1492," *Quaternary Science Reviews* 207, no. 1 (March 2019): 13–36, https://doi.org/10.1016/j.quascirev.2018.12.004.

22 **Old World diseases, combined with colonial violence:** Lizzie Wade, "From Black Death to Fatal Flu, Past Pandemics Show Why People on the Margins Suffer Most," *Science*, May 14, 2020, https://www.science.org/content/article/black-death-fatal-flu-past-pandemics-show-why -people-margins-suffer-most.

22 **Some evidence of this distortion lay:** Katie Bacon, "The Pristine Myth," *Atlantic*, March 7, 2002, https://www.theatlantic.com/magazine/archive/2002/03/the-pristine-myth/303062/.

22 **lingered in many fields, including ecology:** Jon E. Keeley, "Native American Impacts on Fire Regimes of the California Coastal Ranges," *Journal of Biogeography* 29, no. 3 (March 2002): 303–20, https://doi.org/10.1046/j.1365-2699.2002.00676.x.

22 **But over at least the past couple of decades, some historians:** William Cronon, "The Trouble with Wilderness: Or, Getting Back to the Wrong Nature," *Environmental History* 1, no. 1 (January 1996): 7–28; William M. Denevan, "The Pristine Myth: The Landscape of the Americas in 1492," *Annals of the Association of American Geographers* 82, no. 3 (September 1, 1992): 369–85, https://doi.org/10.1111/j.1467-8306.1992.tb01965.x.

22 **Before Europeans arrived in North America, Indigenous communities:** Robin Kimmerer and Frank Kanawha Lake, "Maintaining the Mosaic: The Role of Indigenous Burning in Land Management," *Journal of Forestry* 99, no. 11 (November 2001): 36–41.

22 **In 2011, for instance, researchers from the University of Washington:** Brendan D. O'Fallon and Lars Fehren-Schmitz, "Native Americans Experienced a Strong Population Bottleneck Coincident with European Contact," *Proceedings of the National Academy of Sciences* 108, no. 51 (December 20, 2011): 20444–48, https://doi.org/10.1073/pnas.1112563108.

23 **In the 1950s, the famous wildlife biologist and policy advisor A. Starker Leopold:** Kiki Leigh Rydell, "A Public Face for Science: A. Starker Leopold and the Leopold Report," *George Wright Forum* 15, no. 4 (1998): 50–63; Stephen J. Pyne, "Vignettes of Primitive America: The Leopold Report and Fire Policy," *Forest History Today*, Spring 2017, https://foresthistory.org/wp -content/uploads/2017/10/Pyne_Vignettes.pdf.

23 **some of the oldest federally run prescribed fire programs in the West:** Pyne, "Vignettes of Primitive America."

23 **In the words of Kimmerer and Lake:** Kimmerer and Lake, "Maintaining the Mosaic."

23 **Susan and her colleagues ran an analysis:** Susan J. Prichard, David L. Peterson, and Kyle Jacobson, "Fuel Treatments Reduce the Severity of Wildfire Effects in Dry Mixed Conifer Forest, Washington, USA," *Canadian Journal of Forest Research* 40, no. 8 (August 1, 2010): 1615–26, https://doi.org/10.1139/X10-109.

24 **Then she and another scientist used satellite data:** Prichard and Kennedy, "Fuel Treatments and Landform."

24 **On Monday, July 14, lightning lit four fires:** Jim Kershner, "Carlton Complex Fire,"

HistoryLink.org, December 10, 2014, accessed October 19, 2021, http://www.historylink.org/File/10989.

26 **two years later, at the age of twenty:** William D. Moody, *History of the North Cascades Smoke-jumper Base* (Missoula, MT: National Smokejumper Association, 2019), https://dc.ewu.edu/smokejumping_pubs/4.

26 **across the West, the ski season:** Xubin Zeng, Patrick Broxton, and Nicholas Dawson, "Snow-pack Change from 1982 to 2016 over Conterminous United States," *Geophysical Research Letters* 45, no. 23 (2018): 12940–47, https://doi.org/10.1029/2018GL079621.

27 **Four separate fires lit:** Ann McCreary, "WEDNESDAY UPDATE: More Evacuation Orders as Fires, Smoke Spread," *Methow Valley News*, July 16, 2014, https://methowvalleynews.com/2014/07/16/wednesday-update-more-evacuation-orders-as-fires-smoke-spread/; Ryan Maye Handy, "Carlton Complex Fire Largest in Washington State History," *Wildfire Today*, July 22, 2014, https://wildfiretoday.com/2014/07/22/carlton-complex-fire-largest-in-washington-state-history/.

27 **more than a hundred firefighters:** Kershner, "Carlton Complex Fire."

30 **President Barack Obama arrived in Seattle:** Jim Brunner, "Obama Hits 2 Fundraisers, Gets Word from Inslee on Wildfires," *Seattle Times*, July 22, 2014, https://www.seattletimes.com/seattle-news/obama-hits-2-fundraisers-gets-word-from-inslee-on-wildfires/; "Obama Declares Emergency as Huge Fires Burn in Washington State," Colorado Public Radio, July 23, 2014, https://www.cpr.org/2014/07/23/obama-declares-emergency-as-huge-fires-burn-in-washington-state/.

30 **After all the damages had been assessed:** Kershner, "Carlton Complex Fire."

31 **In mid-August, another fire lit:** Washington State Department of Natural Resources, *Wildland Fire Investigation Report: Rising Eagle Road Fire*, August 1, 2014, https://img1.wsimg.com/blobby/go/9169440b-3652-44f1-8b64-9c12f03c446b/downloads/Dangerous_Trailers.org_FINAL_14-V-AEU_Rising_Eagle_Road_WFIR.pdf; Marcy Stamper, "DNR Investigation: Improperly Maintained Trailer Caused Rising Eagle Road Fire," *Methow Valley News*, March 11, 2015, https://methowvalleynews.com/2015/03/11/dnr-investigation-improperly-maintained-trailer-caused-rising-eagle-road-fire/.

31 **One of these belonged:** Kelli Scott, "Starting Over: A Methow Valley Artist Lost Everything in Last Summer's Wildfires. A Year Later, She's Found It Again," *Wenatchee World*, August 20, 2015, https://www.wenatcheeworld.com/go/starting-over-a-methow-valley-artist-lost-everything-in-last-summers-wildfires-a-year-later/article_d59f67c2-ae89-52fe-8134-1738dbb2ad9b.html.

31 **At the end of August, rains arrived again:** "Rain After Wildfires Triggers Mudslides in Washington State," CBS News, August 22, 2014, https://www.cbsnews.com/news/rain-after-wildfires-triggers-mudslides-in-washington-state/; Kershner, "Carlton Complex Fire."

33 **a common trouble after wildfires loosen earth and liberate:** Kevin D. Bladon, Monica B. Emelko, Uldis Silins, and Micheal Stone, "Wildfire and the Future of Water Supply," *Environmental Science and Technology* 48, no. 16 (August 19, 2014): 8936–43, https://doi.org/10.1021/es500130g.

33 **Fire and Ice, a Robert Frost allusion:** Robert Frost, "Fire and Ice," Poetry Foundation, accessed October 19, 2021, https://www.poetryfoundation.org/poems/44263/fire-and-ice.

34 **Three-fourths of the people living in Pateros:** K. C. Mehaffey, "Fire's Aftermath: Uninsured and Under-Insured Finding Strength to Move Forward," *Wenatchee World*, October 1, 2014, https://www.wenatcheeworld.com/news/local/fire-s-aftermath-uninsured-and-under-insured-finding-strength-to-move-forward/article-6e36c519-59f1-5800-842a-aae50a2491ad.html.

35 **The couple whose broken-down trailer:** Marcy Stamper, "Rising Eagle Road Fire Lawsuit Settled; Couple Cleared," *Methow Valley News*, January 18, 2018, https://methowvalleynews.com/2018/01/18/rising-eagle-road-fire-lawsuit-settled-couple-cleared/.

2: Homesick

I first interviewed Glenn Albrecht for an article in *YES!* magazine, "When Words Fail: Does a Warming World Need a New Vocabulary?" September 19, 2011, and have done several interviews with him in the years since. In 2013, I traveled to Perth, Australia, as a visiting scholar and met with him in person. Many of the quotes and biographical information in this chapter are from these personal communications.

38 **After the English invaded Wales in the thirteenth century:** Pamela Petro, "Dreaming in Welsh," *Paris Review* (blog), September 18, 2012, https://www.theparisreview.org/blog/2012 /09/18/dreaming-in-welsh/.

38 **In 1688, Johannes Hofer, then a medical student:** Thomas Dodman, *What Nostalgia Was: War, Empire, and the Time of a Deadly Emotion* (Chicago: University of Chicago Press, 2018).

39 **He called it *nostalgia*, derived from Greek:** "Medical Dissertation on Nostalgia by Johannes Hofer, 1688," trans. Carolyn Kiser Anspach, *Bulletin of the Institute of the History of Medicine* 2, no. 6 (1934): 376–91.

39 **Modern nostalgia can be pleasant or cloying but also dangerous:** Rhaina Cohen, Shankar Vedantam, Tara Boyle, and Renee Klahr, "Nostalgia Isn't Just a Fixation on the Past—It Can Be About the Future, Too," *Hidden Brain*, National Public Radio, podcast audio, October 16, 2017, https://www.npr.org/2017/10/16/558055384/nostalgia-isnt-just-a-fixation-on-the-past -it-can-be-about-the-future-too.

40 **German word *Heimat*, which translates roughly:** Julia Metzger-Traber, *If the Body Politic Could Breathe in the Age of the Refugee: An Embodied Philosophy of Interconnection* (Wiesbaden, Germany: Springer, 2018).

40 **That meaning still haunts Germany:** Jill Petzinger, "Germany Is Adding a Ministry for 'Heimat,' a Word Loaded with Negative Undertones," *Quartz*, February 20, 2018, https://qz.com /1209547/germany-is-adding-a-ministry-for-heimat-a-word-loaded-with-negative-undertones/.

40 **when the German Ministry of the Interior formally renamed itself:** Ceyda Nurtsch, "'Your Homeland Is Our Nightmare,'" *Qantara.de*, March 29, 2019, https://en.qantara.de/content /germanys-integration-debate-your-homeland-is-our-nightmare.

40 **To them the word still evoked:** "Your Homeland Is Our Nightmare," YouTube video, 1:24:21, panel discussion moderated by Marc Reugebrink with Mithu Sanyal, Fatma Aydemir, and Max Czollek, Goethe-Institut Brüssel, May 27, 2020, posted by "Literaturhaus Berlin," https://www.youtube.com/watch?v=PTsxsOzgB3o.

40 **Social psychiatrist Mindy Fullilove has described:** Mindy Thompson Fullilove, *Root Shock: How Tearing Up City Neighborhoods Hurts America, and What We Can Do About It* (New York: New Village Press, 2016), 11.

40 **As a child in the 1950s and 1960s, he kept a secret aviary:** Glenn A. Albrecht, chap. 1 in *Earth Emotions: New Words for a New World* (Ithaca, NY: Cornell University Press, 2019).

41 **Coal mining in the Hunter dates to the late eighteenth century:** Drew Cottle and Angela Keys, "Open-Cut Coal Mining in Australia's Hunter Valley: Sustainability and the Industry's Economic, Ecological and Social Implications," *International Journal of Rural Law and Policy*, no. 1 (September 10, 2014): 1–7, https://doi.org/10.5130/ijrlp.i1.2014.3844.

41 **Between 1981 and 2012, the amount of land occupied:** Hydrocology Consulting, *Unfair Shares: How Coal Mines Bought the Hunter River*, commissioned by the Lock the Gate Alliance, July 2014, https://d3n8a8pro7vhmx.cloudfront.net/lockthegate/pages/2122/attachments /original/1438071706/LTG_CoalMinesHunterRiver_SML-1.pdf?1438071706.

41 **coal produces the highest carbon emissions:** U.S. Environmental Protection Agency, "Sources of Greenhouse Gas Emissions," December 29, 2015, https://www.epa.gov/ghgemissions /sources-greenhouse-gas-emissions.

42 **Children living in the region have high rates:** Fiona Armstrong, *Coal and Health in the Hunter: Lessons from One Valley for the World* (Melbourne: Climate and Health Alliance, February 23, 2015).

42 **Albrecht was confronted with such a scene:** Albrecht, chap. 1 in *Earth Emotions*.

42 **"In their attempts to halt the expansion":** Glenn A. Albrecht, "'Solastalgia': A New Concept in Health and Identity," *PAN: Philosophy Activism Nature*, no. 3 (2005): 41–55.

42 **In the 1990s, a group of psychologists:** Mark A. Schroll, "Remembering Ecopsychology's

Origins: A Chronicle of Meetings, Conversations, and Significant Publications," *Gatherings: Journal of the International Community for Ecopsychology*, accessed October 20, 2021, https:// www.ecopsychology.org/journal/ezine/ep_origins.html.

43 **In September 2021, more than two hundred medical journals:** Lukoye Atwoli, Abdullah H. Baqui, Thomas Benfield, Raffaella Bosurgi, Fiona Godlee, Stephen Hancocks, Richard Horton, et al., "Call for Emergency Action to Limit Global Temperature Increases, Restore Biodiversity, and Protect Health," *New England Journal of Medicine* 385, no. 12 (September 16, 2021): 1134–37, https://doi.org/10.1056/NEJMe2113200.

43 **There are words in Indigenous languages:** S. Fox, *When the Weather Is Uggianaqtuq: Inuit Observations of Environmental Change* (Boulder: University of Colorado Geography Department Cartography Lab, 2003), summary; Montserrat Madariaga, "Of How a Hopi Ancient Word Became a Famous Experimental Film," *Not Even Past*, May 15, 2018, https://notevenpast .org/of-how-a-hopi-ancient-word-became-a-famous-experimental-film/.

43 **A British trip-hop band:** Daniel B. Smith, "Is There an Ecological Unconscious?" *New York Times Magazine*, January 27, 2010, https://www.nytimes.com/2010/01/31/magazine /31ecopsych-t.html.

44 **A team of social scientists identified feelings:** Petra Tschakert, Raymond Tutu, and Anna Alcaro, "Embodied Experiences of Environmental and Climatic Changes in Landscapes of Everyday Life in Ghana," *Emotion, Space and Society* 7 (May 1, 2013): 13–25, https://doi.org /10.1016/j.emospa.2011.11.001.

44 **A collaboration of environmental scientists and public health researchers observed:** Paul A. Sandifer, Landon C. Knapp, Tracy K. Collier, Amanda L. Jones, Robert-Paul Juster, Christopher R. Kelble, Richard K. Kwok, et al., "A Conceptual Model to Assess Stress-Associated Health Effects of Multiple Ecosystem Services Degraded by Disaster Events in the Gulf of Mexico and Elsewhere," *GeoHealth* 1, no. 1 (2017): 17–36, https://doi.org/10.1002/2016GH000038.

44 **A Los Angeles physician named David Eisenman:** David Eisenman, Sarah McCaffrey, Ian Donatello, and Grant Marshal, "An Ecosystems and Vulnerable Populations Perspective on Solastalgia and Psychological Distress After a Wildfire," *EcoHealth* 12, no. 4 (December 2015): 602–10, https://doi.org/10.1007/s10393-015-1052-1.

44 **It's now becoming more common:** Nylah Burton, "Experts Say Grief Is a Major Response to the Climate Crisis—Here's How to Cope," *Bustle*, August 14, 2019, https://www.bustle.com/p /climate-grief-is-a-more-common-emotion-than-youd-think-18656107; Chris Taylor and Ira Flatow, "You Aren't Alone in Grieving the Climate Crisis," Science Friday, podcast audio, April 17, 2020, https://www.sciencefriday.com/segments/climate-crisis-grief/.

44 **We have even more ways to name:** Ashlee Cunsolo and Neville R. Ellis, "Ecological Grief as a Mental Health Response to Climate Change-Related Loss," *Nature Climate Change* 8, no. 4 (April 2018): 275–81, https://doi.org/10.1038/s41558-018-0092-2; Renee Lertzman, *Environmental Melancholia: Psychoanalytic Dimensions of Engagement* (London: Routledge, 2015).

44 **In a 2020 survey by the American Psychological Association:** "Majority of US Adults Believe Climate Change Is Most Important Issue Today," American Psychological Association, February 6, 2020, https://www.apa.org/news/press/releases/2020/02/climate-change.

44 **In 2019 alone:** Internal Displacement Monitoring Centre, *Global Report on Internal Displacement 2020*, accessed October 20, 2021, https://www.internal-displacement.org/global-report /grid2020/.

45 **A record-breaking monsoon season:** "2019 Saw More Deaths in India Despite Less Number of Extreme Events: State of Environment," Weather Channel, February 10, 2020, https:// weather.com/en-IN/india/news/news/2020-02-10-state-of-environment-more-deaths-despite -less-number-of-extreme-events; "Monsoon Floods Displace Millions in India," BBC News, July 15, 2019, https://www.bbc.com/news/world-asia-48979668; International Federation of Red Cross and Red Crescent Societies, *Final Report, India: Monsoon Rains and Floods*, July 23, 2020, https://reliefweb.int/report/india/india-monsoon-rains-and-floods-final-report-dref-n -mdrin024.

45 **Various economists have tried:** Juan-Carlos Ciscar, Ana Iglesias, Luc Feyen, László Szabó, Denise Van Regemorter, Bas Amelung, Robert Nicholls, et al., "Physical and Economic Consequences of Climate Change in Europe," *Proceedings of the National Academy of Sciences* 108, no. 7 (February 15, 2011): 2678–83, https://doi.org/10.1073/pnas.1011612108.

45 **Others have made calculations:** Juan-Carlos Ciscar, James Rising, Robert E. Kopp, and Luc Feyen, "Assessing Future Climate Change Impacts in the EU and the USA: Insights and Lessons from Two Continental-Scale Projects," *Environmental Research Letters* 14, no. 8 (August 1, 2019): 084010, https://doi.org/10.1088/1748-9326/ab281e; "World Economy Set to Lose up to 18% GDP from Climate Change If No Action Taken, Reveals Swiss Re Institute's Stress-Test Analysis," Swiss Re, April 22, 2021, https://www.swissre.com/media/news-releases/nr-20210422-economics-of-climate-change-risks.html.

45 **Former vice president Al Gore has called:** Ryan Lizza, "Al Gore on the Failure of Climate-Change Legislation," *New Yorker*, October 5, 2010, http://www.newyorker.com/news/news-desk/al-gore-on-the-failure-of-climate-change-legislation.

45 **The journalist David Wallace-Wells:** David Wallace-Wells, "Storytelling," in *The Uninhabitable Earth: Life After Warming* (New York: Crown, 2019).

46 **Because of climate change, the amount of regional rainfall has dropped:** "Climate and Southern WA," Water Corporation, accessed October 20, 2021, https://www.watercorporation.com.au/Our-water/Climate-change-and-WA/Climate-and-Southern-WA.

46 **In the exurbs, entire forests of jarrah trees:** ""Experts Warn WA's Northern Jarrah Forest Under Threat," *PerthNow*, February 21, 2012, https://www.perthnow.com.au/news/wa/experts-warn-was-northern-jarrah-forest-under-threat-ng-9c131c5a78c8bedb5c72897e942472a8.

46 **the concept had been used in a 2013 court ruling:** Amanda Kennedy, "A Case of Place: Solastalgia Comes Before the Court," *PAN: Philosophy Activism Nature* 12 (2016): 23–33.

47 **Albrecht had testified:** Glenn A. Albrecht, *Report for Warkworth Mount Thorley PAC Hearing*, December 2014, https://www.ipcn.nsw.gov.au/resources/pac/media/files/pac/projects/2014/11/mt-thorley-continuation-project/edo-on-behalf-of-bulga-milbrodale-progress-association/3-albrecht-social-impacts-supplementarypdf.pdf.

3: The Flood

Much of the reporting here and in chapter 9 was originally done for a story in *Hakai Magazine*, "The Sea Versus St. Augustine," May 19, 2020. Some also appeared in a 2015 article I wrote for *Al Jazeera America*, "Climate Change Threatens to Wash Away Cultural History."

49 **home, n.:** *home*, n.A.5., *OED Online*.

50 **a town on a group of islands:** "Cedar Key History," Cedar Key Historical Society Museum, November 11, 2018, https://cedarkeyhistory.org/cedarkey/history/.

50 **Her thesis offered a nearly exhaustive review:** Jennifer Marie Wolfe, "Historic Context at Risk: Planning for Tropical Cyclone Events in Historic Cedar Key" (master's thesis, University of Florida, 2006).

50 **Established in the sixteenth century by the Spanish:** "St. Augustine: America's Ancient City," Florida Museum, accessed October 21, 2021, https://www.floridamuseum.ufl.edu/staugustine/.

51 **Her new desk stood inside city hall:** "The Hotel Alcazar," Lightner Museum, accessed October 21, 2021, https://lightnermuseum.org/history-alcazar/.

51 **The city's most iconic structure:** "The Founding of Castillo de San Marcos," Castillo de San Marcos National Monument, U.S. National Park Service, accessed October 21, 2021, https://www.nps.gov/casa/learn/the-founding-of-castillo-de-san-marcos.htm.

51 **built from coquina, a stone formed by the compression of piles of tiny clamshells:** "Geologic Activity," Fort Matanzas National Monument, U.S. National Park Service, accessed October 21, 2021, https://www.nps.gov/foma/learn/nature/geologicactivity.htm; "General Information," Florida Atlantic University Department of Geosciences, accessed October 21, 2021, http://www.geosciences.fau.edu/events/virtual-field-trips/anastasia/general-information.php.

51 **And to the south lies a neighborhood called Lincolnville:** "Florida, St. Augustine, Lincoln-ville Historic District," U.S. National Park Service, accessed October 21, 2021, https://www .nps.gov/nr/travel/geo-flor/28.htm.

51 **In 2009, the U.S. government released:** U.S. Global Change Research Program, ed., *Global Climate Change Impacts in the United States: A State of Knowledge Report* (New York: Cambridge University Press, 2009).

52 **A few months later, four counties down the coast from St. Augustine:** "Southeast Florida Regional Climate Change Compact," accessed October 21, 2021, https://southeastfloridaclimatecompact .org/wp-content/uploads/2014/09/compact.pdf.

52 **but didn't prevent Governor Rick Scott:** Tristram Korten, "In Florida, Officials Ban Term 'Climate Change,'" *Miami Herald*, March 11, 2015, https://www.miamiherald.com/news/state /florida/article12983720.html.

52 **Nor has it prevented real estate tycoons:** Jeff Goodell, *The Water Will Come: Rising Seas, Sinking Cities, and the Remaking of the Civilized World* (New York: Little, Brown, 2017); Megan Mayhew Bergman, "Florida Is Drowning. Condos Are Still Being Built. Can't Humans See the Writing on the Wall?" *Guardian*, February 15, 2019, http://www.theguardian.com/environment/2019/feb /15/florida-climate-change-coastal-real-estate-rising-seas; Greg Allen, "South Florida Real Estate Boom Not Dampened by Sea Level Rise," National Public Radio, December 5, 2017, https:// www.npr.org/2017/12/05/567264841/south-florida-real-estate-boom-not-dampened-by-sea -level-rise.

52 **the Obama White House also directed agencies:** Jane A. Leggett, *Climate Change Adaptation by Federal Agencies: An Analysis of Plans and Issues for Congress* (Washington, DC: Congressional Research Service, February 23, 2015).

52 **Between 2012 and 2014, St. Augustine added a new seawall:** "Avenida Menendez Sea-wall," City of St. Augustine, accessed May 3, 2021, https://www.citystaug.com/357/Avenida -Menendez-Seawall; "Avenida Menendez FAQs," City of St. Augustine, accessed October 21, 2021, https://www.citystaug.com/Faq.aspx?QID=88; "Seawall Helps Preserve History in St. Augustine," *Daytona Beach News-Journal*, December 28, 2016, https://www.news-journalonline .com/news/20161228/seawall-helps-preserve-history-in-st-augustine; Peter Guinta, "St. Augus-tine's Seawall Project Finished," *St. Augustine Record*, January 4, 2014, https://www.staugustine .com/story/news/local/2014/01/05/st-augustines-seawall-project-finished/16072452007/.

53 **It had been platted in the 1920s:** Rodney Kite-Powell, "In Search of D. P. Davis: A Biographi-cal Study of One of Florida's Premier Real Estate Promoters," *Sunland Tribune* 29, no. 7 (2003): 77–114, https://scholarcommons.usf.edu/sunlandtribune/vol29/iss1/7.

53 **St. Augustine's sea level rise report came back:** National Oceanic and Atmospheric Admin-istration, Florida Department of Environmental Protection, and Florida Coastal Management Program, *Coastal Vulnerability Assessment: City of St. Augustine, Florida*, June 24, 2016, https:// www.citystaug.com/DocumentCenter/View/323/Coastal-Vulnerability-Assessment-PDF.

53 **Hurricane Matthew rose up from the Caribbean:** "Hurricane Matthew in the Carolinas: Octo-ber 8, 2016," National Weather Service, October 8, 2016, https://www.weather.gov/ilm/Matthew.

53 **here it produced a horrifying disaster:** Joseph Guyler Delva and Scott Malone, "Hurricane Matthew Kills Almost 900 in Haiti Before Hitting U.S.," Reuters, October 5, 2016, https:// www.reuters.com/article/us-storm-matthew-idUSKCN1250G2; Azam Ahmed, "Hurricane Matthew Makes Old Problems Worse for Haitians," *New York Times*, October 6, 2016, https:// www.nytimes.com/2016/10/07/world/americas/hurricane-matthew-haiti.html.

54 **Then it struck Cuba:** Sam Jones and Nicky Woolf, "Hurricane Matthew Makes Landfall in Cuba After Ripping Through Haiti," *Guardian*, October 5, 2016, https://www.theguardian.com /world/2016/oct/04/hurricane-matthew-haitians-flee-inland-as-violent-storm-approaches; Pam Wright and Rachel Delia Benaim, "Crowdfunding Campaigns Underway to Aid Cuba as Country Grapples with Hurricane Matthew Destruction," Weather Channel, October 8, 2016, https://weather.com/news/news/hurricane-matthew-cuba.

54 **The evacuation order for St. John's County:** "St. Johns County Issues Mandatory Evacuation

Order," First Coast News, October 6, 2016, https://www.firstcoastnews.com/article/news/st
-johns-county-issues-evacuation-order-for-coastal-areas/329221386.

54 **"the western edge of Matthew's eyewall":** Stacy R. Stewart, *National Hurricane Center, Tropical Cyclone Report, Hurricane Matthew*, National Hurricane Center, National Weather Service, April 7, 2017, https://www.nhc.noaa.gov/data/tcr/AL142016_Matthew.pdf.

54 **In video footage from the storm, St. Augustine resembled an underwater ruin:** "St. Augustine Streets Fill with Water from Hurricane Matthew," Action News JAX, October 8, 2016, https://www.actionnewsjax.com/news/local/st-augustine-streets-filled-with-water-from -hurricane-matthew/454618481/.

54 **Downed fences:** Jake Martin, "Hurricane Matthew: Surveying Damage in St. Augustine the Morning After," *St. Augustine Record*, October 8, 2016, https://www.staugustine .com/story/news/local/2016/10/08/hurricane-matthew-surveying-damage-st-augustine -morning/16296294007/.

55 **Beginning at least five thousand years ago, the ancestors:** Kenneth E. Sassaman, Paulette S. McFadden, Micah P. Monés, Andrea Palmiotto, and Asa R. Randall, "North Gulf Coastal Archaeology of the Here and Now," in *New Histories of Pre-Columbian Florida*, ed. Neill J. Wallis and Asa R. Randall (Gainesville: University Press of Florida, 2014), 143–62, https://doi .org/10.5744/florida/9780813049366.003.0008.

56 **Shell mounds may have had various functions:** Margo Schwadron, "Landscapes of Maritime Complexity: Prehistoric Shell Work Sites of the Ten Thousand Islands, Florida" (Ph.D. diss., University of Leicester, 2010); Kenneth E. Sassaman and Friends of the Lower Suwannee and Cedar Keys National Wildlife Refuges, "Shell Mound: A Portal into Another World," interpretive signs, University of Florida, https://www.friendsofrefuges.org/uploads/2/2/9/2/22922110 /shell_mound-printable.pdf; Kenneth E. Sassaman, Meggan E. Blessing, Joshua M. Goodwin, Jessica A. Jenkins, Ginessa J. Mahar, Anthony Boucher, Terry E. Barbour, et al., "Maritime Ritual Economies of Cosmic Synchronicity: Summer Solstice Events at a Civic-Ceremonial Center on the Northern Gulf Coast of Florida," *American Antiquity* 85, no. 1 (January 2020): 22–50, https://doi.org/10.1017/aaq.2019.68.

56 **At some mounds, there is evidence of regular feasts:** Kenneth E. Sassaman, "Summer Solstice Feasts and Other Gatherings at Shell Mound on the Northern Gulf Coast of Florida," *Adventures in Florida Archaeology* (2019): 20–26.

56 **In some locations, people buried their dead inside the mounds:** Kenneth E. Sassaman, John S. Krigbaum, Ginessa J. Mahar, and Andrea Palmiotto, *Archaeological Investigations at McClamory Key (8LV288), Levy County, Florida* (Gainesville: University of Florida, March 2015).

56 **But there were sometimes sudden disruptions:** Miriam C. Jones, G. Lynn Wingard, Bethany Stackhouse, Katherine Keller, Debra Willard, Marci Marot, Bryan Landacre, et al., "Rapid Inundation of Southern Florida Coastline Despite Low Relative Sea-Level Rise Rates During the Late-Holocene," *Nature Communications* 10, no. 1 (July 19, 2019): 3231, https://doi.org /10.1038/s41467-019-11138-4.

57 **In about two and a half decades:** Union of Concerned Scientists, *Underwater: Rising Seas, Chronic Floods, and the Implications for US Coastal Real Estate*, June 18, 2018, https://www .ucsusa.org/resources/underwater.

57 **In 2012, real estate developers:** Erin Durkin, "North Carolina Didn't Like Science on Sea Levels . . . So Passed a Law Against It," *Guardian*, September 12, 2018, https://www.theguardian .com/us-news/2018/sep/12/north-carolina-didnt-like-science-on-sea-levels-so-passed-a-law -against-it.

58 **Since at least the 1970s and 1980s, scientists:** J. H. Mercer, "West Antarctic Ice Sheet and CO_2 Greenhouse Effect: A Threat of Disaster," *Nature* 271, no. 5643 (January 1, 1978): 321–25. https://doi.org/10.1038/271321a0.

59 **For instance, a roughly decade-long Dutch project:** ClimateWire, "How the Dutch Make 'Room for the River' by Redesigning Cities," *Scientific American*, January 20, 2012, https:// www.scientificamerican.com/article/how-the-dutch-make-room-for-the-river/; "Room for the

River Programme," Dutch Water Sector, April 15, 2019, https://www.dutchwatersector.com /news/room-for-the-river-programme.

59 **"Room for the River," or *Ruimte voor de Rivier*, is described:** Elizabeth Kolbert, *Field Notes from a Catastrophe: Man, Nature, and Climate Change* (New York: Bloomsbury USA, 2007), 126–27.

59 **infamous because of cost overruns:** "Venice Flooded as New $8 Billion Dam System Fails to Activate," CBS News, December 9, 2020, https://www.cbsnews.com/news/venice-flooded-as -new-8-billion-dam-system-fails-to-activate/.

59 **Nearly half of Venice can still flood:** Julia Buckley, "Venice Flood Causes 'Serious' Damage Two Months After Flood Barriers Were Introduced," CNN, December 9, 2020, https://www .cnn.com/travel/article/venice-flood-basilica-mose/index.html.

60 **In October 2020, the project:** Julia Buckley, "Venice Holds Back the Water for First Time in 1,200 Years," CNN, October 5, 2020, https://www.cnn.com/travel/article/venice-flood -barrier/index.html; Sarah Cascone, "Offering a Glimmer of Hope for Venice, the City's New Flood Barriers Successfully Prevented a Deluge This Weekend," Artnet News, October 5, 2020, https://news.artnet.com/art-world/venice-flood-barriers-1912983.

60 **Two months later, when the forecast failed:** Buckley, "Venice Flood Causes"; Rebecca Ann Hughes, "Why Isn't Venice's MOSE Holding Back the Tide?" *Forbes*, December 10, 2020, https://www.forbes.com/sites/rebeccahughes/2020/12/10/why-isnt-venices-mose-holding -back-the-tide/.

60 **In 2015, Lisa Craig, the chief of historic preservation of Annapolis, Maryland:** Lisa M. Craig and Leslee F. Keys, "A Tale of Two Cities: Annapolis and St. Augustine Balancing Preservation and Community Values in an Era of Rising Seas," *Parks Stewardship Forum*, January 6, 2020, https://parks.berkeley.edu/psf/?p=1717.

60 **the frequency of chronic flooding:** "NOAA: 'Nuisance Flooding' an Increasing Problem as Coastal Sea Levels Rise," National Oceanic and Atmospheric Administration, July 28, 2014, https://www.noaa.gov/media-release/noaa-nuisance-flooding-increasing-problem-as-coastal -sea-levels-rise; National Oceanic and Atmospheric Administration, *Sea Level Rise and Nuisance Flood Frequency Changes Around the United States* (Silver Spring, MD: NOAA, June 2014), https://tidesandcurrents.noaa.gov/publications/NOAA_Technical_Report_NOS_COOPS _073.pdf.

60 **But when a new mayor was elected in 2017:** Danielle Ohl, "Annapolis Chief of Historic Preservation Quits, Citing Worries About Buckley Administration," *Capital Gazette*, November 15, 2017, https://www.capitalgazette.com/maryland/annapolis/ac-cn-chief-historic-preservation -resigns-20171114-story.html.

61 **The next year, a citizen advisory committee would come together:** Jeremy Cox, "Can Make-over Save Annapolis City Dock from Sea Level Rise?" *Bay Journal*, August 9, 2021, https://www .bayjournal.com/news/climate_change/can-makeover-save-annapolis-city-dock-from-sea-level -rise/article_5b14ee3c-d827-11eb-ac82-4772366f7e6a.html; City Dock Action Committee, "Transforming City Dock," City of Annapolis and Historic Annapolis, March 2019–January 14, 2020, https://www.preservationmaryland.org/wp-content/uploads/2020/03/city-dock-action -committee-transforming-city-dock-report-2020-annapolis-maryland.pdf.

62 **In 1999, after Hurricane Floyd:** David Royce, "Oldest City Protected from Hurricanes," *Tallahassee Democrat*, September 16, 1999, accessed at newspapers.com

62 **the costliest disaster in the United States:** "Hurricane Andrew (1992)," U.S. National Park Service, accessed October 21, 2021, https://www.nps.gov/articles/hurricane-andrew-1992 .htm; "Costliest U.S. Tropical Cyclones," National Centers for Environmental Information and National Oceanic and Atmospheric Administration, last updated October 8, 2021, https:// www.ncdc.noaa.gov/billions/dcmi.pdf.

63 **Leslee Keys would eventually write:** Leslee F. Keys, *Hotel Ponce de Leon: The Rise, Fall, and Rebirth of Flagler's Gilded Age Palace* (Gainesville: University Press of Florida, 2015), 1, 8.

63 **In 1870, he cofounded Standard Oil:** "Henry Flagler," *American Experience*, PBS, accessed

October 21, 2021, https://www.pbs.org/wgbh/americanexperience/features/miami-henry-flagler/; "Florida East Coast Railway," Henry Morrison Flagler Museum, accessed October 21, 2021, https://www.flaglermuseum.us/history/florida-east-coast-railway.

64 **The fantasy of Florida as a paradise:** Michael Grunwald, "A Requiem for Florida, the Paradise That Should Never Have Been," *POLITICO Magazine*, September 8, 2017, https://www.politico.com/magazine/story/2017/09/08/hurricane-irma-florida-215586.

64 **Hurricane Sandy ransacked the museum buildings on Ellis Island:** Debra Holtz, Adam Markham, Kate Cell, and Brenda Ekwurzel, *National Landmarks at Risk: How Rising Seas, Floods, and Wildfires Are Threatening the United States' Most Cherished Historic Sites* (Union of Concerned Scientists, May 2014), 7–9.

66 **The state and NOAA also sponsored:** Florida Department of Environmental Protection, Florida Coastal Management Program, and National Oceanic and Atmospheric Administration, "Florida Community Resiliency Initiative Pilot Project: Adaptation Plan for St. Augustine, Florida," May 2017, https://www.citystaug.com/DocumentCenter/View/321/Strategic-Adaption-Plan-PDF.

66 **Then in September 2017:** Jess Bidgood, "After Irma, a Grim Sense of Déjà Vu in St. Augustine," *New York Times*, September 14, 2017, https://www.nytimes.com/2017/09/13/us/st-augustine-irma-flood.html.

67 **Leslee's presentation, which she delivered:** Clay Henderson and Leslee F. Keys, "Preserving Paradise: There Is a Little of Florida in All of US" (presentation, Keeping History Above Water, Annapolis, 2017).

4: The First Home

70 **In the late 1970s, a young anthropologist named Richard Potts:** Ruth O. Selig, "Human Origins: One Man's Search for the Causes in Time," *AnthroNotes* 21, no. 1 (1999): 1, https://doi.org/10.5479/10088/22375.

70 **In the mid-twentieth century, Mary and Louis Leakey found:** Roger Lewin, "The Old Man of Olduvai Gorge," *Smithsonian Magazine*, October 2002, https://www.smithsonianmag.com/history/the-old-man-of-olduvai-gorge-69246530/.

70 **Potts quite literally followed:** Richard Potts, "Home Bases and Early Hominids," *American Scientist* 72, no. 4 (July–August 1984): 338–47.

70 ***Homo habilis,*** **the "handy man," the long-armed, strong-jawed human ancestor:** "*Homo habilis,*" Smithsonian National Museum of Natural History, accessed October 22, 2021, http://humanorigins.si.edu/evidence/human-fossils/species/homo-habilis.

71 **The bones revealed:** Richard Potts and Pat Shipman, "Cutmarks Made by Stone Tools on Bones from Olduvai Gorge, Tanzania," *Nature* 291, no. 5816 (June 1981): 577–80, https://doi.org/10.1038/291577a0.

72 ***Homo habilis*** **had various descendants or maybe cousins:** "Human Family Tree," Smithsonian National Museum of Natural History, accessed October 22, 2021, http://humanorigins.si.edu/evidence/human-family-tree; Casey Luskin, "Paleoanthropologists Disown *Homo habilis* from Our Direct Family Tree," *Evolution News and Science Today*, August 9, 2007, https://evolutionnews.org/2007/08/paleoanthropologists_disown_ho/.

72 **In the mid-2000s, a team of archaeologists:** Naama Goren-Inbar, Craig S. Feibel, Kenneth L. Verosub, Yoel Melamed, Mordechai E. Kislev, Eitan Tchernov, and Idit Saragusti, "Pleistocene Milestones on the Out-of-Africa Corridor at Gesher Benot Ya'aqov, Israel," *Science* 289, no. 5481 (2000): 944–47; Rivka Rabinovich, Sabine Gaudzinski-Windheuser, and Naama Goren-Inbar, "Systematic Butchering of Fallow Deer (*Dama*) at the Early Middle Pleistocene Acheulian Site of Gesher Benot Ya'aqov (Israel)," *Journal of Human Evolution* 54 (2008): 134–49; Naama Goren-Inbar, Nira Alperson, Mordechai E. Kislev, Orit Simchoni, Yoel Melamed, Adi Ben-Nun, and Ella Werker, "Evidence of Hominin Control of Fire at Gesher Benot Ya'aqov, Israel," *Science* 304, no. 5671 (April 30, 2004): 725–27, https://doi.org/10.1126/science.1095443; James Randerson, "Charred Remains May Be Earliest Human Fires," *New Scientist*, April 29, 2004, https://www.newscientist.com/article/dn4944-charred-remains-may-be-earliest-human-fires/.

72 **Eventually the scientists meticulously sifted through:** Nira Alperson-Afil, Daniel Richter, and Naama Goren-Inbar, "Evaluating the Intensity of Fire at the Acheulian Site of Gesher Benot Ya'aqov—Spatial and Thermoluminescence Analyses," *PloS One* 12, no. 11 (November 16, 2017): e0188091, https://doi.org/10.1371/journal.pone.0188091.

72 **stocky, big-browed people with flat faces:** "*Homo heidelbergensis*," Smithsonian National Museum of Natural History, accessed October 22, 2021, http://humanorigins.si.edu/evidence /human-fossils/species/homo-heidelbergensis.

73 **what looked like the division of space into various tasks:** Mati Milstein, "*Homo erectus* Invented 'Modern' Living?" *National Geographic News*, January 13, 2010, https://www .nationalgeographic.com/culture/article/100112-modern-human-behavior.

73 **you arrive in the French Riviera:** Henry de Lumley, "A Paleolithic Camp at Nice," *Scientific American*, May 1969.

74 **placed a museum therein:** "Présentation du musée de préhistoire de Terra Amata," Ville de Nice, accessed October 22, 2021, https://www.nice.fr/fr/culture/musees-et-galeries/presentation -du-musee-terra-amata.

74 **the first *Homo sapiens* (our species, the "thinking" or "wise people") who appeared:** John Noble Wilford, "When Humans Became Human," *New York Times*, February 26, 2002, https://www.nytimes.com/2002/02/26/science/when-humans-became-human.html; Richard G. Klein, "Why Anatomically Modern People Did Not Disperse from Africa 100,000 Years Ago," in *Neandertals and Modern Humans in Western Asia*, ed. Takeru Akazawa, Kenichi Aoki, and Ofer Bar-Yosef (Boston: Springer US, 1998), 509–21, https://doi.org/10.1007 /0-306-47153-1_33.

76 **we transferred our "individual loyalty":** Ian Tattersall, "In Search of the First Human Home," *Nautilus*, December 2, 2013, http://nautil.us/issue/8/home/in-search-of-the-first-human-home.

75 **Scientist and popular author Jared Diamond:** Jared M. Diamond, *The Third Chimpanzee: The Evolution and Future of the Human Animal* (New York: HarperCollins, 2006). (The first edition was published in 1992.)

76 **a pair of modern human skulls from Ethiopia:** Michael Hopkin, "Ethiopia Is Top Choice for Cradle of *Homo sapiens*," *Nature*, February 16, 2005, https://doi.org/10.1038/news050214-10.

76 **In 2000, two scientists, including a colleague:** Sally McBrearty and Alison S. Brooks, "The Revolution That Wasn't: A New Interpretation of the Origin of Modern Human Behavior," *Journal of Human Evolution* 39, no. 5 (November 2000): 453–563; https://doi.org/10.1006 /jhev.2000.0435.

76 **At one archaeological site, in 2007, an archaeologist named Curtis Marean:** Curtis W. Marean, Miryam Bar-Matthews, Jocelyn Bernatchez, Erich Fisher, Paul Goldberg, Andy I. R. Herries, Zenobia Jacobs, et al., "Early Human Use of Marine Resources and Pigment in South Africa During the Middle Pleistocene," *Nature* 449, no. 7164 (October 18, 2007): 905–8, https://doi.org/10.1038/nature06204.

77 **They had been nosing into caves:** Curtis W. Marean, "When the Sea Saved Humanity," *Scientific American*, November 1, 2012, https://doi.org/10.1038/scientificamericanhuman1112-52.

77 **There were also sharp and sophisticated pieces of stone blades:** Pete Spotts, "When Did Humans Get Smart? Maybe a Lot Earlier Than Some Thought," *Christian Science Monitor*, November 7, 2012, https://www.csmonitor.com/Science/2012/1107/When-did-humans-get -smart-Maybe-a-lot-earlier-than-some-thought.

77 **the first-known seafood-eaters:** Heather Pringle, "The Brine Revolution," *Hakai Magazine*, April 22, 2015, https://www.hakaimagazine.com/features/brine-revolution/.

77 **Meanwhile in Kenya:** Richard Potts, "Turbulent Environment Set the Stage for Leaps in Human Evolution and Technology 320,000 Years Ago," The Conversation, October 21, 2020, http://theconversation.com/turbulent-environment-set-the-stage-for-leaps-in-human -evolution-and-technology-320–000-years-ago-148381.

77 **In the exposed sediment layers, the scientists found one set of artifacts:** Richard Potts, René Dommain, Jessica W. Moerman, Anna K. Behrensmeyer, Alan L. Deino, Simon Riedl, Emily J.

Beverly, et al., "Increased Ecological Resource Variability During a Critical Transition in Hominin Evolution," *Science Advances* 6, no. 43 (October 21, 2020): eabc8975, https://doi.org/10.1126/sciadv.abc8975.

78 **you would probably need to trade:** Alison S. Brooks, John E. Yellen, Richard Potts, Anna K. Behrensmeyer, Alan L. Deino, David E. Leslie, Stanley H. Ambrose, et al., "Long-Distance Stone Transport and Pigment Use in the Earliest Middle Stone Age," *Science* 360, no. 6384 (March 15, 2018): 90–94, https://doi.org/10.1126/science.aao2646; Richard Potts, Anna K. Behrensmeyer, J. Tyler Faith, Christian A. Tryon, Alison S. Brooks, John E. Yellen, Alan L. Deino, et al., "Environmental Dynamics During the Onset of the Middle Stone Age in Eastern Africa," *Science* 360, no. 6384 (March 15, 2018): 86–90, https://doi.org/10.1126/science.aao2200.

78 **This was *Homo sapiens*:** Alan L. Deino, Anna K. Behrensmeyer, Alison S. Brooks, John E. Yellen, Warren D. Sharp, and Richard Potts, "Chronology of the Acheulean to Middle Stone Age Transition in Eastern Africa," *Science* 360, no. 6384 (March 15, 2018): 95–98, https://doi.org/10.1126/science.aao2216.

78 **what seemed like a fully developed sense of human imagination:** Ann Gibbons, "Signs of Symbolic Behavior Emerged at the Dawn of Our Species in Africa," *Science*, March 15, 2018, https://www.science.org/content/article/signs-symbolic-behavior-emerged-dawn-our-species-africa.

78 **In 2017, another skull:** Ann Gibbons, "Oldest Members of Our Species Discovered in Morocco," *Science,* June 7, 2017, https://www.sciencemag.org/news/2017/06/world-s-oldest-homo-sapiens-fossils-found-morocco.

78 **We had used these smarts:** Richard Potts, "Adaptability and the Continuation of Human Origins" (unpublished manuscript, January 11, 2021), typescript.

78 **Climate studies show that in the early existence of *Homo sapiens*:** Marean, "When the Sea Saved Humanity."

79 **One of the oldest-known examples of early *Homo sapiens* architecture:** Brian Handwerk, "A Mysterious 25,000-Year-Old Structure Built of the Bones of 60 Mammoths," *Smithsonian Magazine*, March 16, 2020, https://www.smithsonianmag.com/science-nature/60-mammoths-house-russia-180974426/; Alexander J. E. Pryor, David G. Beresford-Jones, Alexander E. Dudin, Ekaterina M. Ikonnikova, John F. Hoffecker, and Clive Gamble, "The Chronology and Function of a New Circular Mammoth-Bone Structure at Kostenki 11," *Antiquity* 94, no. 374 (April 2020): 323–41, https://doi.org/10.15184/aqy.2020.7.

79 **"Home is a place or places on the landscape":** Jude Isabella, "The Caveman's Home Was Not a Cave," *Nautilus*, December 5, 2013, https://nautil.us/the-cavemans-home-was-not-a-cave-1390/.

80 **"We call the country mother":** Michael Jackson, *At Home in the World* (Durham, NC: Duke University Press, 2000), 22.

79 **The Indigenous San people of the Kalahari Desert:** Jerald Kralik, "Core High-Level Cognitive Abilities Derived from Hunter-Gatherer Shelter Building" (paper presented at the Sixteenth International Conference on Cognitive Modeling, Madison, Wisconsin, July 2018).

80 **In May 2021, the observatory at Mauna Loa:** "Carbon Dioxide Peaks Near 420 Parts per Million at Mauna Loa Observatory," *NOAA Research News*, June 7, 2021, https://research.noaa.gov/article/ArtMID/587/ArticleID/2764/Coronavirus-response-barely-slows-rising-carbon-dioxide.

80 **According to the IPCC, even with the most ambitious:** "Summary for Policymakers," in *Climate Change 2021: The Physical Science Basis; Working Group I Contribution to the Sixth Assessment Report of the Intergovernmental Panel on Climate Change*, ed. Valérie Masson-Delmotte, Panmao Zhai, Anna Pirani, Sarah L. Connors, Clotilde Péan, Yang Chen, Leah Goldfarb, et al., 18.

5: The Thaw

Much of the reporting in this chapter and chapter 10 was initially undertaken for two magazine articles: "Evicted by Climate Change," *Hakai Magazine*, June 1, 2016, and "The Village at the Edge of the Anthropocene," *Sierra* magazine, February 27, 2020. Some of the details on permafrost were also from reporting my story, "Tunnel Vision: Lessons in the Impermanence of Permafrost," in *Undark* magazine, April 28, 2020.

Oral history still holds an important place in Newtok, and documentation of past events in this remote region is sometimes sparse. I relied on oral accounts of the village's past to tell parts of this story.

Yup'ik scholar Alice Rearden provided a review of both chapters on Newtok and verified and supplemented some details about Yup'ik language and culture.

82 **home, n.:** *home*, A.1.4., *OED Online*.

82 **The early 1990s was a period:** Northern Economics, *The Anchorage Economy from 1980 to the Present*, July 2004, https://aedcweb.com/wp-content/uploads/2014/10/History-of-the-Anchorage -Economy-from-1980-2004.pdf.

83 **including some Alaska Natives displaced:** Alaska Natives Commission, *Final Report*, vol. 1. 1996, http://www.alaskool.org/resources/anc/anc00.htm#dedication.

83 **Niugtaq, refers to the rustling of grass:** Ann Fienup-Riordan, ed., and Alice Rearden, trans., *Qaluyaarmiuni Nunamtenek Qanemciput: Our Nelson Island Stories* (Seattle: University of Washington Press, 2013), xx.

84 **In 1984, engineering consultants:** Woodward-Clyde Consultants, *Ninglick River Erosion Assessment: Addendum*, November 29, 1984.

85 **By 1994, the Newtok council had started planning:** Newtok Planning Group, "Newtok Village Relocation History, Part Two: Early Efforts to Address Erosion," Alaska Department of Commerce, Community, and Economic Development, Division of Community and Regional Affairs, accessed October 22, 2021, https://www.commerce.alaska.gov/web /dcra/PlanningLandManagement/NewtokPlanningGroup/NewtokVillageRelocationHistory /NewtokHistoryPartTwo.aspx.

85 **In 1996, during a flood, the massive Ningliq River ate away:** Newtok Planning Group, "Newtok Village Relocation History, Part Three: Progressive Erosion Brings New Problems," Alaska Department of Commerce, Community, and Economic Development, Division of Community and Regional Affairs, accessed October 22, 2021, https://www.commerce.alaska.gov/web /dcra/PlanningLandManagement/NewtokPlanningGroup/NewtokVillageRelocationHistory /NewtokHistoryPartThree.aspx.

86 **Some parts of the Arctic are, according to a civil engineer:** Based on a presentation given by Tom Ravens at the University of Alaska Anchorage, at a side event during the Obama administration's 2015 GLACIER conference. It was first quoted in my January/February 2016 *Audubon* story, "How One Alaskan Community Is Attempting to Adapt to Climate Change."

86 **breathing back out hundreds of millions of tons of carbon:** J. Richter-Menge, M. L. Druckenmiller, and M. Jeffries, eds., *Arctic Report Card 2019*, National Oceanic and Atmospheric Administration, https://arctic.noaa.gov/Portals/7/ArcticReportCard/Documents/Arctic ReportCard_full_report2019.pdf.

86 **Moreover, arctic regions are warming:** Arctic Monitoring and Assessment Programme, *Arctic Climate Change Update 2021: Key Trends and Impacts; Summary for Policy-Makers*, 2021, Tromsø, Norway.

86 **possibly because sea ice:** Aiguo Dai, Dehai Luo, Mirong Song, and Jiping Liu, "Arctic Amplification Is Caused by Sea-Ice Loss Under Increasing CO_2," *Nature Communications* 10, no. 1 (January 10, 2019): 121, https://doi.org/10.1038/s41467-018-07954-9.

87 **An excess of heat in the Arctic can destabilize the global weather system:** Yereth Rosen, "A Changing Bering Sea Is Influencing Weather Far to the South, Scientists Say," ArcticToday, February 19, 2021, https://www.arctictoday.com/a-changing-bering-sea-is-influencing-weather -far-to-the-south-scientists-say/; D. Coumou, G. Di Capua, S. Vavrus, L. Wang, and S. Wang, "The Influence of Arctic Amplification on Mid-Latitude Summer Circulation," *Nature Communications* 9, no. 1 (August 20, 2018): 2959, https://doi.org/10.1038/s41467-018-05256-8.

88 **Beyond lay the Yukon-Kuskokwim:** "About the YK Delta," Yukon-Kuskokwim Health Corporation, accessed October 22, 2021, https://www.ykhc.org/story/about-yk/.

88 **with a population of twenty-seven thousand people:** Anna Rose MacArthur, "Y-K Delta Population Grew About 9% in Past Decade, According to US Census," KYUK, August 16, 2021, https://www.kyuk.org/economy/2021-08-16/y-k-delta-population-grew-about-9-in-past -decade-according-to-us-census; "2020 Census Data for Redistricting," Alaska Department of Labor and Workforce Development Research and Analysis, accessed October 22, 2021, https:// live.laborstats.alaska.gov/cen/2020/downloads.

88 **Between spring and fall, hundreds of millions of migrating birds:** Jim Williams and Paul J. Baicich, "Unbirded Alaska," *BirdWatching,* updated October 2, 2018, https://www .birdwatchingdaily.com/locations-travel/featured-destinations/unbirded-alaska/.

91 **Between 1954 and 2003, the average rate of erosion:** Newtok, Alaska; ASCG Incorporated of Alaska; and Bechtol Planning and Development, *Village of Newtok, Alaska: Local Hazards Mitigation Plan,* March 12, 2008, https://www.commerce.alaska.gov/web/portals/4/pub/2008 _Newtok_HMP.pdf.

91 **as much as 113 linear feet of earth collapsed:** U.S. Army Corps of Engineers, Alaska District, *Alaska Baseline Erosion Assessment, AVETA Report Summary—Newtok, Alaska,* 2006.

91 **In 2003, the U.S. Congress passed a law:** Newtok Planning Group, "Newtok Village Relocation History, Part Two"; Pub. L. No. 108–129, 117 Stat. 1358 (2003), https://uscode.house.gov /statutes/pl/108/129.pdf.

91 **In the fall of 2005, such a storm surrounded:** Newtok Planning Group, "Newtok Village Relocation History, Part Three."

91 **The only place that could provide the "barest shelter":** U.S. Army Corps of Engineers, *Revised Environmental Assessment Finding of No Significant Impact: Newtok Evacuation Center; Mertarvik, Nelson Island, Alaska,* July 2008, https://www.poa.usace.army.mil/Portals/34 /docs/civilworks/reports/Newtok%20Evacuation%20Center%20EA%20&%20FONSI%20 July%2008.pdf.

91 **But Newtok's prospects began to improve:** Newtok Planning Group, "Newtok Village Relocation History, Part Four: The Newtok Planning Group," Alaska Department of Commerce, Community, and Economic Development, Division of Community and Regional Affairs, accessed October 22, 2021, https://www.commerce.alaska.gov/web/dcra/PlanningLandManagement /NewtokPlanningGroup/NewtokVillageRelocationHistory/NewtokHistoryPartFour.aspx.

93 **The closest word to the English *home*:** Steven A. Jacobson, comp., *Yup'ik Eskimo Dictionary,* vols. 1 and 2 (Fairbanks: Alaska Native Language Center, 2012), 360.

95 **Yup'ik people started to camp:** Newtok Planning Group, "Newtok Village Relocation History, Part One."

97 **Once they passed through the cone:** "Fish Trap," Smithsonian Institution, Alaska Native Collections, accessed October 22, 2021, https://alaska.si.edu/record.asp?id=319; "Black fish Traps," YouTube video, 4:00, posted by "Maurice Nanalook," January 8, 2010, https://www.youtube .com/watch?v=GQ1E9ZBLA2E.

97 **Found only in Alaska and Siberia, blackfish:** "Alaska Blackfish," Alaska Department of Fish and Game, 1994, http://www.adfg.alaska.gov/static/education/wns/alaska_blackfish.pdf.

97 **"The world is changing following its people":** Ann Fienup-Riordan and Alice Rearden, *Ellavut: Our Yup'ik World and Weather; Continuity and Change on the Bering Sea Coast* (Seattle: University of Washington Press, 2013), 42.

98 **In 2007, the *New York Times* named:** William Yardley, "Victim of Climate Change, a Town Seeks a Lifeline," *New York Times,* May 27, 2007, https://www.nytimes.com/2007/05/27/us /27newtok.html.

98 **Reuters also published:** Yereth Rosen, "Climate Change Endangers Alaska's Coastal Villages," Reuters, November 9, 2007, https://www.reuters.com/article/environment-alaska-erosion-dc -idUSN0745014320071109.

98 **"We're United States citizens":** Elizabeth Arnold, "Tale of Two Alaskan Villages," National Public Radio, July 29, 2008, https://www.npr.org/templates/story/story.php?storyId=93029431.

99 **And there's the extensive power grid:** Harold D. Wallace, Jr., "Power from the People: Rural Electrification Brought More Than Lights," National Museum of American History, Behring Center, February 11, 2016, https://americanhistory.si.edu/blog/rural-electrification.

99 **In 2008, the community requested help from the Department of Defense:** Newtok Planning Group, "Innovative Readiness Training (IRT) Program Mertarvik," Alaska Department of Commerce, Community, and Economic Development, Division of Community and Regional Affairs, accessed October 22, 2021, https://www.commerce.alaska.gov/web/dcra/PlanningLand Management//NewtokPlanningGroup/IRTMertarvik.aspx.

99 **The conditions were difficult:** Alex Demarban, "Newtok's Opening Move: Military Lays Groundwork to Shift Flood-Threatened Village," *First Alaskans*, October/November 2009, https://irt.defense.gov/Portals/57/Documents/news/First_Alaskans_Nov2009.pdf.

99 **In 2011, the community and the Newtok Planning Group:** Community of Newtok and Newtok Planning Group, *Relocation Report: Newtok to Mertarvik*, August 2011, https://www .commerce.alaska.gov/WEB/PORTALS/4/PUB/MERTARVIK_RELOCATION_REPORT _FINAL.PDF.

100 **So in 2011 and 2012, Newtok made a second attempt to build at Mertarvik:** Newtok Planning Group, "Mertarvik Housing," Alaska Department of Commerce, Community, and Economic Development, Division of Community and Regional Affairs, accessed October 23, 2021, https:// www.commerce.alaska.gov/web/dcra/PlanningLandManagement/NewtokPlanningGroup /MertarvikHousing.aspx.

100 **camping out in a mostly finished house:** Suzanne Goldenberg, "One Family's Great Escape," *Guardian*, May 13, 2013, http://www.theguardian.com/environment/interactive/2013/may /13/alaskan-family-newtok-mertarvik.

101 **One article in the *Guardian*:** Suzanne Goldenberg, "'It's Happening Now . . . the Village Is Sinking'," *Guardian*, May 15, 2013, http://www.theguardian.com/environment/interactive /2013/may/15/newtok-safer-ground-villagers-nervous.

101 **After a new council (which called itself the Newtok Village Council) won:** Charles Enoch, "With a New Tribal Council, Newtok Re-Establishes Efforts for Relocation," KYUK, September 22, 2015, https://www.ktoo.org/2015/09/21/new-tribal-council-newtok-re-establishes-efforts -relocation/; Suzanne Goldenberg, "Relocation of Alaska's Sinking Newtok Village Halted," *Guardian*, August 5, 2013, https://www.theguardian.com/environment/2013/aug/05/alaska -newtok-climate-change; Rachel D'Oro, "Newtok Power Struggle Heats Up as New Leaders Try to Evict Old Ones," *Anchorage Daily News*, December 5, 2015, updated May 31, 2016, https:// www.adn.com/rural-alaska/article/newtok-power-struggle-heats-new-leaders-try-evict-old-ones /2015/12/05/.

101 **Work on the houses and on a new Mertarvik Evacuation Center:** Lisa Demer, "Audits Question Spending for Still-Unfinished Evacuation Center," *Anchorage Daily News*, August 30, 2015, updated May 31, 2016, https://www.adn.com/rural-alaska/article/audits-question -spending-still-unfinished-evacuation-center/2015/08/30/.

102 **A particularly monstrous one:** Alana Semuels, "The Village That Will Be Swept Away," *Atlantic*, August 30, 2015, https://www.theatlantic.com/business/archive/2015/08/alaska-village-climate -change/402604/.

102 **In May 2014, a legal ruling from the U.S. Department of the Interior:** Newtok Village v. Patrick, Order Denying Motion to Set Aside Default Judgment (Docket 65), National Indian Law Library, February 25, 2021, https://narf.org/nill/bulletins/federal/documents/newtok_v _patrick.html.

104 **"The White Seal," one of the stories:** Rudyard Kipling, "The White Seal," in *The Jungle Book* (New York: Century, 1894), 135–71.

104 **An animated version of the story:** "The White Seal (1975)," Chuck Jones Enterprises, Daily-Motion video, 24:26, posted by "Film Gorillas," https://www.dailymotion.com/video/x244acq.

104 **But the people of St. Paul, the Unangan:** Nathaniel Herz, "For Decades, the Government Stood Between the Unangan People and the Seals They Subsist On. Now, That's Changing," *Alaska Public Media*, March 7, 2019, https://www.alaskapublic.org/2019/03/06/for-decades -the-government-stood-between-the-unangan-people-and-the-seals-they-subsist-on-now -thats-changing/; Larry Merculieff, "The Key to Conflict Resolution: Reconnection with the Sacred; The Pribilof Aleut Case Study," *Cultural Survival Quarterly*, September 1995, http:// www.culturalsurvival.org/publications/cultural-survival-quarterly/key-conflict-resolution -reconnection-sacred-pribilof-aleut; E. J. Guarino, "Slaves of the Northern Fur Trade: An American Tragedy," King Galleries, January 1, 2015, https://kinggalleries.com/slaves-northern-fur -trade-american-tragedy/.

104 **In some traditional stories:** Ann Fienup-Riordan, *The Nelson Island Eskimo: Social Structure and Ritual Distribution* (Anchorage: Alaska Pacific University Press, 1983), 177–81; Fienup-Riordan and Rearden, *Our Nelson Island Stories*, xvi.

108 **Romy Cadiente, who had been hired:** Rachel Waldholz, "Newtok Asks: Can the U.S. Deal with Slow-Motion Climate Disasters?" KTOO, January 6, 2017, https://www.ktoo.org/2017 /01/06/newtok-asks-can-u-s-deal-slow-motion-climate-disasters/.

108 **In January 2017, in the last days:** U.S. Federal Emergency Management Agency, *Preliminary Damage Assessment Report: Newtok Village—Flooding, Persistent Erosion, and Permafrost Degradation; Denial*, January 18, 2017, https://www.fema.gov/sites/default/files/2020-03/PDAReportDenial -NewtokVillage.pdf.

108 **On a Sunday morning in late March 2017, Tom John:** Anna Rose MacArthur, "Search Continues into Its Third Week for Missing Newtok Seal Hunter Tom John," KYUK, April 12, 2017, https://www.kyuk.org/post/search-continues-its-third-week-missing-newtok-seal-hunter-tom -john; "Tom John," Charley Project, last updated June 8, 2018, http://charleyproject.org/case /tom-john; Jon-Paul Rios, "Coast Guard Suspends Search for Overdue Hunter Near Newtok," *Alaska Native News*, March 30, 2017, https://alaska-native-news.com/coast-guard-suspends -search-for-overdue-hunter-near-newtok/27482/.

6: The Explosion

Some of the reporting in Chapters 6 and 12 originally appeared in my article "From Vacant City Lots to Food on the Table," *YES!* magazine, Fall 2010.

109 **home, n.:** *home*, n.2., and *home*, v., *Webster's New World Dictionary of the American Language, Second College Edition*, ed. David B. Guralnik (n.p.: World, 1978), 670–71.

109 **The city of Richmond grew up:** "History," Chevron Richmond, accessed October 23, 2021, https://richmond.chevron.com/about/history; Shirley Ann Wilson Moore, *To Place Our Deeds: The African American Community in Richmond, California, 1910–1963* (Berkeley: University of California Press, 2001).

109 **"Wonder City," the port and shipyard humming:** Frederick J. Hulaniski, *The History of Contra Costa County, California* (Berkeley, CA: Elms, 1917), 327.

110 **part of the Great Migration:** Mahlia Posey, "A New Great Migration: The Disappearance of the Black Middle Class," *Richmond Confidential*, December 10, 2015, https://richmondconfidential .org/2015/12/10/a-new-great-migration-the-disappearance-of-the-black-middle-class/.

110 **many of the same troubles in California as in the South:** Richard Rothstein, chap. 1 in *The Color of Law: A Forgotten History of How Our Government Segregated America* (New York: Liveright, 2017); "Climate Safe Neighborhoods," Groundwork Richmond, accessed February 3, 2021, https://gwmke.maps.arcgis.com/apps/Cascade/index.html?appid=720a1dca15ec426 5a94d012cf06fbbf4; Gary Kamiya, "When WWII Brought Blacks to the East Bay, Whites Fought for Segregation," *San Francisco Chronicle*, November 23, 2018, https://www.sfchronicle .com/chronicle_vault/article/When-WWII-brought-blacks-to-the-East-Bay-whites -13417228.php.

110 **"They settled in the foothills":** Isabel Wilkerson, *The Warmth of Other Suns: The Epic Story of America's Great Migration* (New York: Random House, 2010), 236.

110 **And its downtown restaurants:** Jeffrey Callen, "History of Richmond's Iron Triangle," February 3, 2016, https://www.slideshare.net/JeffreyCallenPhD/history-of-richmonds-iron-triangle?from _action=save.

110 **Some chroniclers of Richmond:** Rosie the Riveter WWII Home Front National Historical Park, California, outdoor exhibit and interpretive signs, April 15, 2014; "Eddie Eaton: In Search of the California Dream; From Houston, Texas, to Richmond, California, 1943," interview by Judith K. Dunning in *On the Waterfront: An Oral History of Richmond, California*, 1986, https:// oac.cdlib.org/view?docId=hb0b69n7gm&brand=default&doc.view=entire_text.

111 **This kind of discrimination:** Nikole Hannah-Jones, "From the Magazine: 'It Is Time for Reparations,'" *New York Times Magazine*, June 24, 2020, https://www.nytimes.com/interactive /2020/06/24/magazine/reparations-slavery.html.

111 **Even the company's seemingly benign:** Tom Lochner, "Richmond's Hilltop Mall Rebrands on Way to Hoped-For Revival," *East Bay Times*, September 5, 2017, https://www.eastbaytimes .com/2017/09/05/richmonds-hilltop-mall-rebrands-on-way-to-hoped-for-revival/; Steve Early, *Refinery Town: Big Oil, Big Money, and the Remaking of an American City* (Boston: Beacon Press, 2017), 25.

112 **A half block away and around the corner was a local history museum:** "Richmond, Contra Costa County: San Francisco Bay/Delta/Sacramento Area, San Francisco Bay Area Region; Richmond Museum," Carnegie Libraries of California, accessed June 30, 2020, https://www .carnegie-libraries.org/california/richmond.html.

113 **But in the mid-1980s, the crack cocaine epidemic:** Roland G. Fryer, Jr., Paul S. Heaton, Steven D. Levitt, and Kevin M. Murphy, "Measuring Crack Cocaine and Its Impact," *Economic Inquiry* 51, no. 3 (July 2013): 1651–81, https://doi.org/10.1111/j.1465-7295.2012.00506.x; Deborah J. Vagins and Jesselyn McCurdy, *Cracks in the System: Twenty Years of the Unjust Federal Crack Cocaine Law* (New York: American Civil Liberties Union, October 2006).

114 **Then, on an April day in 1989:** Patrick Lee, "Chevron Fined $877,000 for Refinery Fire," *Los Angeles Times*, September 27, 1989, https://www.latimes.com/archives/la-xpm-1989-09-27-fi-266 -story.html; "Chevron Refinery Incidents," SFGATE, March 26, 1999, https://www.sfgate.com /news/article/CHEVRON-REFINERY-INCIDENTS-3091498.php; "Chevron Oil Refinery Richmond; Richmond, CA," NOAA Incident News, April 10, 1989, accessed October 23, 2021, https://incidentnews.noaa.gov/incident/6689#!.

117 **In a 2009 poll, 57 percent of voters of color:** David Metz and Lori Weigel, "Key Findings from National Voter Survey on Conservation Among Voters of Color" (memorandum from Fairbank, Maslin, Maullin & Associates and Public Opinion Strategies, October 6, 2009); Matthew Ballew, Edward Maibach, John Kotcher, Parrish Bergquist, Seth Rosenthal, Jennifer Marlon, and Anthony Leiserowitz, "Which Racial/Ethnic Groups Care Most About Climate Change?" Yale Program on Climate Change Communication, April 16, 2020, https:// climatecommunication.yale.edu/publications/race-and-climate-change/.

117 **But Hurricane Katrina offered a prelude:** Andrea Thompson, "10 Years Later: Was Warming to Blame for Katrina?" Climate Central, August 27, 2015, https://www.climatecentral.org /news/katrina-was-climate-change-to-blame-19377; Scott Gold, "Trapped in the Superdome: Refuge Becomes a Hellhole," *Seattle Times*, September 1, 2005, https://www.seattletimes.com /nation-world/trapped-in-the-superdome-refuge-becomes-a-hellhole/.

117 **When the levees that were supposed to protect the city broke:** "Lower Ninth Ward Levee Breaches in 2005," New Orleans Historical, accessed October 7, 2021, https://neworleanshistorical .org/items/show/288; "Lower Ninth Ward Statistical Area," Data Center, updated February 24, 2021, https://www.datacenterresearch.org/data-resources/neighborhood-data/district-8/lower -ninth-ward/.

118 **"The income disparity between rich and poor is so great":** Gary Rivlin, "White New Orleans Has Recovered from Hurricane Katrina. Black New Orleans Has Not," TalkPoverty, August 29, 2016, https://talkpoverty.org/2016/08/29/white-new-orleans-recovered-hurricane-katrina -black-new-orleans-not/.

118 **Already, people of color and anyone living in poverty:** Michael Mascarenhas, Ryken Grattet, and Kathleen Mege, "Toxic Waste and Race in Twenty-First Century America: Neighborhood Poverty and Racial Composition in the Siting of Hazardous Waste Facilities," *Environment and Society* 12, no. 1 (September 1, 2021): 108–26, https://doi.org/10.3167/ares.2021.120107.

118 **A 2009 study found that in Los Angeles:** Rachel Morello-Frosch, Manuel Pastor, James Sadd, and Seth B. Shonkoff, *The Climate Gap: Inequalities in How Climate Change Hurts Americans and How to Close the Gap*, May 2009, https://dornsife.usc.edu/assets/sites/242/docs/The _Climate_Gap_Full_Report_FINAL.pdf.

118 **Already, Black American communities face greater risks from flooding:** Jack Graham, "Black Neighborhoods in the US Face Higher Flooding Risks Due to Climate Change," Global Citizen, March 16, 2021, https://www.globalcitizen.org/en/content/black-neighborhoods-flooding -climate-change/.

119 **In sea level rise projections, these landscapes:** Jean Tepperman, "Oakland's Poorest Neighborhoods Will Be the Most Susceptible to Flooding Due to Climate Change and Sea-Level Rise," *East Bay Express*, April 19, 2017, https://eastbayexpress.com/oaklands-poorest -neighborhoods-will-be-the-most-susceptible-to-flooding-due-to-climate-change-and-sea -level-rise-2-1/.

120 **about two-thirds of global carbon emissions:** IPCC, *Climate Change 2014: Mitigation of Climate Change; Working Group III Contribution to the Fifth Assessment Report of the Intergovernmental Panel on Climate Change* (Cambridge: Cambridge University Press, 2015), https://doi .org/10.1017/CBO9781107415416.

120 **home to 55 percent of the global population:** "Urban Development: Overview," World Bank, accessed July 15, 2020, https://www.worldbank.org/en/topic/urbandevelopment/overview.

120 **more than 60 percent of carbon emissions:** "Cities and Pollution," United Nations, accessed October 24, 2021, https://www.un.org/en/climatechange/climate-solutions/cities-pollution.

121 **The dictionary definition of** *front line*: *front line*, n.A.1.b., *OED Online*.

122 **In 1991, poet Adrienne Rich wrote:** Adrienne Rich, "An Atlas of the Difficult World," in *An Atlas of the Difficult World: Poems, 1988–1991* (New York: W. W. Norton, 1991).

122 **full of both toxicity and "human wreckage":** Early, *Refinery Town*.

125 **According to an extensive 2011 investigation:** Jim Morris, Chris Hamby, and M. B. Pell, "Regulatory Flaws, Repeated Violations Put Oil Refinery Workers at Risk," Center for Public Integrity, published February 28, 2011, updated May 19, 2014, https://publicintegrity.org /inequality-poverty-opportunity/workers-rights/regulatory-flaws-repeated-violations-put-oil -refinery-workers-at-risk/.

125 **between 2001 and 2003, the EPA noted:** Antonia Juhasz, *The Tyranny of Oil: The World's Most Powerful Industry—and What We Must Do to Stop It* (New York: HarperCollins, 2009).

125 **Violent crime had been decreasing:** Steve Spiker, Junious Williams, Rachel Diggs, Bill Heiser, and Nic Jay Aulston, *Violent Crime in Richmond: An Analysis of Violent Crime in Richmond, California from January 1, 2005 to December 31, 2006*, Urban Strategies Council, Oakland, February 26, 2007, revised March 29, 2007.

125 **from Black Panthers offering free breakfast to schoolchildren in the 1960s:** Nancy DeVille, "New Exhibit Highlights Richmond's Connection to Black Panthers," *Richmond Pulse*, January 31, 2016, https://richmondpulse.org/2016/01/31/new-exhibit-highlights-richmonds -connection-to-black-panthers/; Early, *Refinery Town*.

125 **A homegrown Richmond environmental group:** Jacob Soiffer, "Emergence of Environmental Justice in Richmond," FoundSF, 2015, http://www.foundsf.org/index.php?title=Emergence_of _Environmental_Justice_in_Richmond; Sara Bernard, "Henry Clark and Three Decades of Environmental Justice," *Richmond Confidential*, December 6, 2012, https://richmondconfidential .org/2012/12/06/henry-clark-and-three-decades-of-environmental-justice/.

125 **The Asian Pacific Environmental Network began organizing:** "Our History," Asian Pacific Environmental Network, accessed October 24, 2021, https://apen4ej.org/our-history/.

126 **A group of local activists:** Early, *Refinery Town*.

126 **That same autumn, locals set up a sit-in:** Chip Johnson, "Making a Bold Stand for Peace: In Richmond's Tough Iron Triangle Neighborhood, Residents Frustrated with a Spate of Killings Erect an Encampment to Help Stem the Violence," SFGATE, October 11, 2006, https://www .sfgate.com/bayarea/johnson/article/Making-a-bold-stand-for-peace-In-Richmond-s-2468468 .php.

127 **The word *tilth* refers:** *tilth*, n., *OED Online*.

128 **In 2007, a thunderous bang:** Bay Area Air Quality Monitoring District, *Incident Report: Chevron Richmond Refinery (Site #AA0010), Richmond, CA*, January 15, 2007, accessed October 24, 2021, https://www.baaqmd.gov/-/media/files/compliance-and-enforcement/incident-reports /i011507_chevron_update.pdf.

132 **To borrow the words of Adrienne Rich:** Rich, "An Atlas of the Difficult World."

133 **In the fall of 2009, a teenage girl:** "Police: As Many as 20 Present at Gang Rape Outside School Dance," CNN, October 28, 2009, http://www.cnn.com/2009/CRIME/10/27/california.gang .rape.investigation/index.html.

135 **In 2009, the old Ford assembly plant:** Bridgette Meinhold, "Historic Ford Factory Transformed into SunPower Photovoltaics Headquarters," *Inhabitat*, February 21, 2011, https:// inhabitat.com/historic-ford-factory-transformed-into-sunpower-photovoltaics-headquarters/.

135 **Meanwhile, a Safeway closed:** Mitzi Mock, "Why Few Grocery Stores Come to Richmond," *Richmond Confidential*, November 17, 2011, https://richmondconfidential.org/2011/11/17 /why-few-grocery-stores-come-to-richmond/.

135 **Meanwhile, a new study said:** Hasan Dudar, "Childhood Obesity in Contra Costa on the Rise," *Richmond Confidential*, November 10, 2011, https://richmondconfidential.org/2011/11 /10/childhood-obesity-in-contra-costa-on-the-rise/.

135 **Meanwhile, a graffiti artist:** Robert Rogers, "'Nacho' Defaces City, Leaves Few Clues," *Richmond Confidential*, February 25, 2011, https://richmondconfidential.org/2011/02/25/nacho -defaces-city-leaves-bitter-taste/.

135 **A small Richmond seedling company sold:** William Harless, "A New Source of Fertilizer in Richmond—Koi Fish," *Richmond Confidential*, September 21, 2011, https://richmondconfidential .org/2011/09/21/a-new-source-of-fertilizer-in-richmond-koi-fish/.

135 **A world-famous submarine designer:** Hannah Dreier, "Richmond Sub in Million-Dollar Race to the Bottom of the Sea," *Mercury News*, August 17, 2011, updated August 13, 2016, https://www.mercurynews.com/2011/08/17/richmond-sub-in-million-dollar-race-to-the -bottom-of-the-sea/.

135 **A congregation in the Iron Triangle:** Christopher Connelly, "Efforts to Save Iron Triangle Church End in Arrests," *Richmond Confidential*, March 18, 2011, https://richmondconfidential .org/2011/03/18/efforts-to-save-iron-triangle-church-end-in-arrests/.

136 **That summer was also the 110th anniversary:** Chris Treadway, "'Other Days, Other Ways: A Refinery Saga' Opens Tuesday in the Seaver Gallery at the Richmond Museum of History, 400 Nevin Ave.," *Contra Costa Times*, August 6, 2012.

137 **And at about 6:30 that evening:** U.S. Chemical Safety and Hazard Investigation Board, *Final Investigation Report: Chevron Richmond Refinery; Pipe Rupture and Fire*, January 2015, https:// www.csb.gov/chevron-refinery-fire/; Robert J. Lopez, "Huge Fire Continues to Rage at Chevron Oil Refinery in Richmond," *Los Angeles Times*, August 6, 2012, https://latimesblogs.latimes .com/lanow/2012/08/crews-battle-richmond-oil-refinery-fire-richmond.html.

7: The Home Fires Burning

141 **home, v.:** *home*, v., *Merriam-Webster*, accessed September 11, 2021, https://www.merriam -webster.com/dictionary/home; the Britannica Library Reference Center accessed through the Seattle Public Library.

142 **You could call it a karmic adjustment:** Carlene Anders told me her personal details during interviews, but much about her experience with the Carlton Complex was also artfully reported here: Michelle Nijhuis, "As Wildfires Get Bigger, Is There Any Way to Be Ready?" *High Coun-*

try News, August 3, 2015, https://www.hcn.org/issues/47.13/after-a-record-setting-wildfire-a
-washington-county-prepares-for-the-next-one.

142 **It's useful to have what's called a "long-term recovery group":** National Voluntary Organizations Active in Disaster, *Long Term Recovery Guide*, 2012, https://www.nvoad.org/wp-content
/uploads/longtermrecoveryguide-final2012.pdf.

142 **The U.S. Federal Emergency Management Agency eventually supplied:** Jim Carlton, "FEMA
Rejects Aid for Residents Who Lost Homes to Washington Wildfires," *Wall Street Journal*,
August 13, 2014, https://www.wsj.com/articles/fema-rejects-aid-for-residents-who-lost-homes
-to-washington-wildfires-1407974651.

143 **teams of volunteers:** Sydney Brownstone, "The 'New Normal' in Washington State," *Stranger*,
September 2, 2015, https://www.thestranger.com/news/feature/2015/09/02/22795813/the-new
-normal-in-washington-state.

143 **"Disaster chaplains":** Disaster Chaplain Code of Ethics and Guiding Principles, Nebraska
Disaster Chaplain Network of Interchurch Ministries of Nebraska, accessed June 4, 2021,
http://cretscmhd.psych.ucla.edu/nola/Video/Clergy/Articles/Ecumenical/Disaster_Chaplain
_Code_of_Ethics_and_Guiding_Principles.pdf.

144 **The Pateros mayor stepped down shortly:** Lisa Cowan, "Pateros Mayor Steps Down After
Losing House to Wildfire," *Seattle Times*, August 12, 2014, https://www.seattletimes.com/news
/pateros-mayor-steps-down-after-losing-house-to-wildfire/.

144 **fire-resistant siding and metal roofs, which are not generally combustible:** David
Bueche and Tim Foley, *FireWise Construction: Site Design and Building Materials; Based
on the 2009 International Wildland-Urban Interface Code*, Colorado State Forest Service, accessed October 24, 2021, https://static.colostate.edu/client-files/csfs/pdfs/firewise
-construction2012.pdf.

145 **The first home would go:** K. C. Mehaffey, "House Blessings Celebrate Recovery from Wildfire," *Wenatchee World*, January 25, 2016, https://www.wenatcheeworld.com/news/local/house
-blessings-celebrate-recovery-from-wildfire/article_1bebc57c-9dc4-5a4b-855f-4b6c0409c5a5
.html.

145 **A heat wave hit the Pacific Northwest in June:** Andrea Thompson, "Soaring Temps in Pacific
Northwest Shattered Records," Climate Central, July 1, 2015, https://www.climatecentral.org
/news/record-temps-pacific-northwest-19179.

145 **A group of fires had lit and were spreading:** Jim Kershner, "Lightning Storms Ignite Okanogan Complex Fires, Which Will Soon Grow into Some of the Biggest in Washington's
Worst-Ever Wildfire Year, on August 14, 2015," HistoryLink.org, April 19, 2016, https://www
.historylink.org/file/11218.

146 **another wildfire, called the Twisp River Fire:** Washington State Department of Natural Resources and U.S. Forest Service, *Twisp River Fire Fatalities and Entrapments: Interagency Learning Review Status Report*, November 18, 2105, https://www.wildfirelessons.net
/HigherLogic/System/DownloadDocumentFile.ashx?DocumentFileKey=77159beb-18bd
-bdbc-57ad-12fe11d38cd2&forceDialog=0.

146 **news reports said the Okanogan Complex:** Ryan Maye Handy, "Okanogan Complex Largest Fire in Washington History," *Wildfire Today*, August 25, 2015, https://wildfiretoday.com
/tag/okanagan-fire/; Evan Bush and Hal Bernton, "Okanogan Complex Wildfire Now Biggest
in State History," *Seattle Times*, August 24, 2015, https://www.seattletimes.com/seattle-news
/northwest/monday-fire-update/.

148 **Susan and her colleagues had decided to develop a series of computer simulations:** "The
Reburn Project: Evaluation of Burn Mosaics on Subsequent Wildfire Behavior, Severity and
Fire Management Strategies," University of Washington, accessed October 24, 2021, https://
depts.washington.edu/nwfire/reburn/index.html.

149 **the Rocky Hull Fire, which had burned down more than thirty houses:** Havillah Community,
"Havillah Community Wildfire Protection Plan Update," accessed October 24, 2021, https://
www.dnr.wa.gov/publications/rp_burn_cwpphavillah.pdf.

150 **In one volume, he ran across an anecdote:** Robert T. Boyd, *Indians, Fire, and the Land in the Pacific Northwest* (Corvallis: Oregon State University Press, 1999), preview at https://osupress .oregonstate.edu/book/indians-fire-and-land-in-pacific-northwest.

150 **eventually collaborating with a local fire ecologist:** Richard Schellhaas, Anne Conway, and Don Spurbeck, *A Report to the Nature Conservancy on the Historical and Current Stand Structure in the Sinlahekin Wildlife Area*, Schellhaas Forestry, December 13, 2009; Richard Schellhaas, Don Spurbeck, and Anne Conway, *Sinlahekin Wildlife Area Phase III Fire History: A Report to the Washington Department of Fish and Wildlife*, Schellhaas Forestry, June 28, 2019.

153 **Lime Belt Fire, one of the five that would merge into the complex:** Central Washington BAER, *Lime Belt Fire: BAER Briefing*, Okanogan-Wenatchee National Forest, November 15, 2015, http:// centralwashingtonfirerecovery.info/2015/wp-content/uploads/2015/10/LimeBeltBriefing.pdf.

154 **the Washington Department of Fish and Wildlife would describe this fire:** Washington Department of Fish and Wildlife, *Scotch Creek and Sinlahekin Wildlife Areas Management Plan*, April 2017.

159 **In mid-2019, it was done:** Justus Caudell, "Recovery Group Hands Keys to Bests, Staffords," *Tribal Tribune*, July 1, 2019, http://www.tribaltribune.com/news/article_d9cccfbc-9c13-11e9 -b574-4fe20b0d6988.html.

160 **"The indigenous worldview emphasizes":** Robin Kimmerer and Frank Kanawha Lake, "Maintaining the Mosaic: The Role of Indigenous Burning in Land Management," *Journal of Forestry* 99, no. 11 (November 2001): 36–41.

161 **It had burned furiously three years previously:** U.S. Forest Service, *Norse Peak Fire 2017: Burned-Area Report*, October 19, 2017, http://centralwashingtonfirerecovery.info/2017/wp -content/uploads/2017/11/Norse-Peak-Report.pdf.

161 **the Palmer Fire scorched nearly eighteen thousand acres:** "Palmer Fire Information," Inci-Web, Incident Information System, accessed June 10, 2021, https://inciweb.nwcg.gov/incident /7029/.

161 **The *Seattle Times* alluded:** Hal Bernton, "Northwest Fire Season: Plenty of Blazes but Total Area Burned Much Smaller Than Recent Years," *Seattle Times*, August 21, 2020, https://www .seattletimes.com/seattle-news/northwest-fire-season-plenty-of-blazes-but-total-area-burned -much-smaller-than-recent-years/.

162 **As Labor Day approached, a local television news website:** Erin Mayovsky, "Warming Up for the Holiday with Record Temps Mid Week. Plus, Red Flag Warnings Across the State," Q13 FOX, September 6, 2020, https://www.q13fox.com/weather/warming-up-for-the-holiday-with -record-temps-mid-week-plus-red-flag-warnings-across-the-state.

162 **In Okanogan County, a fire called the Cold Springs:** "Cold Springs Fire Information," Inci-Web, Incident Information System, accessed October 24, 2021, https://inciweb.nwcg.gov /incident/7161/; Jennifer Forsmann, Mariah Valles, and Katherine Barner, "UPDATE: Cold Springs and Pearl Hill Fires Burns 337K Acres Collectively," KHQ, September 9, 2020, https:// www.khq.com/fires/update-cold-springs-and-pearl-hill-fires-burns-337k-acres-collectively /article_a0bd3b38-f114-11ea-b830-4bdb05c15e42.html; Marcy Stamper, "Progress Reported on Containing Cold Springs, Pearl Hill Fires," *Methow Valley News*, September 16, 2020, https://methowvalleynews.com/2020/09/16/progress-reported-on-containing-cold-springs -pearl-hill-fires/.

162 **A young couple who had been camping:** Chris Thomas, "Citrus Heights Couple's 1-Year-Old Son Dies as Family Tried to Outrun Cold Springs Fire," ABC10, September 11, 2020, https:// www.abc10.com/article/news/local/citrus-heights/citrus-heights-native-burned-while-fleeing -cold-springs-fire/103-f42a8099-8255-4e75-af4c-cf404dee0e57.

162 **a 15,000-acre fire ate up:** Rebecca Moss, "A Year After Fire Destroyed Malden, a Grieving Town Slowly Rebuilds," *Seattle Times*, September 5, 2021, https://www.seattletimes.com/seattle -news/the-day-a-wildfire-took-malden/.

163 **On Tuesday, a fire:** Megan Farmer, "Photos: Wildfires Burning Near Tacoma Suburbs," KUOW, September 9, 2020, https://www.kuow.org/stories/fires-continue-across-washington

-state; Drew Mikkelsen, "'Catastrophic': Gov. Inslee Tours Sumner Grade Fire in Bonney Lake," KING5, September 9, 2020, https://www.king5.com/article/news/local/wildfire/gov-inslee-calls-sumner-grade-fire-among-most-catastrophic-in-washington-history/281-132ec1ec-6ee9-4503-bf9e-86cfc3df57c6; Shelby Miller and Graham Johnson, "Sumner Grade Fire Slowing After Destroying Four Homes," KIRO7, September 10, 2020, https://www.kiro7.com/news/local/sumner-grade-fire-slowing-after-destroying-four-homes/EL2QD4CQMJCYVKMJD7IZAQJAFQ/.

163 **In Oregon, "firefighters fought at least thirty-five large blazes":** Sharon Bernstein and Andrew Hay, "Oregon Wildfires Destroy Five Towns, as Three Fatalities Confirmed in California," Reuters, September 9, 2020, https://www.reuters.com/article/uk-usa-wildfires-idUKKBN26036O; Andrew Freedman, Jason Samenow, Kim Bellware, and Emily Wax-Thibodeaux, "Western Wildfires: Evacuations in California and Oregon as Destructive Fire Outbreak Engulfs Region," *Washington Post*, September 9, 2020, https://www.washingtonpost.com/weather/2020/09/09/western-fires-live-updates/; Jay Barmann, "Massive Evacuation Orders Come as Oregon Wildfire Nears Portland Suburb," SFist, September 11, 2020, https://sfist.com/2020/09/11/massive-evacuation-orders-come-as-oregon-wildfire-nears-portland-suburb/.

164 **By the middle of that week:** Farmer, "Wildfires Burning"; Aimee Green, "Air Quality in Portland and Parts of Oregon Is Worse Than in Beijing or Mexico City, Due to Wildfire Smoke," *Oregonian*, September 10, 2020, https://www.oregonlive.com/news/2020/09/air-quality-in-portland-is-worse-than-in-beijing-mumbai-or-mexico-city-due-to-wildfire-smoke.html; Pat Dooris, "More Than 800,000 Acres of Oregon Burned So Far in Historic Wildfires This Week," KGW8, September 9, 2020, https://www.kgw.com/article/news/local/wildfire/mega-fires-rage-across-oregon/283-a6bfb00c-a037-4251-95c3-f7285d168ea3.

164 **The governor of Oregon reported that five towns:** "Gov. Brown: Towns of Detroit, Blue River, Vida, Phoenix and Talent Are 'Substantially Destroyed,'" KGW8, September 9, 2020, https://www.kgw.com/article/news/local/wildfire/gov-brown-press-conference-on-wildfires-in-oregon/283-d8014ecc-cf04-4ba5-bb05-3609141542ab.

164 **At least a couple of the fire starts:** Jamie Parfitt, "FireWatch: One Year Later, There's Still No Suspect in Almeda Fire Investigation," KDRV, September 7, 2021, https://www.kdrv.com/content/news/FireWatch-One-year-later-theres-still-no-suspect-in-Almeda-Fire-investigation-575259561.html; Tony Buhr, "Sheriff Says It's Unlikely Antifa Started Cold Springs Fire," *Wenatchee World*, September 13, 2020, https://www.wenatcheeworld.com/paywalloff/sheriff-says-its-unlikely-antifa-started-cold-springs-fire/article_07305fd8-f5e0-11ea-b5ba-0b8f62692c57.html.

164 **Rumors surged on social media:** Madeleine Carlisle, "FBI Calls Rumors of Extremists Starting Oregon Wildfires 'Untrue,'" *Time*, September 12, 2020, https://time.com/5888371/oregon-wildfires-antifa-proud-boys-rumors-false/.

164 **In Northern California, the August Complex Fire:** J. D. Morris, "California's New Largest-Ever Wildfire: North Coast's August Complex Shatters Record Set Two Years Ago," *San Francisco Chronicle*, September 10, 2020, https://www.sfchronicle.com/california-wildfires/article/North-Coast-complex-is-now-California-s-second-15554767.php.

164 **Three other California fires:** Alex Meier, "SCU, LNU Lightning Complex Fires Become Top 5 Largest Wildfires in CA History," ABC7 San Francisco, October 3, 2020, https://abc7news.com/lnu-lightning-complex-scu-size-largest-fire-california-history-cal/6383626/; "Top 20 Largest California Wildfires," California Department of Forestry and Fire Protection, accessed February 16, 2022, https://www.fire.ca.gov/media/4jandlhh/top20_acres.pdf.

164 **National Guard helicopters airlifted:** Alex Wigglesworth, "As Fire 'Engulfed Everything' Around Campers, an Air Rescue Like No Other in the Sierra," *Los Angeles Times*, September 6, 2020, https://www.latimes.com/california/story/2020-09-06/dramatic-night-airlift-rescues-scores-of-victims-trapped-by-creek-fire-at-mammoth-pool.

165 **The smoke rose into the atmosphere:** Anna Buchmann, "See West Coast Wildfire Smoke Get Sucked into a Cyclone over the Pacific Ocean," *San Francisco Chronicle*, September 14,

2020, https://www.sfchronicle.com/california-wildfires/article/See-West-Coast-wildfire-smoke -get-sucked-into-a-15565759.php.

166 **A few hundred evacuees:** Enrique Pérez de la Rosa, "Fire-Evacuated Farmworkers From Bridgeport Slept in a Brewster Park as the Pearl Hill Fire Burned," Northwest Public Broadcasting, September 18, 2020, https://www.nwpb.org/2020/09/18/fire-evacuated-farmworkers -from-bridgeport-slept-in-a-brewster-park-as-the-pearl-hill-fire-burned/.

166 **In the November 2021 mayoral election:** "Election 2021: Getting to Know the Candidates," *Omak-Okanogan County Chronicle*, October 13, 2021, https://www.omakchronicle .com/free/election-2021-getting-to-know-the-candidates/article_3f71274c-2b81-11ec-b4be -3bcd03ead03f.html.

8: Finding Home Ground

168 **In her childhood in Kentucky, Black essayist and scholar bell hooks:** bell hooks, *Belonging: A Culture of Place* (New York: Routledge, 2009), 121–24.

169 *Localism* **is a** *preference: localism*, n.1.a., *OED Online*.

169 **As a student at Stanford University, bell hooks worked:** hooks, *Belonging*, 6–24, 53–58.

170 **In the rural South, for instance, discrimination in property law:** Summer Sewell, "There Were Nearly a Million Black Farmers in 1920. Why Have They Disappeared?" *Guardian*, April 29, 2019, http://www.theguardian.com/environment/2019/apr/29/why-have-americas-black -farmers-disappeared.

171 **The acronym NIMBY probably first appeared:** Danny De Vaal, "What Is a Nimby, What Does the Acronym Stand For and Where Did It Come From?" *Sun* (U.K.), March 5, 2018, https://www.thesun.co.uk/news/5727581/nimby-acronym-where-did-it-come-from-building/; "NIMBY," Online Etymology Dictionary, accessed October 24, 2021, https://www.etymonline .com/word/nimby.

171 **With a family fortune derived from shipbuilding, coal, and steel, Ridley:** Patrick Cosgrave, "Obituary: Lord Ridley of Liddesdale," *Independent*, March 6, 1993, https://www.independent .co.uk/news/people/obituary-lord-ridley-of-liddesdale-1495860.html.

171 **When Margaret Thatcher was elected prime minister in 1979:** Margaret Thatcher, "Nicholas Ridley Memorial Lecture," November 22, 1996, full transcript published by Margaret Thatcher Foundation, https://www.margaretthatcher.org/document/108368; Christine Berry, "Thatcher Had a Battle Plan for Her Economic Revolution—Now the Left Needs One Too," openDemocracy, October 28, 2019, https://www.opendemocracy.net/en/oureconomy/thatcher-had -a-battle-plan-for-her-economic-revolution-now-the-left-needs-one-too/.

171 **Initially, neoliberalism was partly a reaction against communism:** George Monbiot, "Neoliberalism—the Ideology at the Root of All Our Problems," *Guardian*, April 15, 2016, http:// www.theguardian.com/books/2016/apr/15/neoliberalism-ideology-problem-george-monbiot.

171 **I am simplifying and narrowing this discussion:** Naomi Klein, *This Changes Everything: Capitalism vs. the Climate* (New York: Simon & Schuster, 2014); Naomi Klein, *The Shock Doctrine: The Rise of Disaster Capitalism* (New York: Picador, 2010).

171 **the British government was pushing for new nuclear power development:** Ian Welsh, "The NIMBY Syndrome: Its Significance in the History of the Nuclear Debate in Britain," *British Journal for the History of Science* 26, no. 1 (1993): 15–32.

172 **Around the same time in the United States, Wendell Berry:** Wendell Berry, "Higher Education and Home Defense," in *Home Economics: Fourteen Essays* (Berkeley, CA: Counterpoint, 2009), 49–53.

172 **In 1995, Friends of the Earth revealed:** "Nirex and Their Nuclear Waste Dump," *Nuclear Monitor*, June 16, 1995, https://wiseinternational.org/nuclear-monitor/433–434/nirex-and -their-nuclear-waste-dump.

172 **In 1997, the group leaked a memo:** Rob Edwards, "Nirex Betrays Nerves over Nuclear Dump," *New Scientist*, January 25, 1997, https://www.newscientist.com/article/mg15320660-900 -nirex-betrays-nerves-over-nuclear-dump/.

173 **Ridley himself was later labeled:** "Would YOU Live Next to a Nimby?" May 21, 2002, BBC News, http://news.bbc.co.uk/2/hi/uk_news/2000000.stm.

173 **In 2011, *Time* magazine named NIMBYism:** Bryan Walsh, "Top 10 Green Trends: 5. NIM-BYism," December 7, 2011, http://content.time.com/time/specials/packages/article/0,28804,2101344_2101393_2101398,00.html.

173 **Cape Wind, a proposed offshore wind power:** Amanda Little, "RFK Jr. and Other Prominent Enviros Face Off over Cape Cod Wind Farm," *Grist*, January 13, 2006, https://grist.org/article/capecod/; Katharine Q. Seelye, "After 16 Years, Hopes for Cape Cod Wind Farm Float Away," *New York Times*, December 19, 2017, https://www.nytimes.com/2017/12/19/us/offshore-cape-wind-farm.html.

173 **Kennedy insisted his primary concern:** Robert F. Kennedy, Jr., "An Ill Wind Off Cape Cod," *New York Times*, December 16, 2005, https://www.nytimes.com/2005/12/16/opinion/an-ill-wind-off-cape-cod.html.

173 **though some evidence from the U.K. and Belgium:** "Offshore Renewable Energy Improves Habitat, Increases Fish," Rhode Island Sea Grant, June 26, 2020, https://seagrant.gso.uri.edu/offshore-renewable-energy-improves-habitat-increases-fish/.

173 **In response, a group of prominent environmentalists:** "Over 150 Activists Send Letter Asking Kennedy to Reconsider Position," *Grist*, January 7, 2006, https://grist.org/article/enviros-call-on-rfk-jr-to-support-cape-wind-project/.

174 **For one thing, there is no clear definition:** Kate Burningham, Julie Barnett, and Diana Thrush, "The Limitations of the NIMBY Concept for Understanding Public Engagement with Renewable Energy Technologies: A Literature Review," University of Surrey, Working Paper 1.3, August 2006.

174 **many opponents of windmills simply distrust:** Eric R. A. N. Smith and Holly Klick, "Explaining NIMBY Opposition to Wind Power" (paper presented at the annual meeting of the American Political Science Association, Boston, Massachusetts, August 29, 2007).

174 **After the world's first major tidal energy generator:** Patrick Devine-Wright, "Place Attachment and Public Acceptance of Renewable Energy: A Tidal Energy Case Study," *Journal of Environmental Psychology* 31, no. 4 (December 2011): 336–43, https://doi.org/10.1016/j.jenvp.2011.07.001.

174 **According to a 2021 report from the real estate company Redfin:** Lily Katz, Sebastian Sandoval-Olascoaga, and Sheharyar Bokhari, "39% of Utah Homes Face High Fire Risk—a Bigger Share Than Other Western States," Redfin Real Estate News, June 30, 2021, https://www.redfin.com/news/wildfire-real-estate-risk-2021/.

175 **Today, the United States already has $20 billion in expected real estate losses:** "Highlights from 'The Cost of Climate: America's Growing Flood Risk,'" First Street Foundation, February 22, 2021, https://firststreet.org/research-lab/published-research/highlights-from-the-cost-of-climate-americas-growing-flood-risk/.

175 **He calls this phenomenon *soliphilia*:** Glenn A. Albrecht, "Soliphilia," *Psychoterratica* (blog), September 8, 2019, https://glennaalbrecht.wordpress.com/2019/09/08/soliphilia/.

175 **Chinese American geographer Yi-Fu Tuan came up:** Yi-Fu Tuan, *Topophilia: A Study of Environmental Perception, Attitudes, and Values* (New York: Columbia University Press, 1990).

176 **Consider one study, for instance, from the Indian state of Odisha:** Sasmita Mishra, Sanjoy Mazumdar, and Damodar Suar, "Place Attachment and Flood Preparedness," *Journal of Environmental Psychology* 30, no. 2 (June 2010): 187–97, https://doi.org/10.1016/j.jenvp.2009.11.005.

176 **In other studies in the western United States, Canada, and Australia:** Menka Bihari and Robert Ryan, "Influence of Social Capital on Community Preparedness for Wildfires," *Landscape and Urban Planning* 106, no. 3 (June 15, 2012): 253–61, https://doi.org/10.1016/j.landurbplan.2012.03.011; Charis E. Anton and Carmen Lawrence, "Does Place Attachment Predict Wildfire Mitigation and Preparedness? A Comparison of Wildland–Urban Interface and Rural Communities," *Environmental Management* 57, no. 1 (January 1, 2016): 148–62, https://doi.org/10.1007/s00267-015-0597-7; Robin S. Cox and Karen-Marie Elah Perry, "Like

a Fish out of Water: Reconsidering Disaster Recovery and the Role of Place and Social Capital in Community Disaster Resilience," *American Journal of Community Psychology* 48, no. 3–4 (December 2011): 395–411, https://onlinelibrary.wiley.com/doi/abs/10.1007/s10464-011 -9427-0.

176 **Hurricane (or Superstorm) Sandy, which slammed against the eastern United States:** Ross Toro, "Hurricane Sandy's Impact (Infographic)," Live Science, October 29, 2013, https://www .livescience.com/40774-hurricane-sandy-s-impact-infographic.html.

176 **One set of places that became important after Hurricane Sandy:** Joana Chan, Bryce DuBois, and Keith G. Tidball, "Refuges of Local Resilience: Community Gardens in Post-Sandy New York City," *Urban Forestry and Urban Greening* 14, no. 3 (2015): 625–35, https://doi.org/10 .1016/j.ufug.2015.06.005.

177 **Take the case of Staten Island:** Elizabeth Rush, *Rising: Dispatches from the New American Shore* (Minneapolis: Milkweed Editions, 2018), 113–32.

177 **In a survey led by a psychology Ph.D. student:** Sherri Brokopp Binder, "Resilience and Postdisaster Relocation: A Study of New York's Home Buyout Plan in the Wake of Hurricane Sandy" (Ph.D. diss., University of Hawaiʻi, 2014).

178 **When bell hooks was a girl:** hooks, *Belonging* 46, 69–88, 143–52, 203.

9: Living with Water

180 **home waters, n.,** and **home wind, n.:** *OED Online.*

180 **On a warming planet, these are already becoming:** Thomas R. Knutson, Maya V. Chung, Gabriel Vecchi, Jingru Sun, Tsung-Lin Hsieh, and Adam J. P. Smith, "Climate Change Is Probably Increasing the Intensity of Tropical Cyclones," *ScienceBrief Review*, March 2021.

180 **Other kinds of storms are also growing more severe:** Peter Ciurczak, "It's Not Just You: Nor'Easters Really Have Gotten More Frequent and More Intense," Boston Indicators, March 22, 2018, https://www.bostonindicators.org/article-pages/2018/march/nor-easters; John Walsh, Donald Wuebbles, Katharine Hayhoe, James Kossin, Kenneth Kunkel, Graeme Stephens, Peter Thorne, et al., "Our Changing Climate," in *Climate Change Impacts in the United States: The Third National Climate Assessment*, ed. Jerry M. Melillo, Terese (T. C.) Richmond, and Gary Yohe (U.S. Global Change Research Program, 2014), 19–67, https://doi.org/10.7930 /J0KW5CXT.

180 **A second is via high tides:** "What Is a Perigean Spring Tide?" U.S. National Oceanic and Atmospheric Administration, accessed April 30, 2021, https://oceanservice.noaa.gov/facts/perigean -spring-tide.html.

181 ***nuisance flooding:*** William V. Sweet and John J. Marra, *2015 State of U.S. "Nuisance" Tidal Flooding*, National Oceanic and Atmospheric Administration's Center for Operational Oceanographic Products and Services and National Centers for Environmental Information, June 8, 2016, https://www.ncdc.noaa.gov/monitoring-content/sotc/national/2016/may/sweet-marra -nuisance-flooding-2015.pdf.

181 **Early on a Monday in May 2019, Andrea Dutton:** Andrea Dutton, "The Past, Present, and Future of Sea Level Rise Along the Florida Coast" (presentation at Keeping History Above Water, St. Augustine, FL, May 6, 2019).

182 **In this part of the twenty-first century, the global rate of sea level rise:** Joseph F. Donoghue, "Sea Level History of the Northern Gulf of Mexico Coast and Sea Level Rise Scenarios for the Near Future," *Climatic Change* 107, no. 1–2 (July 2011): 17–33, https://doi.org/10.1007 /s10584-011-0077-x; Andrea D. Hawkes, Andrew C. Kemp, Jeffrey P. Donnelly, Benjamin P. Horton, W. Richard Peltier, Niamh Cahill, David F. Hill, et al., "Relative Sea-Level Change in Northeastern Florida (USA) During the Last ~8.0 Ka," *Quaternary Science Reviews* 142 (June 2016): 90–101, https://doi.org/10.1016/j.quascirev.2016.04.016; Miriam C. Jones, G. Lynn Wingard, Bethany Stackhouse, Katherine Keller, Debra Willard, Marci Marot, Bryan Landacre, et al., "Rapid Inundation of Southern Florida Coastline Despite Low Relative Sea-Level Rise

Rates During the Late-Holocene," *Nature Communications* 10, no. 1 (July 19, 2019): 3231, https://doi.org/10.1038/s41467-019-11138-4.

182 **A few years previously, Andrea Dutton and some of her colleagues:** Arnoldo Valle-Levinson, Andrea Dutton, and Jonathan B. Martin, "Spatial and Temporal Variability of Sea Level Rise Hot Spots over the Eastern United States," *Geophysical Research Letters* 44, no. 15 (2017): 7876–82, https://doi.org/10.1002/2017GL073926; Arnoldo Valle-Levinson and Andrea Dutton, "An X-Factor in Coastal Flooding: Natural Climate Patterns Create Hot Spots of Rapid Sea Level Rise," The Conversation, January 1, 2018, http://theconversation.com/an-x-factor-in-coastal -flooding-natural-climate-patterns-create-hot-spots-of-rapid-sea-level-rise-82628; Justin Gillis, "The Sea Level Did, in Fact, Rise Faster in the Southeast U.S.," *New York Times*, August 9, 2017, https://www.nytimes.com/2017/08/09/climate/the-sea-level-did-in-fact-rise-faster-in-the -southeast-us.html.

187 **Some spots along that trail are already at risk:** Debra Holtz, Adam Markham, Kate Cell, and Brenda Ekwurzel, *National Landmarks at Risk: How Rising Seas, Floods, and Wildfires Are Threatening the United States' Most Cherished Historic Sites* (Union of Concerned Scientists: May 2014), 7–9.

187 **"We have always been a pluralist nation":** Kathryn Schulz, "Citizen Khan," *New Yorker*, May 30, 2016, https://www.newyorker.com/magazine/2016/06/06/zarif-khans-tamales-and-the -muslims-of-sheridan-wyoming.

188 **A few years ago, beneath the central plaza:** Marcia Lane, "Plaza Gives Up Its Secrets," *St. Augustine Record*, June 20, 2010, https://www.staugustine.com/article/20100620/NEWS/306209935.

189 **Spain treated slavery as an "unnatural condition":** Jane Landers, "Transforming Bondsmen into Vassals: Arming Slaves in Colonial Spanish America," in *Arming Slaves: From Classical Times to the Modern Age*, ed. Christopher Leslie Brown and Philip D. Morgan (New Haven, CT: Yale University Press, 2006), 120–45, https://doi.org/10.12987/yale/9780300109009.003.0006.

189 **When Pedro Menéndez and his crew:** Jane Landers, "Spanish Sanctuary: Fugitives in Florida, 1687–1790," *Florida Historical Quarterly* 62, no. 3 (1984): 296–313; "African Americans in St. Augustine 1565–1821," Castillo de San Marcos National Monument, U.S. National Park Service, accessed September 23, 2021, https://www.nps.gov/casa/learn/historyculture/african -americans-in-st-augustine-1565-1821.htm.

189 **Fort Mose might never have existed:** Jane Landers, "Filling in the Missing Pieces: The Extraordinary Life of Captain Francisco Menendez, Leader of the Free Black Town of Gracia Real de Santa Theresa de Mose" (paper presented at the Florida Conference of Historians for the Florida Lecture Series at Florida Southern College, February 14, 2015).

190 **which was established in 1866:** "Florida, St. Augustine, Lincolnville Historic District," National Register of Historic Places, U.S. National Park Service, accessed October 21, 2021, https://www.nps.gov/nr/travel/geo-flor/28.htm.

191 **In the nineteenth century, a civil engineer hired by Standard Oil founder Henry Flagler:** "From Creek to Lake: The Maria Sanchez," Governor's House Library, June 16, 2021, https:// governorshouselibrary.wordpress.com/2021/06/16/from-creek-to-lake-the-maria-sanchez/.

191 **But it wasn't fully excavated until the mid-1980s:** Kathleen A. Deagan and Jane Landers, "Fort Mosé: Earliest Free African-American Town in the United States," in *"I, Too, Am America": Archaeological Studies of African-American Life*, ed. Theresa A. Singleton (Charlottesville: University Press of Virginia, 1999); "Fort Mose," St. Johns County Government, accessed May 17, 2021, http://www.co.st-johns.fl.us/LAMP/Projects/FLct_mose.aspx.

193 **This hurricane dampened:** Sheldon Gardner, "Dorian Grazes City," *St. Augustine Record*, September 4, 2019, https://www.staugustine.com/news/20190904/hurricane-dorian-grazes-city -brings-some-minor-flooding; Lena Pringle, Ashley Harding, Jennifer Ready, Vic Micolucci, and Francine Frazier, "St. Johns County Mostly Unscathed by Hurricane Dorian," WJXT, September 6, 2019, https://www.news4jax.com/news/2019/09/06/st-johns-county-mostly-unscathed -by-hurricane-dorian/.

194 **Still, the city government:** Aaron London, "Moving History: Florida Ag Museum Hauls 1940s-Era Cottage to Flagler County," *Daytona Beach News-Journal*, November 12, 2019.

194 **Miami Beach—with six times the population of St. Augustine:** Greg Allen, "As Waters Rise, Miami Beach Builds Higher Streets and Political Willpower," NPR, May 10, 2016, https://www.npr.org/2016/05/10/476071206/as-waters-rise-miami-beach-builds-higher-streets-and-political-willpower; Mario Ariza and Alex Harris, "Miles of Florida Roads Face 'Major Problem' from Sea Rise. Is State Moving Fast Enough?" *South Florida Sun-Sentinel*, March 19, 2021, https://www.sun-sentinel.com/news/environment/fl-ne-sea-level-rise-threatens-florida-roads-20210319-lcheqk6p4rcb5ivprpzfqg3wfq-story.html.

195 **A previous St. Augustine mayor had traveled to the Netherlands:** Jessica Clark, "St. Augustine Mayor Travels to the Netherlands to Learn About Sea Level Rise Solutions," First Coast News, June 11, 2018, https://www.firstcoastnews.com/article/news/local/st-augustine/st-augustine-mayor-travels-to-the-netherlands-to-learn-about-sea-level-rise-solutions/77-563346222; Sheldon Gardner, "City Taps Dutch Researchers on Flooding," *St. Augustine Record*, August 31, 2018, https://www.staugustine.com/news/20180831/st-augustine-officials-in-talks-with-dutch-researchers-on-sea-level-rise-solutions.

195 **But she had to resign:** Sheldon Gardner, "St. Augustine Mayor Steps Down After Stroke," *Florida Times-Union*, February 28, 2019, https://www.jacksonville.com/news/20190228/st-augustine-mayor-steps-down-after-stroke.

195 **In September 2020, a nor'easter arrived:** Sheldon Gardner, "City Sees Worst Nor'easter in Decades," September 21, 2020, https://www.staugustine.com/story/news/2020/09/21/noreaster-causes-flooding-st-augustine/5855398002/.

198 **The mayor who visited:** Ryan Benk, "St. Augustine Mayor: You Can Address Sea Level Rise Without Talking Climate Change," WJCT News, June 29, 2017, https://news.wjct.org/first-coast/2017-06-29/st-augustine-mayor-you-can-address-sea-level-rise-without-talking-climate-change.

198 **In early 2021:** Mary Ellen Klas, "Florida Lawmakers Advance Bills to Halt Local Clean Energy Efforts," *Tampa Bay Times*, March 9, 2021, https://www.tampabay.com/news/florida-politics/2021/03/09/florida-lawmakers-advance-bills-to-halt-local-clean-energy-efforts/; Emily Pontecorvo and Brendan Rivers, "A Florida City Wanted to Move Away from Fossil Fuels. The State Just Made Sure It Couldn't," *Grist*, July 29, 2021, https://grist.org/cities/tampa-wanted-renewable-energy-resolution-florida-lawmakers-made-sure-it-couldnt-gas-ban-preemption/.

10: A Safe Space

200 **The first of these studies, published in 2008:** James Hansen, Makiko Sato, Pushker Kharecha, David Beerling, Robert Berner, Valerie Masson-Delmotte, Mark Pagani, et al., "Target Atmospheric CO_2: Where Should Humanity Aim?" *Open Atmospheric Science Journal* 2, no. 1 (October 31, 2008): 217–31, https://doi.org/10.2174/1874282300802010217.

200 **The second, in 2009—with the evocative title "A Safe Operating Space for Humanity":** Johan Rockström, Will Steffen, Kevin Noone, Åsa Persson, F. Stuart Chapin III, Eric F. Lambin, Timothy M. Lenton, et al., "A Safe Operating Space for Humanity," *Nature* 461, no. 7263 (September 2009): 472–75, https://doi.org/10.1038/461472a.

200 **The original definition of *threshold*:** *threshold*, n.1.a., *OED Online*.

200 **There are those, such as Elon Musk:** Nick Statt, "Elon Musk Still Thinks a Mars Colony Will Save Us from a Future Dark Age," *Verge*, March 11, 2018, https://www.theverge.com/2018/3/11/17106910/elon-musk-ai-threat-mars-moon-colonization-nukes-sxsw-2018; Bill McKibben, epilogue to *Falter: Has the Human Game Begun to Play Itself Out?* (New York: Henry Holt, 2019).

201 **In 1833, the British economist and professor William Forster Lloyd:** W. F. Lloyd, *Two Lectures on the Checks to Population, Delivered Before the University of Oxford, in Michaelmas Term 1832* (S. Collingewood, 1833), accessed on Google Books.

202 **echoed the influential (though also controversial) scholar Thomas Malthus:** Morgan Rose,

"In Defense of Malthus," Econlib, September 16, 2002, https://www.econlib.org/library/Columns/Teachers/defendmalthus.html.

202 **In 1968, writing in the journal *Science*, Hardin penned:** Garrett Hardin, "The Tragedy of the Commons," *Science* 162, no. 3859 (December 13, 1968): 1243–48.

203 **Hardin's ideas about human scarcity and the commons:** Matto Mildenberger, "The Tragedy of the *Tragedy of the Commons*," *Scientific American*, April 23, 2019, https://blogs.scientificamerican.com/voices/the-tragedy-of-the-tragedy-of-the-commons/; Matto Mildenberger, "Trump and the Tragedy of the Commons," Commons Network, July 9, 2019, https://www.commonsnetwork.org/news/trump-and-the-tragedy-of-the-commons/; Cathy Gere, "The Drama of the Commons," *Point*, June 12, 2020, https://thepointmag.com/politics/the-drama-of-the-commons/.

203 **Elsewhere in his writing, he wasn't so subtle:** "Garrett Hardin," Southern Poverty Law Center, accessed October 25, 2021, https://www.splcenter.org/fighting-hate/extremist-files/individual/garrett-hardin.

203 **A 1985 essay in *American Zoologist* on environmental education suggested:** John A. Moore, "Science as a Way of Knowing—Human Ecology," *American Zoologist* 25, no. 2 (May 1985): 483–637, https://doi.org/10.1093/icb/25.2.483.

204 **a political scientist named Elinor Scott (better known as Elinor Ostrom):** Vlad Tarko, *Elinor Ostrom: An Intellectual Biography* (London: Rowman and Littlefield, 2016); Emily Langer, "Elinor Ostrom, First Woman to Receive Nobel Prize in Economics, Dies at 78," *Washington Post*, June 13, 2012, https://www.washingtonpost.com/national/elinor-ostrom-first-woman-to-receive-nobel-prize-in-economics-dies-at-78/2012/06/13/gJQAMO2vaV_story.html; Gere, "Drama of the Commons."

204 **Vincent was an expert in Western resources:** *Actual World, Possible Future* (WTIU Documentaries, 2020), 1:26:51, https://www.pbs.org/video/actual-world-possible-future-09rkab/.

205 **the Central and West Coast Groundwater Basins of Los Angeles:** "Coastal Plain of Los Angeles County Groundwater Basin, West Coast Subbasin," *California's Groundwater Bulletin* 118, updated February 27, 2004, https://water.ca.gov/-/media/DWR-Website/Web-Pages/Programs/Groundwater-Management/Bulletin-118/Files/2003-Basin-Descriptions/4_011_03_WestCoastSubbasin.pdf; Ted Johnson, "An Introduction to the Hydrogeology of the Central and West Coast Basins," Water Replenishment District of Southern California, Technical Bulletin no. 1, Fall 2004.

205 **Elinor wanted to understand:** Elinor Ostrom, chap. 4 in *Governing the Commons: The Evolution of Institutions for Collective Action* (New York: Cambridge University Press, 2015); Elinor Ostrom, "Public Entrepreneurships: A Case Study in Ground Water Basin Management" (Ph.D. diss., University of California, Los Angeles, 1964).

206 **Essayist and former city official D. J. Waldie writes:** D. J. Waldie, "Beneath Our Feet: Water and Politics in Southeast L.A.," KCET, September 2, 2016, https://www.kcet.org/shows/lost-la/beneath-our-feet-water-and-politics-in-southeast-l-a.

206 **The Metropolitan Water District of Southern California is now considering:** Sammy Roth, "Climate Change Spells Trouble for the Colorado River. But There's Still Hope," *Los Angeles Times*, December 31, 2020, https://www.latimes.com/environment/newsletter/2020-12-31/boiling-point-colorado-river-in-trouble-still-hope-boiling-point; Jeremy P. Jacobs, "Could LA Water Recycling Be a Miracle for Parched West?" *E&E News*, September 27, 2021, https://www.eenews.net/articles/could-la-water-recycling-be-a-miracle-for-parched-west/; Carl Smith, "California Invests in Recycled Water as Droughts Take a Toll," *Governing*, August 4, 2021, https://www.governing.com/next/california-invests-in-recycled-water-as-droughts-take-a-toll.

206 **"No one 'owns' the basins themselves":** Ostrom, chap. 4 in *Governing the Commons*.

206 **In 1973, five years after the publication:** Tarko, *Elinor Ostrom*.

207 **Business and finance journalist:** John Cassidy, "The Nobel That Should Have Been," *New Yorker*, October 13, 2009, https://www.newyorker.com/news/john-cassidy/the-nobel-that-should-have-been.

207 **Communal resources also have:** Ostrom, chap. 2 in *Governing the Commons*; "Elinor Ostrom

Prize Lecture," video, Sveriges Riksbank Prize in Economic Sciences in Memory of Alfred Nobel 2009, 28:05, December 8, 2009, https://www.nobelprize.org/prizes/economic-sciences /2009/ostrom/lecture/.

207 **But Elinor herself took issue:** Fran Korten, "Elinor Ostrom Wins Nobel for Common(s) Sense," *YES!* magazine, February 27, 2010, https://www.yesmagazine.org/issue/america-remix /2010/02/27/elinor-ostrom-wins-nobel-for-common-s-sense.

207 **Such irrigation systems exist:** Robert Neuwirth, "Centuries-Old Irrigation System Shows How to Manage Scarce Water," *National Geographic*, May 17, 2019, https://www.nationalgeographic .com/environment/article/acequias.

208 **Naomi Klein writes in *On Fire*:** Naomi Klein, "Capitalism vs. the Climate," in *On Fire: The (Burning) Case for a Green New Deal* (New York: Simon and Schuster, 2020).

208 **just ninety companies have borne:** Richard Heede, "Tracing Anthropogenic Carbon Dioxide and Methane Emissions to Fossil Fuel and Cement Producers, 1854–2010," *Climatic Change* 122, no. 1 (January 1, 2014): 229–41, https://doi.org/10.1007/s10584-013-0986-y.

209 **Since at least the late 1970s:** "Exxon: The Road Not Taken," Inside Climate News, nine-part series of stories, accessed October 25, 2021, https://insideclimatenews.org/project/exxon-the -road-not-taken/; Neela Banerjee, "Exxon's Oil Industry Peers Knew About Climate Dangers in the 1970s, Too," Inside Climate News, December 22, 2015, https://insideclimatenews.org /news/22122015/exxon-mobil-oil-industry-peers-knew-about-climate-change-dangers-1970s -american-petroleum-institute-api-shell-chevron-texaco/.

209 **In 2015, the *Los Angeles Times* and Inside Climate News published:** Neela Banerjee, Lisa Song, and David Hasemyer, "Exxon Believed Deep Dive into Climate Research Would Protect Its Business," Inside Climate News, September 17, 2015, https://insideclimatenews.org/news /17092015/exxon-believed-deep-dive-into-climate-research-would-protect-its-business/; Sara Jerving, Katie Jennings, Masako Melissa Hirsch, and Susanne Rust, "What Exxon Knew About the Earth's Melting Arctic," *Los Angeles Times*, October 9, 2015, https://graphics.latimes .com/exxon-arctic/.

209 **In social science, the technical term for a mooch or a thief:** Natalie M. Roy, "Climate Change's Free Rider Problem: Why We Must Relinquish Freedom to Become Free," *William and Mary Environmental Law and Policy Review* 45, no. 3 (2021), https://scholarship.law.wm .edu/wmelpr/vol45/iss3/7.

209 **In the same month that she won the Nobel, Elinor Ostrom:** Elinor Ostrom, "A Polycentric Approach for Coping with Climate Change," Policy Research Working Paper 5095, Background Paper to the 2010 World Development Report, The World Bank, October 2009.

210 **In February 2012, four months before:** Elinor Ostrom, "Nested Externalities and Polycentric Institutions: Must We Wait for Global Solutions to Climate Change Before Taking Actions at Other Scales?" *Economic Theory* 49, no. 2 (February 2012): 353–69, https://doi.org/10.1007 /s00199-010-0558-6.

210 **In 2008, a group of young activists:** Bill McKibben, *Oil and Honey: The Education of an Unlikely Activist* (New York: Henry Holt, 2013).

210 **It has also drawn strength and power:** Rebecca Hersher, "Key Moments in the Dakota Access Pipeline Fight," NPR, February 22, 2017, https://www.npr.org/sections/thetwo-way/2017/02 /22/514988040/key-moments-in-the-dakota-access-pipeline-fight.

210 **In 2020, legal actions:** Lisa Friedman, "Standing Rock Sioux Tribe Wins a Victory in Dakota Access Pipeline Case," *New York Times*, March 25, 2020, https://www.nytimes.com/2020/03 /25/climate/dakota-access-pipeline-sioux.html.

211 **In the summer of 2020, the Sunrise Movement:** "If You Care About the Green New Deal, We Need You to Join the Movement for Black Lives," Sunrise Movement, June 20, 2020, https:// www.sunrisemovement.org/theory-of-change/if-you-care-about-the-green-new-deal-we-need -you-to-join-the-movement-for-black-lives-7d4395918408/.

211 **In September 2019, millions of people joined:** Oliver Milman, "US to Stage Its Largest Ever Climate Strike: 'Somebody Must Sound the Alarm,'" *Guardian*, September 20, 2019, https://

www.theguardian.com/world/2019/sep/20/climate-strikes-us-students-greta-thunberg; Eliza Barclay and Brian Resnick, "How Big Was the Global Climate Strike? 4 Million People, Activists Estimate," Vox, September 20, 2019, https://www.vox.com/energy-and-environment/2019/9/20/20876143/climate-strike-2019-september-20-crowd-estimate.

211 **More than a decade after 350 became:** Bill McKibben, "350—the Most Important Number on the Planet. We Just Need to Get the Politicians to Listen to the Scientists," *Guardian*, October 23, 2009, https://www.theguardian.com/environment/blog/2009/oct/23/350-rally.

11: To Move Home

212 **out of house and home (idiom):** *Out of house and home, The Free Dictionary*, accessed October 25, 2021, https://idioms.thefreedictionary.com/out+of+house+and+home.

215 **They hired the Alaska Native Tribal Health Consortium:** Denali Commission, "ANTHC Chosen as Project Manager for Newtok Village Move," *Delta Discovery*, June 20, 2018, https://deltadiscovery.com/anthc-chosen-as-project-manager-for-newtok-village-move/.

216 **Just in Alaska, dozens of other remote rural communities:** University of Alaska Fairbanks Institute of Northern Engineering, U.S. Army Corps of Engineers Alaska District, and U.S. Army Corps of Engineers Cold Regions Research and Engineering Laboratory, *Statewide Threat Assessment: Identification of Threats from Erosion, Flooding, and Thawing Permafrost in Remote Alaska Communities*, report prepared for the Denali Commission, November 2019, https://www.denali.gov/wp-content/uploads/2019/11/Statewide-Threat-Assessment-Final-Report-20-November-2019.pdf.

216 **Over the last three decades, in more than a thousand counties:** Katharine J. Mach, Caroline M. Kraan, Miyuki Hino, A. R. Siders, Erica M. Johnston, and Christopher B. Field, "Managed Retreat Through Voluntary Buyouts of Flood-Prone Properties," *Science Advances* 5, no. 10 (October 2019): eaax8995, https://doi.org/10.1126/sciadv.aax8995.

216 **But according to an analysis published by the National Institute of Building Sciences:** Keith Porter, Nicole Dash, Charles Huyck, Joost Santos, Charles Scawthorn, Michael Eguchi, Ron Eguchi, et al., *Natural Hazard Mitigation Saves: 2019 Report*, National Institute of Building Sciences, Multihazard Mitigation Council, December 2019, https://www.nibs.org/reports/natural-hazard-mitigation-saves-2019-report.

216 **In Louisiana, an Indigenous community on a rapidly shrinking island:** "The Story of Isle de Jean Charles," Isle de Jean Charles Resettlement, accessed October 25, 2021, https://isledejeancharles.la.gov/.

217 **On the Olympic Peninsula in Washington State, the Quileute Tribe:** Morgan Keith, "The 'Twilight' Saga Used Its People and Traditions. Now, the Quileute Tribe Is Relocating in an Effort to Preserve Them," *Insider*, July 24, 2021, https://www.insider.com/the-quileute-tribe-is-relocating-2021-7.

217 **Between 2008 and 2012, then president of the Maldives:** Ben Doherty, "Climate Change Castaways Consider Move to Australia," *Sydney Morning Herald*, January 7, 2012, https://www.smh.com.au/environment/climate-change/climate-change-castaways-consider-move-to-australia-20120106-1pobf.html; Randeep Ramesh, "Paradise Almost Lost: Maldives Seek to Buy a New Homeland," *Guardian*, November 9, 2008, https://www.theguardian.com/environment/2008/nov/10/maldives-climate-change.

217 **But Nasheed was ousted:** "Maldives Ex-President Mohamed Nasheed Arrested," *Guardian*, October 8, 2012, https://www.theguardian.com/world/2012/oct/08/maldives-mohamed-nasheed-arrested; Krishan Francis and Aniruddha Ghosal, "Maldives Minister: Failure to Limit Warming a Death Sentence," AP News, October 20, 2021, https://apnews.com/article/climate-science-business-sri-lanka-europe-2da12921977d7ef067dfec43028528ed.

217 **According to some estimates, the diaspora that permanently departed:** Maria Godoy, "Tracking the Katrina Diaspora: A Tricky Task," NPR, August 2006, https://legacy.npr.org/news/specials/katrina/oneyearlater/diaspora/index.html; "Hurricane Katrina Migration: Where Did People Go? Where Are They Coming from Now?" *Times-Picayune* (New Orleans, LA), August

27, 2015, updated July 18, 2019, https://www.nola.com/news/article_b84a9b86-e0dc-511a-872d-09fef0012508.html.

217 **In 2020, the Internal Displacement Monitoring Centre:** "2020 Internal Displacement," Internal Displacement Monitoring Centre, accessed October 25, 2021, https://www.internal-displacement.org/database/displacement-data.

217 **but the organization predicts:** Internal Displacement Monitoring Centre, *Global Report on Internal Displacement 2021: Internal Displacement in a Changing Climate*, https://www.internal-displacement.org/global-report/grid2021/.

217 **"There are no reliable estimates":** "A Complex Nexus," International Organization for Migration, accessed October 25, 2021, https://www.iom.int/complex-nexus.

217 **Already, a portion of the vast population of migrants:** Abrahm Lustgarten, "The Great Climate Migration," *New York Times Magazine*, July 23, 2020, https://www.nytimes.com/interactive/2020/07/23/magazine/climate-migration.html.

218 **two men died on the Kuskokwim River:** Krysti Shallenberger, "Melting Ice Is Disrupting Daily Life in the Y-K Delta in the Worst Possible Way," Alaska Public Media, April 12, 2019, https://www.alaskapublic.org/2019/04/12/melting-ice-is-disrupting-daily-life-in-the-y-k-delta-in-the-worst-possible-way/; Matthew Cappucci and Andrew Freedman, "More Freak Weather Comes to Alaska, Which Has Had an Unprecedented Summer," *Washington Post*, August 16, 2019, https://www.washingtonpost.com/weather/2019/08/16/more-freak-weather-comes-alaska-which-has-had-an-unprecedented-summer/.

218 **But that July:** Krysti Shallenberger, "Alaska Heat Wave Hits Yukon-Kuskokwim Delta," Alaska Public Media, July 10, 2019, https://www.alaskapublic.org/2019/07/10/alaska-heat-wave-hits-yukon-kuskokwim-delta/.

219 **The warming of the Arctic and the resulting wrinkles:** "Scientists See Link Between Arctic Warming and Texas Cold Snap," *Yale Environment 360*, September 3, 2021, https://e360.yale.edu/digest/scientists-see-link-between-climate-change-and-the-texas-cold-snap; Judah Cohen, Laurie Agel, Mathew Barlow, Chaim I. Garfinkel, and Ian White, "Linking Arctic Variability and Change with Extreme Winter Weather in the United States," *Science* 373, no. 6559 (September 3, 2021): 1116–21, https://www.science.org/doi/10.1126/science.abi9167.

221 **First, the military returned:** Greg Kim, "Newtok Partners with Military to Escape Coastal Erosion," KYUK, August 7, 2019, https://www.kyuk.org/post/newtok-partners-military-escape-coastal-erosion.

222 **"We can't have very many mistakes":** "The Newtok to Mertarvik Innovative Readiness Training Documentary," Facebook video, 11:33, posted by Innovative Readiness Training, August 2, 2019, https://www.facebook.com/watch/?v=457260611791011.

222 **Linguists at the Alaska Native Language Center:** Lawrence Kaplan, "Inuit or Eskimo: Which Name to Use?" Alaska Native Language Center, accessed October 25, 2021, https://www.uaf.edu/anlc/resources/inuit_or_eskimo.php.

227 **"This is an emergency. Our house is on fire":** "'Our House Is on Fire': Greta Thunberg Addresses Hundreds of Thousands at NYC Climate Strike," *Democracy Now!*, September 23, 2019, https://www.democracynow.org/2019/9/23/new_york_city_climate_strike_greta.

232 **Communities in rural Alaska have not forgotten:** Kyle Hopkins, "An Alaska Hospital Executive Downplayed the COVID-19 Threat in an Email to Staff. She's No Longer on the Job," *Anchorage Daily News*, March 31, 2020, updated April 1, 2020, https://www.adn.com/alaska-news/rural-alaska/2020/04/01/an-alaska-hospital-executive-downplayed-the-covid-19-threat-in-an-email-to-staff-shes-no-longer-on-the-job/.

233 **Then, in the still-eroding village site:** Greg Kim, "Newtok Regains Power After Month-Long Outage," KYUK, September 22, 2020, https://www.kyuk.org/post/newtok-regains-power-after-month-long-outage.

233 **Sadly, Newtok was not able to keep the pandemic out forever:** Greg Kim, "How COVID-19 Is Slowing Down the Relocation of a Southwest Alaska Village," Alaska Public Media,

September 8, 2021, https://www.alaskapublic.org/2021/09/08/how-covid-19-is-slowing-down-the-relocation-of-a-southwest-alaska-village/.

12: To Clean House

236 **bring something home to (idiom):** under entry for *home, Webster's New World Dictionary.*

237 **On Twitter, someone launched:** Ch3vron PR (@ch3vronPR), "In PR school, we had nightmares about refineries blowing up in our cities so we built them all in yours. #ChevronFire #ThoughtsDuringSchool," Twitter, August 7, 2012, 12:11 P.M., https://twitter.com/ch3vronPR /status/232871665917374464.

238 **When the evening arrived, the Urban Tilthers gathered:** Jennifer Baires, "At Town Hall Meeting, Questions and Anger over Chevron Refinery Fire," *Richmond Confidential,* August 8, 2012, https://richmondconfidential.org/2012/08/08/at-town-hall-meeting-questions-and -anger-over-chevron-refinery-fire/.

238 **Doria also spoke to the media:** KQED News Staff, "Chevron Refinery Fire," KQED, August 6, 2012, https://www.kqed.org/news/72350/chevron-refinery-fire-shelter-in-place-for-richmond -north-richmond-and-san-pablo-residents.

239 **This foundation:** "eQuip Richmond," RCF Connects, accessed September 19, 2021, https:// www.rcfconnects.org/community-initiatives/community-growth/equip-richmond/.

240 **Meanwhile, the regional air quality district:** Bay Area Air Quality Management District, "Air District Statement on Chevron Fire Air Quality Samples," news release, August 7, 2012.

240 **Three days later, that agency backpedaled:** Laird Harrison, "Air Quality Agency Detected Contaminant After Chevron Richmond Refinery Fire," August 10, 2012, https://www.kqed .org/news/72878/richmond-air-quality-officials-say-pollution-detected-from-refinery-fire.

240 **Much later, an independent academic analysis:** Linda L. Remy, Ted Clay, Vera Byers, and Paul E. Rosenfeld, "Hospital, Health, and Community Burden After Oil Refinery Fires, Richmond, California 2007 and 2012," *Environmental Health* 18, no. 1 (May 16, 2019): 48, https://doi .org/10.1186/s12940-019-0484-4.

240 **A year after the accident, Chevron:** Mark Andrew Boyer, "Chevron Installs New Air Monitor in North Richmond," *Richmond Confidential,* September 28, 2013, https://richmondconfidential .org/2013/09/27/chevron-installs-new-air-monitor-in-north-richmond/.

241 **"At first the shipyards":** Richard Rothstein, *The Color of Law: A Forgotten History of How Our Government Segregated America* (New York: Liveright, 2017).

241 **According to various estimates:** Eli Moore, The Othering and Belonging Institute, email message, December 3, 2021; Eli Moore and Swati Prakash, "Richmond's Tax Revenue from Chevron," Pacific Institute, October 2008; "Richmond Facing Chevron Decision," *East Bay Times,* March 2, 2008, https://www.eastbaytimes.com/2008/03/02/richmond-facing-chevron -decision/.

242 **However, the company has disputed:** David R. Baker, "Chevron, Richmond End Dispute over Taxes," SFGATE, May 13, 2010, https://www.sfgate.com/business/article/Chevron-Richmond -end-dispute-over-taxes-3188996.php; Conor Dougherty, "California's 40-Year-Old Tax Revolt Survives a Counterattack," *New York Times,* November 10, 2020, https://www.nytimes.com /2020/11/10/business/economy/california-prop-15-property-tax.html.

242 **In the summer of 2012:** Isabella Fertel, "'The World Needs More Najaris,'" *Richmond Confidential,* September 28, 2020, https://richmondconfidential.org/2020/09/28/the-world-needs -more-najaris/.

243 **That year, the city government had also adopted:** City of Richmond, "Community Health and Wellness," *Richmond General Plan 2030,* April 25, 2012.

244 **Harvard University historians Geoffrey Supran and Naomi Oreskes:** Geoffrey Supran and Naomi Oreskes, "Rhetoric and Frame Analysis of ExxonMobil's Climate Change Communications," *One Earth* 4, no. 5 (May 21, 2021): 696–719, https://doi.org/10.1016/j.oneear.2021 .04.014; Naomi Oreskes, "Testimony Before the House Committee on Oversight and Reform, Subcommittee on Civil Rights and Civil Liberties: Examining the Oil Industry's Efforts to

Suppress the Truth About Climate Change," October 23, 2019, https://docs.house.gov/meetings/GO/GO02/20191023/110126/HHRG-116-GO02-Wstate-OreskesN-20191023.pdf.

244 **In response, even long-established:** Lauren Feeney, "Why the Sierra Club Broke Tradition to Protest the Keystone Pipeline," BillMoyers.com, February 14, 2013, https://billmoyers.com/2013/02/14/sierra-club-lifts-120-year-ban-on-civil-disobedience-to-protest-keystone-xl-pipeline/; Sierra Club National, "Sierra Club, Allies Engage in Historic Act Civil Disobedience to Stop Keystone XL," news release, February 13, 2013, https://content.sierraclub.org/press-releases/2013/02/sierra-club-allies-engage-historic-act-civil-disobedience-stop-keystone-xl; Loren Blackford, "Civilly Disobedient; Morally Imperative," Sierra Club, September 17, 2018, https://www.sierraclub.org/change/2018/09/civilly-disobedient-morally-imperative.

244 **Climate activists became willing:** Jamie Henn, "Here's How We Defeated the Keystone XL Pipeline," *Sierra*, January 30, 2021, https://www.sierraclub.org/sierra/here-s-how-we-defeated-keystone-xl-pipeline.

244 **Analysts even warned that activists:** Katrina Rabeler, "Gas Industry Report Calls Anti-Fracking Movement a 'Highly Effective Campaign,'" *YES!* magazine, March 27, 2013, https://www.yesmagazine.org/environment/2013/03/27/gas-industry-report-calls-anti-fracking-movement-highly-effective; Control Risks, *The Global Anti-Fracking Movement: What It Wants, How It Operates and What's Next*, http://epl.org.ua/wp-content/uploads/2015/04/what_is_antishale_movement.pdf.

245 **In August 2013, three days before the one-year anniversary of the fire:** Susie Cagle, "A Year After a Refinery Explosion, Richmond, Calif., Is Fighting Back," *Grist*, August 6, 2013, https://grist.org/climate-energy/a-year-after-a-refinery-explosion-richmond-cali-is-fighting-back/.

245 **She held a bright blue microphone:** Some quotes and details in this section first appeared in this news report: "Chevron to Pay $2 Million for 2012 Refinery Fire in Richmond, CA; 200 Arrested at Protest," *Democracy Now!*, video, 44:04, August 6, 2013, accessed October 26, 2021, http://www.democracynow.org/2013/8/6/chevron_to_pay_2_million_for.

245 **she told the crowd, *The community doesn't deserve*:** Fareed Abdulrahman, "Thousands March to Chevron Days Before Fire Anniversary," *Richmond Confidential*, August 3, 2013, https://richmondconfidential.org/2013/08/03/protestors-march-to-chevron-days-before-fire-anniversary/.

246 **Idle No More activists led them:** "Idle No More Solidarity SF Bay Led the March of over 2,500 People to Chevron in Richmond, California," Idle No More, accessed September 19, 2021, https://idlenomore.ca/idle-no-more-solidarity-sf-bay-led-the-march-of-over-2500-people-to-chevron-in-richmond-california-idle-no-more/.

246 **In the end, the police arrested more than two hundred people:** Jill Tucker and Victoria Colliver, "210 Arrested at Chevron Refinery Protest," *San Francisco Chronicle*, August 3, 2013, updated August 4, 2013, https://www.sfchronicle.com/bayarea/article/210-arrested-at-Chevron-refinery-protest-4705509.php.

246 **Amazonian communities had sued over contamination left behind:** Sharon Lerner, "How the Environmental Lawyer Who Won a Massive Judgment Against Chevron Lost Everything," *Intercept*, January 29, 2020, https://theintercept.com/2020/01/29/chevron-ecuador-lawsuit-steven-donziger/.

246 **The company had also filled hundreds:** Patrick Radden Keefe, "Reversal of Fortune," *New Yorker*, January 1, 2012, https://www.newyorker.com/magazine/2012/01/09/reversal-of-fortune-patrick-radden-keefe; David Feige, "Pursuing the Polluters," *Los Angeles Times*, April 20, 2008, https://www.latimes.com/archives/la-xpm-2008-apr-20-op-feige20-story.html.

247 **According to one report commissioned by a human rights group:** James Brooke, "Pollution of Water Tied to Oil in Ecuador," *New York Times*, March 22, 1994, https://www.nytimes.com/1994/03/22/science/pollution-of-water-tied-to-oil-in-ecuador.html.

247 **In 2014, a U.S. judge declared:** Steven Mufson, "U.S. Judge Rules for Chevron in Ecuador Pollution Case," *Washington Post*, March 4, 2014, https://www.washingtonpost.com/business/economy/2014/03/04/ec828d00-a3bb-11e3-84d4-e59b1709222c_story.html; James North, "Is

Chevron's Vendetta Against Steven Donziger Finally Backfiring?" *Nation*, October 4, 2021, https://www.thenation.com/article/environment/steven-donziger-chevron-sentencing/; Human Rights Council, Working Group on Arbitrary Detention, "Opinions Adopted by the Working Group at Its Ninety-First Session, 6–10 September, 2021," October 1, 2021, https://www.ohchr.org/_layouts/15/WopiFrame.aspx?sourcedoc=/Documents/Issues/Detention/Opinions/Session91/A_HRC_WGAD_2021_24_AdvanceEditedVersion.pdf&action=default&DefaultItemOpen=1; "29 Nobel Laureates Demand That Chevron Face Justice for Amazon Pollution and Call for Freedom for Human Rights Defender Steven Donziger," https://static1.squarespace.com/static/5ac2615b8f5130fda4340fcb/t/5e9890f6d641a53d544792d6/1587056892216/2020–04-nobel-laureates-statement.pdf; Isabella Grullón Paz, "Lawyer Who Won $9.5 Billion Judgment Against Chevron Reports to Prison," *New York Times*, October 27, 2021, https://www.nytimes.com/2021/10/27/business/energy-environment/steven-donziger-chevron.html.

247 **He would become an even more controversial figure:** John Otis, "Correa's Legacy Leaves a Long Road to Recovery for Ecuador's Journalists," Committee to Protect Journalists, November 2, 2017, https://cpj.org/2017/11/correas-legacy-leaves-a-long-road-to-recovery-for/.

247 **But McLaughlin's intent was clearer:** Gayle McLaughlin, *Winning Richmond: How a Progressive Alliance Won City Hall* (Brooklyn, NY: Hard Ball Press, 2018).

248 **Over a week, they went to Quito:** John Geluardi, "From Richmond to the Rainforest," *East Bay Express*, October 16, 2013, https://eastbayexpress.com/from-richmond-to-the-rainforest-1/.

248 **Doria kept a blog:** Doria Robinson, "The Dirty Hand of Chevron-Texaco," *Richmond, Ecuador* (blog), September 18, 2013, https://richmondecuador.wordpress.com/2013/09/18/the-dirty-hand-of-chevron-texaco/.

249 **Chevron later funded a billboard advertisement:** Steve Early, *Refinery Town: Big Oil, Big Money, and the Remaking of an American City* (Boston: Beacon Press, 2017).

249 **There was also an underlying socioeconomic clash:** Eli Moore, Samir Gambhir, and Phuong Tseng, *Belonging and Community Health in Richmond: An Analysis of Changing Demographics and Housing*, Haas Institute for a Fair and Inclusive Society, University of California, Berkeley, February 20, 2015, https://belonging.berkeley.edu/belonging-and-community-health-richmond.

249 **The phrase *identity politics* originated in the 1970s:** Mychal Denzel Smith, "What Liberals Get Wrong About Identity Politics," *New Republic*, September 11, 2017, https://newrepublic.com/article/144739/liberals-get-wrong-identity-politics; "Combahee River Collective Statement," Combahee River Collective, April 1977, accessed September 20, 2021, http://combaheerivercollective.weebly.com/the-combahee-river-collective-statement.html; Alicia Garza, "Identity Politics: Friend or Foe?" Othering and Belonging Institute, September 24, 2019, https://belonging.berkeley.edu/identity-politics-friend-or-foe.

250 **In the fossil fuel politics of Richmond, some residents:** Norimitsu Onishi, "Together a Century, City and Oil Giant Hit a Rough Patch," *New York Times*, January 2, 2013, https://www.nytimes.com/2013/01/03/us/chevron-hits-rough-patch-in-richmond-calif.html.

250 **In 2008, a previous city council, more sympathetic:** David Helvarg, "Opinion: Chevron Wants to Buy My Vote," *Los Angeles Times*, October 8, 2014, https://www.latimes.com/opinion/opinion-la/la-ol-chevron-richmond-campaign-spending-20141007-story.html.

250 **A judge halted:** Earthjustice, "Chevron Refinery Expansion at Richmond, CA Halted," news release, July 2, 2009, https://earthjustice.org/news/press/2009/chevron-refinery-expansion-at-richmond-ca-halted.

251 **Kamala Harris, then serving as the attorney general of California, wrote:** Amy Standen, "Chevron Refinery Plans in Richmond Pose Risks, Says Attorney General," June 9, 2014, https://www.kqed.org/science/18206/chevron-refinery-plans-in-richmond-pose-risks-says-attorney-general.

251 **The company started its own news site:** John Upton, "Chevron Creates Its Own News Outlet for a Poor City That It Pollutes," *Grist*, March 24, 2014, https://grist.org/climate-energy/chevron-creates-its-own-news-outlet-for-a-poor-city-that-it-pollutes/; Early, *Refinery Town*.

251 **its consultants helped launch:** Jimmy Tobias, "Stealth Chevron Consultants Administer Richmond News Website," *Richmond Confidential*, October 30, 2014, https://richmondconfidential.org/2014/10/30/stealth-chevron-consultants-administer-richmond-news-website/.

251 **the company also spent about $3 million on political campaigning:** Brett Murphy and Elly Schmidt-Hopper, "Accusations and Money Fly as Chevron Spends on Richmond City Council Race," KQED, October 20, 2014, https://www.kqed.org/news/10344454/accusations-and-money-fly-as-chevron-spends-on-richmond-city-council-race.

251 **The Civic Center is Richmond's city hall:** John King, "Richmond Civic Center—Where Past Meets Future," SFGATE, May 26, 2009, https://www.sfgate.com/bayarea/article/Richmond-Civic-Center-where-past-meets-future-3231718.php.

251 **On one curved wall of this building:** "The Extraordinary Ordinary People of Richmond," Judy Baca, accessed September 20, 2021, http://www.judybaca.com/artist/art/extraordinaryordinarypeopleofrichmondca/.

252 **He had also recently fulminated against:** Sara Lafleur Vetter, "The Battle for Bikes in Richmond," *Richmond Confidential*, December 3, 2013, https://richmondconfidential.org/2013/12/03/the-battle-for-bikes-in-richmond/.

252 **But Nat Bates and Doria Robinson:** Betty Marquez Rosales, "Nat Bates, City Council," *Richmond Confidential*, January 31, 2018, https://richmondconfidential.org/2018/01/31/nat-bates-city-council/.

253 **Critics of Bates have raised questions:** Harriet Rowan, "Independence of Independent Expenditure Groups Called into Question," *Richmond Confidential*, December 20, 2014, https://richmondconfidential.org/2014/12/20/independence-of-independent-expenditure-groups-called-into-question/.

255 **In July 2014, the Richmond city council voted to approve:** Molly Samuel, "Richmond Approves Contentious Chevron Project," KQED, July 30, 2014, https://www.kqed.org/science/20001/richmond-approves-contentious-chevron-project.

255 **Doria's mother attended:** Malcolm Marshall, "Richmond Approves Stalled Modernization Plan at Chevron Refinery," Richmond Pulse, July 31, 2014, https://richmondpulse.org/2014/07/31/richmond-approves-stalled-modernization-plan-at-chevron-refinery-2/.

255 **David Horsey, a cartoonist and editorial writer:** David Horsey, "Chevron Funds Brazen Campaign to Buy a City Government," *Los Angeles Times*, October 29, 2014, https://www.latimes.com/opinion/topoftheticket/la-na-tt-chevron-brazen-campaign-20141029-story.html.

255 **Senator Bernie Sanders stopped in Richmond:** Bonnie Chan, Loi Almeron, Semany Gashaw, Martin Totland, and Phil James, "The Best of Senator Bernie Sanders in Richmond," *Richmond Confidential*, October 19, 2014, https://richmondconfidential.org/2014/10/19/the-best-of-senator-bernie-sanders-in-richmond/.

255 **In November, every candidate:** Richard Gonzales, "Chevron Spends Big, and Loses Big, in a City Council Race," *Two-Way* (blog), NPR, November 5, 2014, https://www.npr.org/sections/thetwo-way/2014/11/05/361875792/chevron-spends-big-and-loses-big-in-a-city-council-race.

256 **Soil can also be forgiving:** Borislava Lukić, Antonio Panico, David Huguenot, Massimiliano Fabbricino, Eric D. van Hullebusch, and Giovanni Esposito, "A Review on the Efficiency of Landfarming Integrated with Composting as a Soil Remediation Treatment," *Environmental Technology Reviews* 6, no. 1 (2017): 94–116, https://doi.org/10.1080/21622515.2017.1310310; Zhenyu Wang, Ying Xu, Jian Zhao, Fengmin Li, Dongmei Gao, and Baoshan Xing, "Remediation of Petroleum Contaminated Soils Through Composting and Rhizosphere Degradation," *Journal of Hazardous Materials* 190, no. 1–3 (June 15, 2011): 677–85, https://doi.org/10.1016/j.jhazmat.2011.03.103; Kawina Robichaud, Catherine Girard, Dimitri Dagher, Katherine Stewart, Michel Labrecque, Mohamed Hijri, and Marc Amyot, "Local Fungi, Willow and Municipal Compost Effectively Remediate Petroleum-Contaminated Soil in the Canadian North," *Chemosphere* 220 (April 1, 2019): 47–55, https://doi.org/10.1016/j.chemosphere.2018.12.108.

256 **More than a decade earlier, for instance:** A. A. Meharg, J. Wright, H. Dyke, and D. Osborn, "Polycyclic Aromatic Hydrocarbon (PAH) Dispersion and Deposition to Vegetation and Soil Following a Large Scale Chemical Fire," *Environmental Pollution* 99, no. 1 (January 1, 1998): 29–36, https://doi.org/10.1016/S0269-7491(97)00180-2.

257 **To the north lies a community outside city limits:** Robert Rogers, "Part 2: North Richmond's Inauspicious Beginnings," *Richmond Confidential,* June 8, 2011, https://richmondconfidential.org /2011/06/08/part-2-north-richmonds-inauspicious-beginnings/; Robert Rogers, "Part 6: North Richmond's Unceasing Struggle," *Richmond Confidential,* July 13, 2011, https://richmondconfidential.org /2011/07/13/part-6-north-richmonds-unceasing-battle/.

260 **The Berkeley soil science students:** "Healthy Soil in Richmond's Concrete Jungle," *Just Food,* podcast audio, Berkeley Food Institute, November 2, 2017, https://food.berkeley.edu/resources /just-food-podcast/healthy-soil-richmonds-concrete-jungle/; http://food.berkeley.edu/wp-content /uploads/2017/10/SOILS.pdf.

260 **In the fall of 2016, Doria and the Tilthers:** Reis Thebault, "Urban Tilth Launches New Farm in North Richmond," *Richmond Confidential,* October 19, 2016, https://richmondconfidential .org/2016/10/19/urban-tilth-launches-new-farm-in-north-richmond/.

261 **Around Labor Day that year, the San Francisco Bay Area:** Brendan Weber, "Hottest Temperatures Ever Recorded in the Bay Area," NBC Bay Area, September 3, 2017, https://www .nbcbayarea.com/news/local/hottest-temperatures-ever-recorded-bay-area/30810/.

261 **The following year, heat waves:** "Southern California Heat Wave Breaks Records," CBS News, July 6, 2018, https://www.cbsnews.com/news/southern-california-heat-wave-breaks-records-in -los-angeles-today-2018-07-06/.

261 **In May 2018, the city of Richmond:** Ted Goldberg, "Chevron, Richmond Settle Lawsuit over 2012 Refinery Fire," KQED, May 3, 2018, https://www.kqed.org/news/11665999/chevron -richmond-move-to-settle-lawsuit-over-2012-refinery-fire-that-sickened-thousands.

261 **By comparison, Chevron's CEO, Michael Wirth:** George Avalos, "Chevron Execs Capture Big Pay Raises, CEO Made $33 Million in 2019," *East Bay Times,* April 7, 2020, https://www .eastbaytimes.com/2020/04/07/chevron-execs-capture-big-pay-raises-ceo-made-33-million-in -2019/.

261 **Meanwhile, the Richmond refinery continued:** Denis Cuff, "New Year's Eve Gasp: Chevron Refinery Emits Gases from Flare," *East Bay Times,* January 2, 2018, https://www.eastbaytimes .com/2018/01/02/new-years-eve-gasp-chevron-refinery-emits-gases-from-flare/; Ted Goldberg, "Chevron Flaring Took Place After Four Steam Boilers Malfunctioned," KQED, April 16, 2018, https://www.kqed.org/news/11662610/chevron-flaring-took-place-after-four-steam -boilers-malfunctioned.

261 **Local activists said they were pollution:** Greg Karras, Carla M. Pérez, Jessica Guadalupe Tovar, and Adrienne Bloch, *Flaring Prevention Measures,* Communities for a Better Environment, April 2007.

261 **In October 2018, the Intergovernmental Panel on Climate Change:** IPCC, "Summary for Policymakers," in *Global Warming of 1.5°C: An IPCC Special Report,* ed. Valérie Masson-Delmotte, Panmao Zhai, Hans-Otto Pörtner, Debra C. Roberts, James Skea, Priyadarshi R. Shukla, Anna Pirani, et al., World Meteorological Organization, Geneva, Switzerland.

262 **A pair of earth scientists announced:** Manoochehr Shirzaei and Roland Bürgmann, "Global Climate Change and Local Land Subsidence Exacerbate Inundation Risk to the San Francisco Bay Area," *Science Advances* 4, no. 3 (March 1, 2018): eaap9234, https://doi.org/10.1126 /sciadv.aap9234.

262 **That summer, the Richmond city council:** Resolution of the Council of the City of Richmond, Declaring a Climate Emergency, Res. 69-18, July 24, 2018, http://www.ci.richmond.ca .us/ArchiveCenter/ViewFile/Item/8947.

263 **some experts say they could also fail in the long term:** Greg Karras, *Decommissioning California Refineries: Climate and Health Paths in an Oil State,* Communities for a Better Environment:

Huntington Park, Oakland, Richmond, and Wilmington, CA, 2020, https://www.cbecal.org/wp-content/uploads/2020/07/Decomm-CA-Refineries-July2020.pdf.

263 **California beat its 2020 goals:** Dale Kasler, "California Beats Its 2020 Goals for Cutting Greenhouse Gases," *Sacramento Bee*, July 11, 2018, https://www.sacbee.com/latest-news/article214717585.html.

263 **And as such a facility reduces its carbon output, it also generally gasps out:** Much of this analysis is based on reporting I've done on cap-and-trade and environmental justice over the last ten years, including for an article titled "Will California's Cap and Trade Be Fair?" published on March 20, 2013, in *The Nation*, https://www.thenation.com/article/archive/will-californias-cap-and-trade-be-fair/.

263 **But many experts say the Golden State:** Paul Rogers, "California's Behind on Its 2030 Climate Goals. What's at Stake If It Doesn't Catch Up?" *Mercury News*, January 16, 2020, https://www.mercurynews.com/2020/01/16/new-study-more-renewable-energy-electric-vehicles-needed-for-california-to-hit-greenhouse-gas-targets.

263 **According to an investigation by ProPublica:** Lisa Song, "Cap and Trade Is Supposed to Solve Climate Change, but Oil and Gas Company Emissions Are Up," ProPublica, November 15, 2019, https://www.propublica.org/article/cap-and-trade-is-supposed-to-solve-climate-change-but-oil-and-gas-company-emissions-are-up?token=6OpPu6Rt0wdCykgO57Up0RfvUivWw30A.

264 **In the fall of 2018, the governor:** Mark Hertsgaard, "At the Global Climate Action Summit, Brown and Bloomberg Make Bold New Pledges," *Nation*, September 13, 2018, https://www.thenation.com/article/archive/at-the-global-climate-action-summit-brown-and-bloomberg-make-bold-new-pledges/.

266 **Big green crates of food from Urban Tilth's:** Cameron Nielsen, "Urban Farmers in Richmond Are Helping in the Fight Against Food Insecurity," *Richmond Confidential*, November 5, 2020, https://richmondconfidential.org/2020/11/05/urban-farmers-in-richmond-are-helping-in-the-fight-against-food-insecurity/.

266 **And in late April the price of crude oil:** "US Oil Prices Turn Negative as Demand Dries Up," BBC News, April 21, 2020, https://www.bbc.com/news/business-52350082.

266 **By the summer, investigative journalist:** Antonia Juhasz, "The End of Oil Is Near," *Sierra*, August 24, 2020, https://www.sierraclub.org/sierra/2020-5-september-october/feature/end-oil-near.

267 **His organization had released a report, authored:** Karras, *Decommissioning California Refineries*.

267 **In late July, the oil and gas company Marathon:** Ted Goldberg, "Shutdown of Marathon's Martinez Refinery Prompts Calls for 'Just Transition' for Oil Workers," KQED, August 3, 2020, https://www.kqed.org/news/11831607/shutdown-of-marathons-martinez-refinery-prompts-calls-for-just-transition-for-oil-workers; "Phillips 66 to Close Refinery on Nipomo Mesa, Phase Out Associated Pipelines," *Santa Maria Times*, https://santamariatimes.com/news/san_luis_obispo_county_news/phillips-66-to-close-refinery-on-nipomo-mesa-phase-out-associated-pipelines/article_6b52217c-2a3b-5fe8-a68a-93ae8390b525.html.

271 **Beyond that, a solar company had opened a sixty-acre farm:** Aaron Davis, "New 10.5 Megawatt Solar Farm Opens on Site of Former Chevron Landfill in Richmond," *East Bay Times*, April 18, 2018, https://www.eastbaytimes.com/2018/04/18/new-10-5-megawatt-solar-farm-opens-on-site-of-former-chevron-landfill-in-richmond/.

272 **That same week, an eight-thousand-gallon tanker truck:** "Update: Tanker Truck Fire Shuts Down I-80; Richmond Shelter-in-Place Order Lifted; Westbound Lanes Reopened," October 24, 2020, https://sanfrancisco.cbslocal.com/2020/10/24/tanker-fire-interstate-80-hilltop-emergency/.

273 **On November 2, 2020, the day before the election:** "Flaring Activity Reported at Chevron Richmond Refinery," ABC7, November 2, 2020, https://abc7news.com/richmond-chevron-refinery-flaring-fire/7585937/.

273 **During 2021, community activists:** NRDC, "Groups Recommend Steps for Just Transition Roadmap," news release, February 10, 2021, https://www.nrdc.org/media/2021/210210.

Epilogue

277 **The days were thirty to forty degrees above normal:** Jason Samenow and Ian Livingston, "Canada Sets New All-Time Heat Record of 121 Degrees amid Unprecedented Heat Wave," *Washington Post*, June 29, 2021, https://www.washingtonpost.com/weather/2021/06/27/heat-records-pacific-northwest/.

277 **Just north of Seattle, transportation crews shut down:** "Pavement on I-5 Buckles in Extreme Seattle Heat," KING5, June 29, 2021, https://www.king5.com/article/traffic/traffic-news/i5-pavement-buckles-extreme-seattle-heat/281-5a302ac2-304c-4fe8-9ac1-e0d76cf43c5f.

277 **At the shores of the estuary that curves between Seattle:** John Ryan, "Extreme Heat Cooks Shellfish Alive on Puget Sound Beaches," KUOW, July 7, 2021, https://www.kuow.org/stories/extreme-heat-cooks-shellfish-alive-on-puget-sound-beaches?fbclid=IwAR1QX9JvVIN5dIH1LOF1-G09rb6Km87sC5La5ZlJ9lhNYcd4EC4Fbs2K_Aw.

278 **The estimated human death toll:** Nadja Popovich and Winston Choi-Schagrin, "Hidden Toll of the Northwest Heat Wave: Hundreds of Extra Deaths," *New York Times*, August 11, 2021, https://www.nytimes.com/interactive/2021/08/11/climate/deaths-pacific-northwest-heat-wave.html.

ACKNOWLEDGMENTS

It is a kind of myth that a book is the product of one individual's imagination and labor. In truth, it is a manifestation of the generosity and creativity of many. So many people have contributed to the research and writing of *At Home on an Unruly Planet*.

I could not have put this book together without the support and faith of my agents, Stuart Krichevsky and Laura Usselman: from the beginning, they understood that a deep conversation about home in an era of tumult was powerful and necessary and that the stories herein needed to be told. My editor, Conor Mintzer, worked tirelessly and diligently to help me bring this book into being. I am also grateful to my previous editor, Barbara Jones, for her early support of this project.

The extensive research that formed the bones of these stories would not have been possible without support from the Alfred P. Sloan Foundation books program and a project grant from Artist Trust. I also wrote an early outline of the book during a retreat at the Mesa Refuge that was funded by a Jonathan Rowe Memorial Fellowship. With the help and sponsorship of poet and professor Glen Phillips, Edith Cowan University funded my research trip to Australia as a visiting scholar several years ago, a trip that influenced several pieces of this book, most obviously the chapter on homesickness. (Glen also taught me years before, when I was an undergraduate exchange student in Perth, about the importance of creative persistence.)

The Fund for Investigative Journalism supported two projects that ultimately helped bolster the reporting for this book: an investigation into cap and trade policies and the impacts on frontline communities in California, and a report on black carbon pollution in Alaska, which led to my explorations of arctic climate change impacts. I'm also grateful to my editors at *Hakai, Sierra, Audubon, The Nation,* and *Undark* and my former colleagues at *YES!* magazine for letting me take on projects that helped fuel the research of this book.

Numerous other friends, colleagues, and fellow travelers and journalists contributed in ways large and small. Robert McLaughlin read early drafts of the manuscript—which gave me an extra dose of courage to push ahead with the writing and revision—and also contributed financially to the book. (Early in my adult life, when I was a student in his literature classes at Illinois State University, Bob M. also led me to understand better how words can be revolutionary.) My friends, Robert and Teri Stephens, also made a financial gift to support the book's research expenses. Ebonye Gussine Wilkins and Alice Rearden, my sensitivity readers, helped me identify important cultural nuances in the draft and bring greater authenticity to descriptions of communities on the front lines of climate change. Doug Pibel, Heather Purser, Lizzie Wade, Valerie Schloredt, Mark Kramer, Jessica Bruder, Deborah Blum, Annalee Newitz, Jen Marlowe, Erica Howard, and Jackie Varriano read early pieces of either the proposal or the manuscript and offered astute feedback. Sarah Neilson and Palmer Stroup lent their time and energy as research assistants. Austin Price, James Steinbauer, and James Gaines helped me rigorously fact-check the manuscript.

Researching this book required extensive travel, the vast majority done in the years before the pandemic. Kit and Phyllis Barnett, Rebekah Chapman, and Brittany Retherford offered me a roof to sleep under during travels in Alaska (when I was initially just a stranger to them). Kyle Hopkins and Rebecca Palsha also loaned me a bed (and once a car) during trips through Anchorage. Ash Adams quite literally allowed me to cry on her shoulder. Though I focused my Alaskan narratives on Newtok, many others in communities like Utqiagvik and Wainwright invited me into their homes and shared stories and wisdom. Linda and Ransom Agnasagga especially offered their time, insight, and kindness and fed me delicious Iñupiaq food.

In addition to painting the beautiful forest-house that graces the cover, Sarah Gilman put me up during a trip through the Methow Valley and read an early draft of the full manuscript with keen eyes, offering much valuable feedback. Mark Fritzel gave me a place to stay in Oakland numerous times and let me borrow his car to drive around Richmond. Rebecca Kim loaned me a set of wheels to take myself to a writing retreat in a remote part of Washington State, where I could concentrate deeply and write multiple chapter drafts.

My coterie (or perhaps coven) of brilliant and wise women science writers, the League of Extraordinary Writing Dames, provided ongoing moral support through some rocky moments in both the writing process and in my personal life. Antonia Juhasz, Alexandra Witze, and Julia Rosen were my writing and goal-setting partners at various moments.

This book also exists because of the generosity of hundreds of people who were willing to share their expertise, ideas, and personal histories, and I'm profoundly grateful to all the people who made time for both formal and informal conversations and interviews. One of my deepest sources of hope comes from knowing so many individuals who have dedicated themselves to protecting both our planetary home and our particular communities and beloved places. From them I have learned that optimism is not a passive feeling but an active fighting stance necessary for survival.

Finally, thank you to the people who have shaped my career. Especially to my mom, who made sure my childhood room was full of beautiful books, who read me not just bedtime stories but chapters and sections of novels and fanciful tales, and who taught me to love poetry and music. To my late grandfather, Harvey Clure Barnes, who instilled in me a love of nature and who knew how to get lost in the woods, to sit in stillness, and to attend to the landscape and the plants and animals.

To my stepdad, who pushed me, early in adulthood, to balance out the artistic and analytical parts of my brain by taking classes in organic chemistry and calculus, so that later I would better understand the scientific process and the forces that are changing the atmosphere.

To my dad, who taught me how to use a computer when I was about five years old, and has, in my adulthood, helped equip me with the technological tools every good journalist needs.

To my brother and my sister-in-law, who have trod some of these bumpy roads with me and have offered their unqualified enthusiasm for this project.

To my husband for sharing with me a life of exploration, adventure, and wonder, and for his unwavering faith in me.

To my husband's family, who inspire me with their passion for sustainability and for building a better future.

To Seattle, a city full of activists and rabble-rousers, politicians, literati, artists, backyard gardeners, scientists, hikers, adventurers, and so many people who are passionate about nature, thank you for the ongoing inspiration and for everything you do to protect this place and the magnificence and beauty of the Puget Sound.

And my deep gratitude to the Coast Salish peoples—including the Duwamish, Suquamish, Muckleshoot, and Tulalip communities—whose traditional land I live upon and who, for many generations, have nourished what is now my home place.

INDEX

ABOUT THE AUTHOR

Madeline Ostrander is a science journalist and writer whose work has appeared in the NewYorker.com, *The Nation*, *Sierra* magazine, PBS's *NOVA Next*, *Slate*, and numerous other outlets. Her reporting on climate change and environmental justice has taken her to locations such as the Alaskan Arctic and the Australian outback. She's received grants, fellowships, and residencies from the Alfred P. Sloan Foundation, Artist Trust, the USC Annenberg Center for Health Journalism, the Fund for Investigative Journalism, the Jack Straw Cultural Center, the Mesa Refuge, Hedgebrook, and Edith Cowan University in Australia. She is the former senior editor of *YES!* magazine and holds a master's degree in environmental science from the University of Wisconsin–Madison. She lives in Seattle with her husband.